Key Technologies on New Energy Vehicles

Key Technologies on New Energy Vehicles publishes the latest developments in new energy vehicles - quickly, informally and with high quality. The intent is to cover all the main branches of new energy vehicles, both theoretical and applied, including but not limited:

- Control Technology of Driving System
- Hybrid Electric Vehicle Coupling Technology
- Cross Disciplinary design optimization technology
- Single and Group Battery Technology
- Energy Management Technology
- Lightweight Technology
- New Energy Materials and Device
- Internet of Things (IoT)
- Cloud Computing
- 3D Printing
- Virtual Reality Technologies

Within the scopes of the series are monographs, professional books or graduate textbooks, edited volumes, and reference works purposely devoted to support education in related areas at the graduate and post-graduate levels.

To submit a proposal or request further information, please contact:
Dr. Mengchu Huang, Senior Editor, Applied Sciences
Email: mengchu.huang@springer.com
Tel: +86-21-2422 5094

More information about this series at https://link.springer.com/bookseries/16377

Xi Zhang · Chong Zhu · Haitao Song

Wireless Power Transfer Technologies for Electric Vehicles

華中科技大學出版社

Springer

Xi Zhang
Shanghai Jiao Tong University
Shanghai, China

Chong Zhu
Shanghai Jiao Tong University
Shanghai, China

Haitao Song
Shanghai Jiao Tong University
Shanghai, China

ISSN 2662-2920 ISSN 2662-2939 (electronic)
Key Technologies on New Energy Vehicles
ISBN 978-981-16-8347-3 ISBN 978-981-16-8348-0 (eBook)
https://doi.org/10.1007/978-981-16-8348-0

Jointly published with Huazhong University of Science and Technology Press
The print edition is not for sale in China (Mainland). Customers from China (Mainland) please order the print book from: Huazhong University of Science and Technology Press.

This Springer imprint is published by the registered company Springer Nature Singapore Pte Ltd.
The registered company address is: 152 Beach Road, #21-01/04 Gateway East, Singapore 189721, Singapore

Committee of Reviewing Editors

Chairman of the Board

Ouyang Minggao (Tsinghua University)

Vice Chairman of the Board

Wang Junmin (University of Texas at Austin)

Members

Ma Fangwu (Jilin University)
Wang Jianqiang (Tsinghua University)
Ai Xinping (Wuhan University)
Li Keqiang (Tsinghua University)
Yu Zhuoping (Tongji University)
Chen Yong (Hebei University of Technology)
Yin Chengliang (Shanghai Jiaotong University)
Wang Feiyue (Institute of Automation, Chinese Academy of Sciences)
Deng Weiwen (Beijing University of Aeronautics and Astronautics)
Hua Lin (Wuhan University of Technology)
Wu Chaozhong (Wuhan University of Technology)
Chen Hong (Jilin University)
Yin Guodong (Southeast University)
Huang Yunhui (Huazhong University of Science and Technology)

Foreword: New Energy Vehicles and New Energy Revolution

The past two decades have witnessed the research and development (R&D) and the industrialization of China's new energy vehicles. Reviewing the development of new energy vehicles in China, we can find that the "Tenth Five Year Plan" is the period when China's new energy vehicles began to develop and our nation started to conduct organized R&D of the electric vehicle technology on a large scale; the "Eleventh Five Year Plan" is the period when China's new energy vehicles shifted from basic development to demonstration and examination as the Ministry of Science and Technology carried out the key project themed at "energy saving and new energy vehicles"; the period of the "Twelfth Five Year Plan" is the duration when China's new energy vehicles transitioned from demonstration and examination to the launch of industrialization as the Ministry of Science and Technology organized the key project of "electric vehicles"; the period of the "Thirteenth Five Year Plan" is the stage when China's new energy vehicle industry realized the rapid development and upgrading as the Ministry of Science and Development introduced the layout of the key technological project concerning "new energy vehicles."

The decade between 2009 and 2018 witnessed the development of China's new energy automobile industry starting from scratch. The annual output of new energy vehicles developed from zero to 1.27 million while the holding volume increased from zero to 2.61 million, each of which occupied over 53% in the global market and ranked first worldwide; the energy density of lithium-ion power batteries had more than doubled and the cost reduced by over 80%. In 2018, six Chinese battery companies were among the top 10 global battery businesses, with the first and the third as China's CATL and BYD. In the meanwhile, a number of multinational automobile businesses shifted to develop new energy vehicles. This was the first time for China to succeed in developing high-technology bulk commodities for civic use on a large scale in the world, also leading the trend of the global automobile development. The year of 2020 marked the landmark in the evolution of new energy automobile. Besides, this year was the first year when new energy vehicles entered families on a large scale and the watershed where new energy vehicles shifted from policy-driven to market-driven development. This year also saw the successful wrapping up of the mission in the *Development Plan on Energy Saving and New Energy Vehicle*

Industry (2012–2020) and the official release of *Development Plan on New Energy Vehicle Industry (2021–2035)*. At the end of 2020, in particular, President Xi Jinping proposed that China strove to achieve the goal typified by peak carbon dioxide emissions by 2030 and carbon neutral by 2060, so as to inject great power into the sustainable development of new energy vehicles.

Looking back to the past and looking forward to the future, we can see even more clearly the historical position of the current development of new energy vehicles in the energy and industrial revolution. As is known to us all, each and every energy revolution started from the invention of power installations and transportation vehicles. On the other hand, the progress of power installations and transportation vehicles contributed to the development and exploitation of energy and led to industrial revolution. In the first energy revolution, steam engine was used as the power installation, with coal and train as energy and transportation, respectively. As for the second energy revolution, internal combustion engine was taken as the power installation, oil and natural gas as energy, gasoline and diesel as energy carrier and automobile as the transportation vehicle. At the current stage of the third energy revolution, all kinds of batteries are power installation, the renewable energy as the subject of energy and electricity and hydrogen as energy carrier and electric vehicles as the means of transportation. In fact, the first energy revolution enabled the UK to outperform the Netherlands while the second energy revolution made the USA overtook the UK, both in terms of the economic strength. The present energy revolution may be the opportunity for China to catch up with and surpass other nations. How about the Fourth Industrial Revolution? In my opinion, it is the green revolution based on renewable energy and also the smart revolution on the basis of digital network.

From the perspective of energy and industrial revolution, we can find three revolutions closely related to the new energy vehicles: electrification of power—the revolution of electric vehicles; low-carbon energy—the revolution of new energy; systematic intelligence—the revolution of artificial intelligence (AI).

First—electrification of power and the revolution of electric vehicles.

The invention of lithium-ion battery triggered the technological revolution in the area of storage battery over the past 100 years. Viewed from the development of power battery and power electronic device, the involvement of high specific energy battery and high specific power electric drive system would contribute to the platform development of electric chassis. The volume power of the machine controller based on new-generation power electric technology has more than doubled to 50 kW. In future, the volume power of the high-speed and high-voltage machine can be nearly doubled to 20 kW, and the power volume of the automobile with 100 kW volume power could be no more than 10 L. With the constant decline of the volume of the electric power system, the electrification will lead to the platform and module development of chassis, which will lead to a major change in terms of vehicle design. The platform development of electric chassis and the lightweight of body materials will bring about the diversification and personalization of types of vehicles. Besides, the combination of active collision avoidance technology and body lightweight technology will result in a significant change in automobile manufacturing system. The revolution of power electrification will promote the popularity of new energy electric vehicles and will eventually contribute to the overall electrification of the transportation sector. China Society of Automobile Engineers proposed the development goals of China's new energy vehicles in the *2.0 Technology Road Map of Energy Saving and New Energy Vehicles*: The sales of new energy vehicles would reach 40% of the total sale of vehicles by 2030; new energy vehicles would become the mainstream by 2035 with its sale accounting for over 50% of the total sale of vehicles. In the foreseeable future, electric locomotives, electric ships, electric planes and other types will become a reality.

Second—low-carbon energy and new energy revolution.

In the *Strategy on Energy Production and Consumption Revolution (2016–2030)* jointly issued by National Development and Reform Commission and National Energy Administration, a target was proposed that the non-fossil energy would account for around 20 percent of total energy consumption by 2030 and over 50% by 2050. Actually, there are five pillars aimed to realize the energy revolution: firstly, the transformation of renewable resources and the development of photovoltaic and wind power technologies; secondly, the energy system is shifting from centralized to distributed development, turning every building into a micro-power plant; thirdly, the storage of intermittent energy by way of relevant technologies such as hydrogen and battery; fourth, the development of energy (electric power) Internet technology; fifth, enabling electric vehicles to become the end of energy usage, energy storage and energy feedback. In fact, China's photovoltaic and wind power technologies are fully qualified for large-scale distribution, but energy storage remains a bottleneck which needs to be solved by way of battery, hydrogen and electric vehicles. With the large-scale promotion of electric vehicles, along with the mixture of electric vehicles and renewable energy, electric vehicles will become the "real" new energy vehicles utilizing the entire chain of clean energy. In doing so, it could both solve the pollution and carbon emission problems of the vehicle itself, but could also be conducive to the carbon emission reduction of the entire energy system, thus bringing about a new energy revolution for the entire energy system.

Third—intelligent development of system and AI revolution.

Electric vehicles have three attributes of travel tools, energy devices and intelligent terminals. Intelligent and connected vehicles (ICVs) will restructure the industrial chain and value chain of vehicles. Software defines vehicles while data determine value. The traditional vehicle industry will be transformed into a high-tech industry leading the AI revolution. In the meanwhile, let us take a look at the Internet connection and the feature of sharing regarding vehicles, among "four new attributes," from the perspectives of both intelligent travel revolution and the new energy revolution: For one thing, the connotation of the Internet pays equal attention to the Internet of vehicle information and the Internet of mobile energy. For another, the connotation of sharing lays equal emphasis on sharing travel and energy storage information. And both stationery and running electric vehicles can be connected to the mobile energy Internet, finally realizing a full interaction (V2G, Vehicle to Grid). As long as the energy storage scale of distributed vehicles is large enough, it will become the core hub of intelligent transportation energy, namely the mobile energy Internet.

Intelligent charging and vehicle to grid will meet the demand of absorbing renewable energy fluctuations. By 2035, China's inventory of new energy vehicles will reach about 100 million. At that time, the new energy vehicle-mounted battery power will reach approximately 5 billion kWh (kilowatt-hours) with 2.5–5 billion kWh as the charging and discharging power. By 2035, the maximum installed capacity of wind power and photovoltaic power generation will not surpass 4 billion kW. The combination of vehicle-mounted energy storage battery and hydrogen energy could completely meet the demand of load balance.

All in all, with the accumulation of experience over the past two decades, since 2001, China's electric vehicles have "shifted to another path and led in the sector of new energy vehicles" worldwide. At the same time, China could build its advantage in terms of renewable resources with AI leading the world. It can be predicted that the period between 2020 and 2035 will be a new era when the revolution of new energy electric vehicles, the revolution of renewable energy and the revolution of artificial intelligence will leapfrog and develop in a coordinated manner and create a Chinese miracle featuring the strategic product and industry of new energy intelligent electric vehicles. Focusing on one strategic product and sector, such three technological revolutions and three advantages will release huge power, which could help realize the dream of a strong vehicle nation and play a leading role in all directions. With the help of such advantages, China will create a large industrial cluster with the scale of the main industry exceeding 10 trillion yuan and the scale of related industries reaching tens of trillions of yuan. The development of new energy vehicles at a large scale will result in a new energy revolution, which will bring earthshaking changes to the traditional vehicle, energy and chemical industry, thus truly embracing the great changes unseen in a century since the replacement of carriages by vehicles.

The technology revolution of new energy vehicle is advancing the rapid development of related interdiscipline subjects. From the perspective of technical background, the core technology of energy saving and new energy vehicles—the new energy power system technology–remains the frontier technology at the current stage. In 2019, China Association for Science and Technology released 20 key scientific and engineering problems, two of them (electrochemistry of high energy and density power battery materials, and hydrogen fuel battery power system) belonging to the scope of new energy system technology; the report of *Engineering Fronts 2019* published by Chinese Academy of Engineering mentioned the power battery 4 times, fuel battery 2 times, hydrogen energy and renewable energy 4 times as well as electricity-driven/hybrid electric-driven system 2 times. Over the past two decades, China has accumulated plenty of new knowledge, new experience and so many methods during the research and development regarding new energy vehicles. The "research series of key technologies on energy saving and new energy vehicles" are based on Chinese practice and the international frontier, aiming to review China's research and development achievements on energy saving and new energy vehicles, meet the needs of technological development concerning China's energy saving and new energy vehicles, reflect the key technology research trend of international energy saving and new energy vehicles and promote the transformation and application of key technologies as regards China's energy saving and new energy vehicles. The series involve four modules: vehicle control technology, power battery technology, motor driving technology as well as fuel battery technology. All those books included in the series are research achievements with the support of National Natural Science Foundation of China (NSFC), major national science and technology projects or national key research and development programs. The publish of the series plays a significant role in enhancing the knowledge accumulation of key technologies concerning China's new energy vehicle, improving China's independent innovation capability, coping with climate change and promoting the green development of the vehicle industry. Moreover, it could contribute to China's development into a strong vehicle nation. It is hoped that the series could build a platform for academic and technological communication, and the author and readers could jointly make contributions to reaching the top in the international stage concerning the technological and academic level in terms of China's energy saving and new energy vehicle.

January 2021 (2021年1月)

Ouyang Minggao
Academician of Chinese Academy of Sciences
Professor of Tsinghua University (THU)
Beijing, China

Contents

About the Authors

Dr. Xi Zhang received his Bachelor, Master and Ph.D. degrees from Shanghai Jiao Tong University, Shanghai, China, in 2002, 2004 and 2007, respectively. He joined the University of Michigan-Dearborn, USA, as a postdoctor in 2007 and a research scientist in 2009. He is currently a full professor and a doctoral supervisor in the School of Mechanical Engineering, and the director of the Institute of Intelligent Vehicles, Shanghai Jiao Tong University, China. His research interests include wireless power transfer technologies, EV electric drive systems, battery management systems, etc. He acts as PI in about 30 research projects. Meanwhile, he has published three books and over 100 research papers. He was granted about 30 patents of invention. Besides, he was sponsored by Pujiang Talent Program and won the Chinese Industry-University-Institute Cooperation Promotion Award. He is now an IEEE senior member and New Energy Power Conversion Technology Committee member of the China Power Supply Society. He was the chair, the vice chair or the program committee chair for some international academic conferences.

Dr. Chong Zhu received the B.S. degree in Electrical Engineering from the China University of Mining and Technology, Xuzhou, China, in 2010, and the Ph.D. degree in Electrical Engineering from Zhejiang University, Hangzhou, China, in 2016. He was a postdoctoral researcher with San Diego State University, San Diego, CA, the USA, from 2017 to 2019. He is currently an assistant professor in the School of Mechanical Engineering, Shanghai Jiao Tong University. His research interests include battery management system and wireless power transfer for electric vehicles.

Dr. Haitao Song received his B.Eng. and M.Sc. majoring in Electrical Engineering from Shanghai Jiao Tong University in 2008 and 2013, respectively, and received his Ph.D. degree in Electronic Information Engineering from Shanghai Jiao Tong University in 2021. He is currently the vice dean of Artificial Intelligence Research Institute of Shanghai Jiao Tong University, and the executive president of Shanghai Artificial Intelligence Research Institute Co., Ltd. His current research interests include artificial intelligence, industrial Internet, integrated energy systems and so on.

Chapter 1
Introduction

1.1 Electric Vehicle Development

1.1.1 History of Electric Vehicles

As an important means of transportation, cars greatly facilitate people's travel. With the improvement of people's living standards, the number of cars also continues to grow rapidly. However, due to the emergence of the oil crisis and the intensification of environmental pollution, traditional fuel vehicles do not meet the requirements of sustainable development. In contrast, electric vehicles are powered by electricity and have developed rapidly due to their energy conservation, emission reduction, and environmental protection advantages. Especially in the past two or three decades, governments of various countries have successively introduced various preferential policies to promote the development of electric vehicles. At present, the scale of the global electric vehicle industry is expanding, and sales continue to rise. Electric vehicles are the inevitable trend of the future development of the automobile industry (Jian 2018).

As early as 1832, Scottish Robert Anderson installed batteries and motors on the carriage and transformed it into an electric carriage. Then, in 1834, American Thomas Davenport made the first electric tricycle powered by a dry battery and a DC motor, and since then began the history of electric vehicles. That's half a century earlier than the internal combustion engine (the world's first internal combustion engine was born in 1885) (Fig. 1.1).

In April 1881, Gustave Truvi of France built a three-wheel electric vehicle with lead-acid batteries. Thus, the prototype of electric vehicles was born. Then a series of new electric vehicles came into being, and they were widely used in Europe in the late 19th and early twentieth century. In the 1920s, the development of internal combustion engines gave fuel vehicles an absolute advantage at this stage. In contrast, electric vehicles are slow and short-range; fuel vehicles have strong endurance and

X. Zhang et al., *Wireless Power Transfer Technologies for Electric Vehicles*,
Key Technologies on New Energy Vehicles,
https://doi.org/10.1007/978-981-16-8348-0_1

Fig. 1.1 Davenport with his electric tricycle. *Source* http://www.qcwxjs.com/qichewenhua/19014.html#images-1

gasoline prices are reasonable. Fuel vehicles have surpassed electric vehicles in all aspects, and electric vehicles have gradually withdrawn from the market since then. Until the 1980s, the oil crisis and environmental pollution caused people to re-examine electric vehicles. Various developed countries began to re-study electric vehicles. The highly valued electric vehicles developed rapidly under the vigorous promotion of various countries and gradually returned to people's vision.

1.1.2 Battery Electric Vehicles and Hybrid Electric Vehicles

Battery electric vehicles are powered entirely by rechargeable batteries (lead-acid batteries, nickel–cadmium batteries, nickel-metal hydride batteries, or lithium-ion batteries), such as Tesla Model 3, Chevrolet Menlo, Buick Velite, etc. Battery electric vehicles have little environmental pollution and high efficiency. However, for battery electric vehicles, the development of battery technology is still its bottleneck. Despite the rapid development of battery technology, the driving range of battery electric vehicles has been dramatically improved, and the charging time has been significantly shortened. Compared with fuel vehicles, battery electric vehicles still have disadvantages such as a high price, short life, large size and weight, and short cruising range (Fig. 1.2).

Hybrid electric vehicles generally use traditional internal combustion engines (gasoline engines) and electric motors as power sources, such as BYD Qin proDM, PRIUS Hybrid. Compared with traditional vehicles, the internal combustion engine of hybrid electric vehicles always works in the best condition with very low fuel consumption. In the busy urban areas, the internal combustion engine can be shut

Fig. 1.2 Tesla model3. *Source* http://00imgmini.eastday.com/mobile/20180928/20180928173947_2d63098338624b8add7effdc74fe9d58_1.jpeg

down and driven by the battery alone to achieve "zero" emissions. And hybrid electric vehicles don't have the problem of short cruising range. Under the current technological level and application conditions, the hybrid electric vehicle is the most industrialized and market-oriented model among electric vehicles and is the best choice to improve fuel economy and reduce emissions (Yin et al. 2011). However, hybrid electric vehicles still need a gasoline engine, so they are only the "transition models" of battery electric vehicles in the future (Fig. 1.3).

1.1.3 Development Status of Electric Vehicles at Home and Abroad

China's new energy automobile industry began at the beginning of the twenty-first century. In 2001, the ministry of science and technology set up a major electric vehicle special project in the "tenth five-year" national "863" plan, taking the electric vehicle as the main direction of China's automobile technological innovation, which marked the official launch of China's electric vehicle special project. In 2010, the Ministry of Finance, the Ministry of Science and Technology, the Ministry of Industry and Information Technology, and the National Development and Reform Commission jointly issued the new energy subsidy policy: the maximum subsidy per battery electric vehicle is 60,000 yuan, and the maximum subsidy per plug-in hybrid

Fig. 1.3 PRIUS hybrid. *Source* http://img1.gtimg.com/gd/pics/hv1/228/76/927/60297783.jpg

vehicle is 50,000 yuan. Until now, a series of preferential policies on electric vehicles are continually introduced. Over the past two decades, China's electric vehicle industry has developed rapidly, including BYD, FAW, Geely, Changan, Chery, SAIC, BAIC, and other automobile manufacturers have formulated electric vehicle product development and industrialization plans. At the same time, domestic sales of electric vehicles continue to climb. As shown in Fig. 1.4, from 2010 to 2018, domestic sales of new energy vehicles have continued to increase, especially since 2015, which an annual growth of hundreds of thousands. At present, China's electric vehicles are still in the stage of rapid development, and battery electric vehicles are the main research direction.

In the United States, new energy vehicles are generally called "alternative fuel vehicles." Affected by environmental pollution and the oil crisis, the US government has gradually increased its emphasis on new energy vehicles and supported the

Fig. 1.4 Domestic sales of new energy vehicles from 2010 to 2018

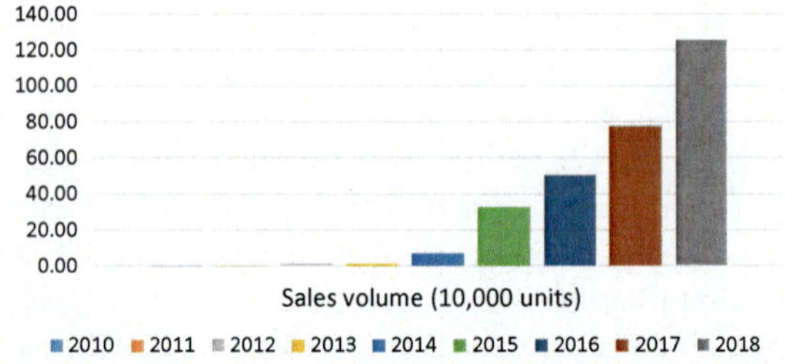

Fig. 1.5 Tesla's factory in Shanghai. *Source* https://ss1.bdstatic.com/70cFuXSh_Q1YnxGkpoWK1HF6hhy/it/u=3196924258,185930641&fm=11&gp=0.jpg

development of the new energy vehicle industry through the use of preferential tax relief policies. In the United States, Tesla is the most representative brand of electric vehicles, and it is a global leader in technological innovation and product design. Driven by the strong sales of Tesla Model 3, the sales volume of the new energy vehicle market in the United States reached 358,000 units in 2018, with a growth rate of up to 79%. In 2019, Tesla established a super factory in Shanghai, which further promoted the development of electric vehicles (Fig. 1.5).

The Japanese government attaches great importance to the development of electric vehicles because of the lack of resources. In the 1970s, Japan began the research and development of electric vehicles and was one of the first countries to start developing electric vehicles. Similarly, the Japanese government has promoted the development of electric vehicles through subsidies and tax breaks. The development of Japanese electric vehicle companies is dominated by Honda, Toyota, and Nissans, whose hybrid technology development is in the world's leading position. The most representative is the Toyota Prius. By the end of July 2015, Toyota's cumulative global sales of hybrid electric vehicles exceeded 8 million, of which the "Prius" sales were 3.527 million, accounting for more than 40%. Japan plans to increase the proportion of electric vehicle sales to total new vehicle sales to 70% by 2030; by 2050, the number of electric vehicles is expected to be 8.8 million.

Europe attaches great importance to carbon emissions, so electric vehicles are also highly valued in Europe. The UK, France, Italy, and Spain are prioritizing electric vehicles and their infrastructure. Norway has introduced a large number of incentives to promote electrified transportation. Germany, as the largest automotive market in Europe, lags behind its electric vehicle development goals. The European Union, on the other hand, constantly enhance the vehicle CO_2 emissions and emissions requirements, objectively promoting the development of electric vehicles. At present, the European automobile market is accelerating the electrification.

Until now, the EVs are not so attractive to consumers even with many government incentive programs (Lu et al. 2015). The long charging time and a short cruising range of electric vehicles are the main factors restricting their development. Therefore, the further development of electric vehicles depends on the innovation of battery technology and charging technology.

1.1.4 Electric Vehicles and Wireless Charging

As the global environment deteriorates and oil resources become scarcer, electric vehicles have emerged in the automobile market. Electric vehicles are more convenient to obtain power and have less impact on the environment. The related technologies of electric vehicles have been developed, making it gradually become the protagonist of the automobile market and the darling of people traveling. There are many aspects of electric vehicle technology research, such as power battery systems, motor drive systems, charging systems, vehicle control, and functional safety. Among them, the power battery as the core of the electric vehicle determines the vital performance of the electric vehicle's cruising range. Meanwhile, due to the existence of power batteries, electric vehicles need to recharge the batteries before they run out of power, so research on electric vehicle recharging systems is crucial.

At present, the charging methods for electric vehicles can be broadly divided into two types: power replacement and charging. The power replacement method refers to removing the power battery that is about to run out and replacing it with a fully charged power battery. Although this method can be realized, it faces a series of problems in design, production, and promotion. For example, the safety problems faced by the easy-to-remove power battery, the uniformity of power battery specifications, and the concerns of car owners. Compared with the power replacement method, the charging method is more feasible. Power battery charging methods include contact charging and non-contact charging. Contact charging, also known as conductive charging. The charging is completed by the metal contact conduction of the wire plug on the charging pile and the electric vehicle battery socket. Non-contact charging, also known as wireless charging, is a new type of charging method. This method uses electromagnetic fields as a medium to achieve electrical energy transmission. When charging an electric vehicle, the power supply equipment on the ground and the vehicle power receiving equipment do not need to be connected by wires.

Contact charging technology is now relatively mature. However, there are still some problems including cumbersome cable plugging and unplugging, risk of electric shock, and severe impacts from extreme weather such as rain and snow. As an emerging charging technology, wireless charging has certain advantages over the current plug-in or contact charging technology in terms of charging station construction, actual use, and later maintenance: (1) high space utilization, especially suitable for future large-scale cluster construction; (2) Strong environmental adaptability,

Fig. 1.6 Pure electric vehicle power system structure. *Source* https://www.d1ev.com/kol/51831

higher reliability of sealed installation; (3) High safety, eliminating the risk of electric shock; (4) Convenient usage and management, especially suitable for future unmanned management development trend (Lu et al. 2015) (Fig. 1.6).

Wireless charging is mainly divided into static charging and dynamic charging according to whether the car is stationary during charging.

1.1.4.1 Static Wireless Charging

The vigorous development of electric vehicles also heralds a broad market for wireless charging technology. In the field of electric vehicles, electric buses and electric cars are the most widely used. The traditional pure electric bus charging technology uses a conductive DC charging system, which is connected to the vehicle charging socket through the charging gun of the non-vehicle charger. Charging stations occupy large spaces. The management cost of charging stations is high. The charging gun and the vehicle charging socket are severely worn, which leads to problems such as poor contact, burning of contact heating, and short service life of charging gun and vehicle socket. Static wireless charging provides a convenient and reliable charging method, which can reduce users' burden and improve users' charging experience.

In the vision of the future city, self-driving vehicles have become a daily mainstream mobility tool for people. In the future parking lot, each parking space is equipped with a static wireless charging device, so that the vehicles parked in the parking lot can be quickly charged in time to ensure the cruising range of the vehicle. In addition, the wireless charging station will also become a conventional form of charging station in the future. Compared with the traditional electric vehicle charging pile and charging station, wireless charging can substantially save floor space, increase the charging rate per unit area, and greatly improve the convenience of electric car charging. Wireless charging can be combined with automated parking technology in the future. In short, static wireless charging is an essential foundation for realizing the fully automatic unmanned driving of electric vehicles.

1.1.4.2 Dynamic Wireless Charging

Because wireless charging technology can solve the connector limitations and safety problems faced by traditional conductive charging, it has gradually developed into the mainstream of electric vehicle charging. However, both static wireless charging and wired charging have the problems of frequent charging, short cruising range, large battery consumption, and high cost. Especially, electric buses have higher requirements for continuous battery life. Static wireless charging places, including parking lots and charging stations, are unable to meet the growing demand for electric vehicles. Urban road charging will play a huge role, and the road charging mode is mainly dynamic wireless charging. Dynamic charging for on-line electric vehicles (OLEVs), that is, the car is charged during driving. It can reduce the on-board battery capacity and the quality of the vehicle, make the electric vehicle equipped with a small number of battery packs, extend its cruising range. Simultaneously, the power supply becomes safer and convenient (Fig. 1.7).

Fig. 1.7 Dynamic wireless charging scenario. *Source* https://www.sohu.com/a/164021220_470482

Dynamic wireless charging is an important driving force and guarantees for electric vehicles to get rid of the shackles of cruising range and truly become the mainstream of the market. Therefore, the development of a dynamic wireless power supply system with high safety, high power, high efficiency, strong lateral adaptability, low electromagnetic radiation, and moderate cost has become the main research focus of major research institutions at home and abroad.

1.1.4.3 Key of Automatic Access for EV to Energy Internet

Due to the mobility of electric vehicles, wireless charging will become the key to the automatic connection of electric vehicles into the energy internet. Especially supported by the bidirectional wireless energy transmission technology, electric vehicles will realize automatic bidirectional wireless charging and discharging, so that energy flows bidirectionally between the electric vehicle and the grid. The electric vehicle will truly become a highly autonomous mobile energy storage device and energy node. At the same time, under the new energy microgrid system, wireless charging and discharging technology can realize the safe and orderly charging and discharging of electric vehicles, and automatically participate in the microgrid energy regulation to achieve peak shifting and valley filling, economical charging, and grid coordination, thereby making electric vehicles a new important part of the energy interactive management system.

Innovative development of electric vehicle wireless charging technology will effectively promote the development of electric vehicles, which largely determines the market application and development prospects of electric vehicles. In the process of developing wireless charging technology, technicians should strictly abide by the safety guidelines, strive to promote technological transformation, and accelerate the development of electric vehicles.

1.2 Lithium-Ion Battery and EV Charging Technologies

1.2.1 The Lithium-Ion Battery Technology Development

In the early 1970s, the research work on lithium-ion batteries was promoted due to the energy crisis. Li or Li–Al alloys were used as anode materials. After that, lithium primary batteries, including Li/MnO_2, Li/I_2, Li/SOC_{12}, and Li/FeS_2, appeared one by one. Panasonic obtained a patent for $Li/(CF)_n$ batteries in the United States in 1970. During the discharge process $(CF)_n$ was lithiated to form C and LiF, while the reaction was not reversible so that the lithium primary battery at the time was disposable. During the 1970s, Sanyo, the world's largest manufacturer of lithium-ion batteries at that time, designed the Li/MnO_2 battery, which was immediately applied to the rechargeable electronic calculator produced by the company and powered

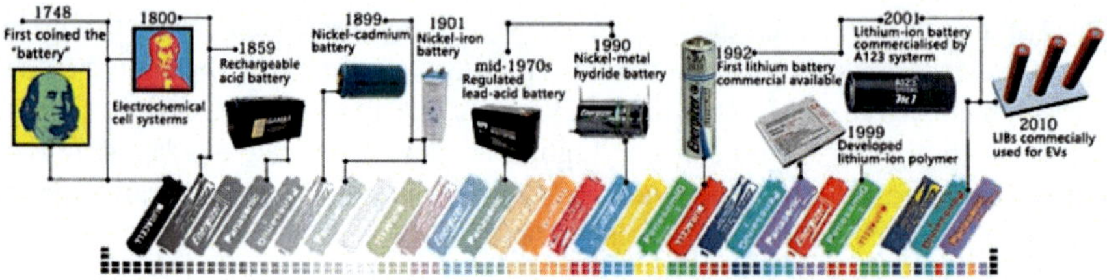

Fig. 1.8 Development of battery systems. *Source* https://www.upsbatterycenter.com/blog/history-batteries-timeline/

by the solar energy. For the initial research, General Motors (GM) and Argonne National Laboratory (ANL) focused their attention on lithium-ion battery systems that use molten salts as electrolytes. The raw materials of the positive and negative electrodes of the molten salt electrolyte battery are molten lithium and sulfur sources. The electrolyte uses a mixed lithium salt melt, such as LiCl and KCl. But this kind of molten salt electrolyte battery requires a high temperature of 45 °C. However, this type of battery cannot suppress the damage to the environment during cycling; thus the capacity quickly deteriorates. In the next study, the researchers gave up lithium and sulfur for the anode and switched their research point to lithium aluminum alloys (LiAl) and iron sulfides (FeS and FeS_2), which improved battery cycle performance. While, due to the rapid development of organic electrolyte lithium-ion batteries, the research work of high-temperature molten salt lithium-ion batteries around 1990 was basically terminated (Li et al. 2014) (Fig. 1.8).

The real lithium-ion battery was first proposed by Whittingham of Exxon company in 1976. The so-called lithium-ion battery means that there is no elemental lithium inside the battery, and the source of lithium is provided by lithium-ion compounds. Whittingham found that at room temperature, layered TiS_2 could electrochemically react with metallic lithium, and organically combined insertion reactions with energy storage for the first time. The operating voltage of the Li/TiS_2 battery is about 2 V, and the cycle performance is very superior. The capacity attenuation is only 0.05% per time within 1000 cycles. Exxon Company adopted LiAl alloy instead of metal lithium as the negative electrode of the lithium-ion battery, which was introduced to the market in 1977 and used in watches and small electronic equipment. Since then, compounds represented by VSe_2, MoO_3, $CuTi_2S_4$, V_2O_5, V_6O_3, and LiV_3O_8, which can reversibly deintercalate with lithium, have been applied as anode materials for lithium-ion battery. In 1980, the MoliEnergy of Canada inserted 1 mol of lithium into 1 mol of MoS_2 by the chemical method to form $LiMoS_2$, which improved the electrochemical cycle performance of MoS_2 and industrialized this product. This is also the closest to the prototype of modern lithium-ion batteries (Haering et al. 1980).

After the 1980s, great breakthroughs have been made in the research of lithium-ion batteries. In 1980, John Goodenough's group introduced $LiCoO_2$ as the cathode material (Mizushima et al. 1980). In 1981, Bell Labs used graphite as the anode material for lithium-ion batteries. In the same year, Japanese scientist Yoshino Akira discovered that conductive polyacetylene can be used as the negative electrode of

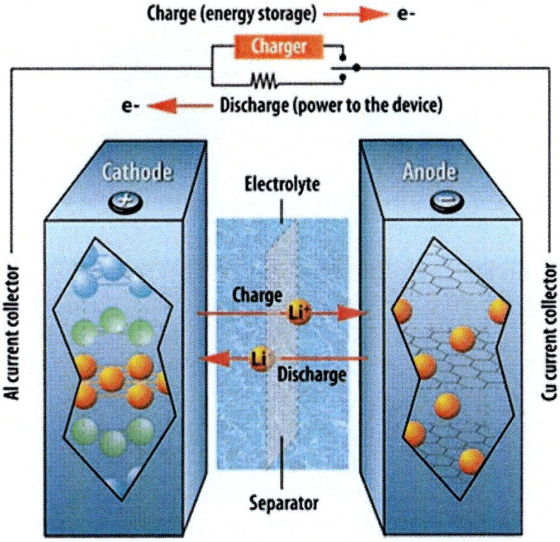

Fig. 1.9 Charge–Discharge mechanism of Li-ion batteries. *Source* https://genesisnanotech.wordpress.com/2018/05/15/predictions-for-the-evolution-of-the-battery-markets-for-evs-and-more-looking-back-to-see-what-is-ahead/

rechargeable batteries using organic solvents. Two years later, Dr. Goodenough's research group made the cathode material $LiMn_2O_4$ for the first time (Goodenough and Mizuchima 1981) (Fig. 1.9).

After that, taking the discovery of Goodenough, Yoshino Akira developed a prototype of a lithium-ion battery in 1983, thereby eliminating metallic lithium in the battery and improving the safety of the battery. However, the polyacetylene performed as the negative electrode has the disadvantages of low specific gravity, inability to increase the battery capacity, and insufficient stability. For this reason, Yoshino Akira continued to develop and used carbon-based materials as the negative electrode, and the positive electrode is still lithium cobalt oxide. New lithium-ion secondary battery (LIB), thus establishing the basic framework of modern lithium-ion batteries in 1985 (Yoshino et al. 1987). In 1989, Manthiram and Goodenough reported that the induction effect of polyanions (such as $SO4^{2-}$) can improve the working voltage of metal oxides. In 1990, Sony's commercial lithium-ion secondary battery ($C/LiCoO_2$) became a true lithium-ion battery. A lithium secondary battery using graphitized carbon material as a negative electrode was realized, and its composition structure is lithium and transition metal composite oxide/electrolyte/graphitized carbon material. Moreover, Tarascon and Guyomard made an electrolyte system based on ethylene carbonate and dimethyl carbonate in 1994. Goodenough reported a cathode material $LiFePO_4$ in 1997. Until then, the lithium-ion battery system has been fully established (Nishi 2001).

Lithium-ion batteries are currently the most available generation of secondary batteries with high energy density and cycle life. They are currently widely used in mobile communications, digital technology, electric vehicles, energy storage, and other fields. The future demand for lithium-ion batteries and their materials is huge, and its supporting upstream and downstream industrial chains also have a huge

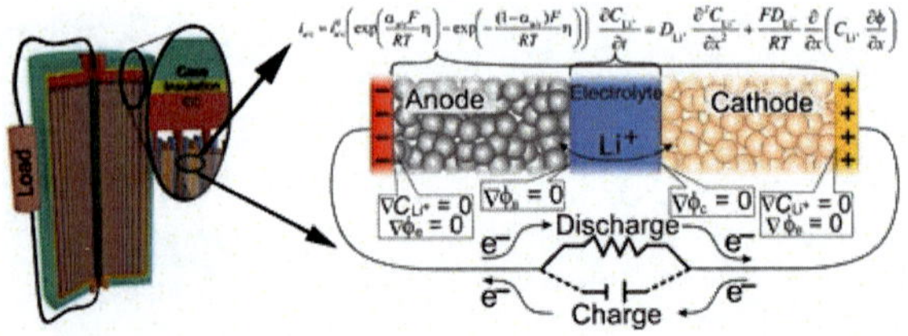

Fig. 1.10 Modeling framework for the simulation of Li-ion batteries. *Source* https://cmtl.uic.edu/research4.html

market. Under the technological development of lithium-ion batteries, pure electric new energy vehicles, hybrid new energy vehicles, and extended range new energy vehicles have sprung up constantly, corresponding to the development of new energy vehicle industry chain all over the world, especially for pure electric. As a kind of high energy density energy storage device, the lithium-ion battery is widely used as the power energy system for various new energy vehicles. It is expected that by the year 2025, the global electric vehicle market will have a demand for lithium-ion batteries of up to 200 million kWh. Therefore, except for the amount of battery production, more accurate understanding and control of power batteries have become an important requirement for the development of electric vehicles. At the same time, the accurate assessment of the aging state of electric vehicle power batteries will help to establish a more accurate battery management model, which will improve battery efficiency and service life (Fig. 1.10).

According to different research directions, battery models can be roughly divided into mechanism models and empirical models. Empirical models are mainly divided into equivalent circuit models and electrochemical impedance models which using electrochemical impedance spectroscopy. The mechanism model is mainly divided into an electrochemical model based on the chemical reaction mechanism of the battery, and a simplified electrochemical model that is simplified for ease of application. In practical electric vehicle battery management systems, most models are battery experience models based on equivalent circuits currently. Because the equivalent circuit model has a simple structure, with fewer parameters involved, and the model can simulate the actual charging and discharging dynamic characteristics of the battery well so that it is widely used in engineering most recently. However, the battery model still has the following problems in practical applications. Firstly, the electrochemical reaction mechanism inside the battery is ignored, and the model parameters do not correspond to a clear electrochemical meaning, it is difficult to reflect the precise actual battery characteristics. Secondly, during the use of the battery, the electrochemical reactions that occur inside the battery are also accompanied by side reactions. Irreversible side reactions will cause permanent aging and degradation of battery performance, which is contrary to the accurate estimation of the battery model. Therefore, the battery electrochemical model based on the internal electrochemical principle of the battery can integrate the fixed battery parameters of each part of the battery and the battery parameters that change with the battery cycles into a set of theoretical models. That founded electrochemical battery model can truly reflect the battery in various actual operating conditions in the vehicle running states.

1.2.2 The EV Charging Technology Development

EV charging has been categorized by the U.S. Department of Energy in three levels: Level 1 is standard charging that has a charge power less than 5 kW, Level 2 is fast-charging that occurs between 5 and 50 kW, and Level 3 is super-fast charging that is greater than 50 kW. Level 3 charging usually conveys the power as DC, while Level 1 and Level 2 charging often contain onboard electronic converters, allowing for AC energy transfer. The need for fast charging exists for consumer convenience as to not limit drivers to the daily average driving distance, while also improving the energy transfer rate to closer values to that of gasoline and diesel refueling stations. However, it should be noted that as the charging process takes place, a certain upper voltage limit should be used to prevent the battery from over-charged. It is known that the acceptable charging current is a monotonically decreasing function of the state-of-charge.

In recent years, the problems of "Range anxiety" associated with electric vehicles (EVs) have been alleviated by the improvements in high energy density batteries. With the increasing popularity of electric vehicles, "Charging anxiety" is now replacing the "Range anxiety". The EV charging system is the equipment required to condition and transfer energy from the constant-frequency, constant-voltage network to the direct-current, variable-voltage EV traction battery bus to charge the battery while connected. The developments of robust and efficient charging technologies are widely considered key requirements for a large-scale transition to electro-mobility. Such methods would not only provide more charging options for drivers but would also promote awareness and range confidence for prospective electric vehicle owners. There are different physical implementations and charging strategies when recharging the EV batteries. The physical implementation of battery charging means how the energy is transferred to the vehicle physically, while the battery charging strategy refers to the curves of the currents/voltages to be used at charging.

In general, the available Li-ion battery charging strategy can be divided into three classes based on the internal mathematical models. The first category is a model-free methodology, including constant-current (CC), constant-current and constant-voltage (CCCV), multi-stage CCCV, and pulse charging techniques. These approaches can be characterized by their predefined charging profiles with fixed current, voltage, and/or power constraints but ignoring the responses of battery dynamics. Considering the operability of model-free methods, the corresponding programs are viewed as a heuristic. Hence, this motivates the necessity to explore advanced charging strategies to meet fast charging requirements and alleviate the impact on battery state-of-health (SOH) meanwhile.

The second category of charging strategies utilizes empirical models, such as equivalent circuit-based models and neural network models. These models predict battery states and calculate electrical elements using past experimental data. By means of different circuit models, Kalman-type filters, recursive least squares, sliding mode observers, and moving horizon estimations were adopted to estimate battery states. Meanwhile, frequency optimization, multi-objective optimization, fuzzy control, linear quadratic control, and model predictive control were formulated to improve charging performances. The empirical models are computationally fast and simple, but unable to reflect physics-based parameters and battery aging. Therefore, an empirical model-oriented charging control protocol may fail to work properly after certain cycles.

The charging algorithms based on electrochemical models governed by kinetics and transport equations are relatively sophisticated. The closed-loop optimization problems can be formulated to minimize charging time and compensate for model uncertainties and disturbances. In addition, temperature variation can also be predicted with thermal-related relations. Thus, the electrochemistry-based control algorithm is close to the actual battery mechanism when used as a state observer. However, the intractable computation complexity associated with full-order nonlinear partial differential equations (PDEs) limits the further application to a real-time charging controller (Gao et al. 2019).

1.2.2.1 Constant-Current Constant Voltage Related Charging Methods

The constant current constant voltage (CC/CV) charging algorithm is widely adopted in charging Li-ion batteries because of its simplicity and easy implementation. Under the CC/CV algorithm, the battery is initially charged with constant current until the battery voltage reaches a preset maximum charging voltage, then the charging voltage is held constant until the current is reduced to a preset minimum value. The charging curve of the CC/CV is shown in Fig. 1.11.

Many variants of the CC/CV charging strategy were developed. A multi-stage fast charging profile is split into three different stages, referred to as CC-I, CC-II, and CV-I in Fig. 1.12. The algorithm is based on the evolution of internal resistance during charging. Due to the cell's smaller resistance in the lower SOC range, the highest current is applied. The last two stages are used as the cell's internal resistance increases rapidly. Experimental results verify that the procedure is useful for avoiding a considerable temperature rise and extending the cycle life.

Another form of the multi-stage CC/CV method is introduced in Fig. 1.13, where a $V_{\max}$ voltage is utilized to charge the battery in the initial CV-mode during a very

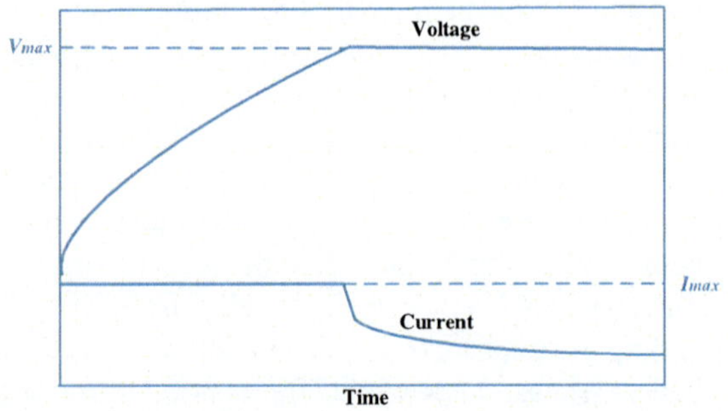

Fig. 1.11 Constant current-Constant voltage curve

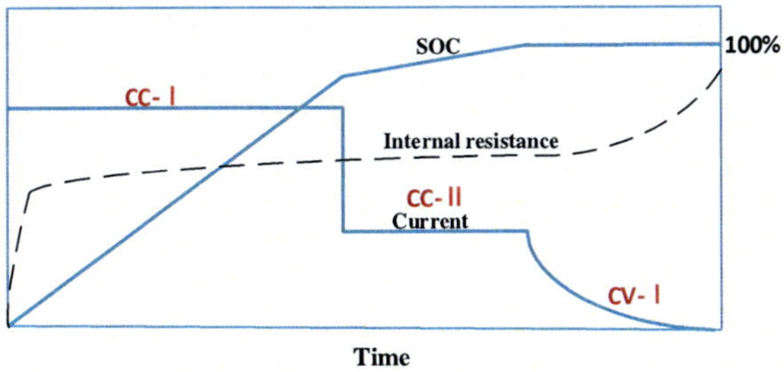

Fig. 1.12 Current, state of charge, resistance profiles of the fast charging technique

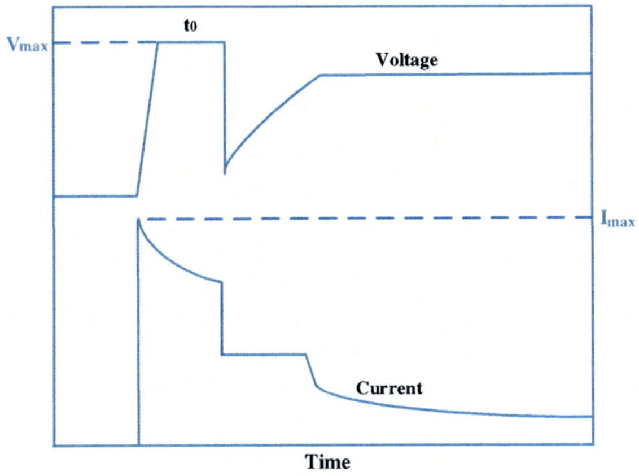

Fig. 1.13 Voltage and current characteristics of the boost-charging technique

short period t_0, subsequently followed by a standard CCCV process. This boost-charging technique ensures that the fully discharged battery can be recharged to one-third of its rated capacity without inducing any extra degradation effects.

1.2.2.2 Pulse Charge Based Charging Curve

The pulse charge has been claimed to be a fast and efficient charging algorithm for lithium-ion batteries. The pulse charging generates pulses of current to the battery in a fashion that optimizes the charging time while considering polarization, battery heating, SOC, and variable battery impedance, as depicted in Fig. 1.14. The rest time of each pulse period allows the ions to diffuse through the electrode materials, increasing the efficiency of the charging process. Negative pulse charging methods, originally developed to enhance the efficiency of charging converters but now extended to lithium-ion batteries, impose small discharges to the battery during the pulse charging rest period. The negative impulse decreases stress in the cell and helps minimize the temperature rise of the cell. Since the negative pulse pulls a small amount of energy from the battery, circuit configurations that recapture that energy

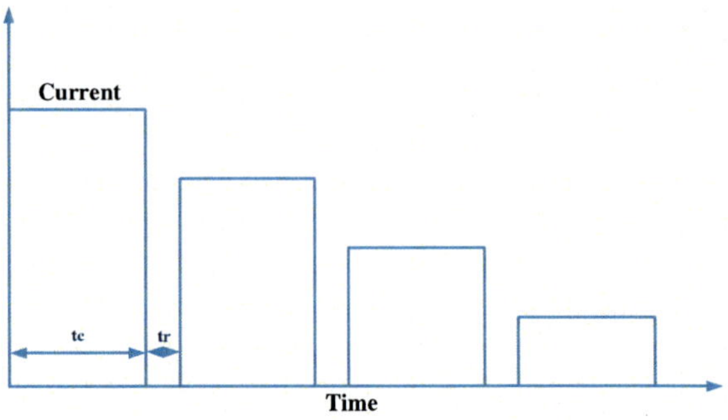

Fig. 1.14 Nonlinearly decreasing pulse charge technique

have been devised. By occasionally depolarizing the cell, high currents can continually be pumped into the battery, enabling a higher charge rate and lower charge time. This method helps the chemical reactions within the battery and can significantly improve the life of the battery (Lin et al. 2019).

1.2.2.3 Generalized Active Optimal Charging Algorithm

The commonly-used active battery charging management structure is often composed of three important elements. They are the battery model, state estimator, and model-based controller. Considering the process noise, a closed-loop control structure for the battery optimal charging strategy is formed, as depicted in Fig. 1.15.

Generally speaking, the battery model is carefully designed to be lower-order and easily implementable for outstanding controllability. The aim of constituting such a battery model is to simulate the real battery system dynamics under the specified loading current profile. The output variables of the model integrated with the noise vector are used as input for the observer. Given that the battery contains a lot of state variables, many of which are unmeasurable, such as concentration and over-potential. Thus, a robust and effective model-based estimator is required to observe internal

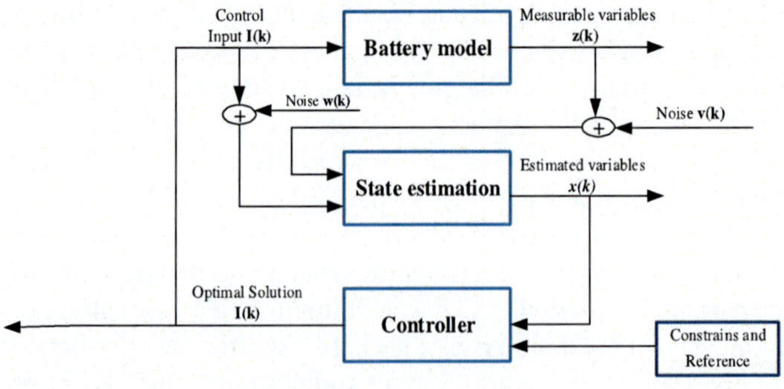

Fig. 1.15 Active optimal charging structure

states of the battery system. Based on the reduced-order battery model and state estimator, an active charging strategy is formulated with an optimal control algorithm. Meanwhile, for better charging performances, constraints of battery health-related variables (temperature rise, side reaction rate, and so on) need to be considered as outputs references (Zou et al. 2018).

The optimal charging targets are often related to the battery performances in two aspects, which are the temperature rise and the aging rate. Therefore, the optimal charging strategy is formulated considering battery performances and charging speed at the same time.

The battery aging-associated properties are selected from part of internal states. The overpotential of side reactions occurring at the cathode electrode or anode electrode and the solid electrolyte interface (SEI) growth are considered as side effects results. The electrolyte concentration is also viewed as an incentive to side effects. Considering the particle mechanical stress induced by diffusion, the lithium concentration at the surface of particles is limited no less than zero. Based on empirical equations between capacity loss and total discharge Ah throughout, the state of health (SOH) is selected as degradation representatives. The sharp temperature rise may lead to a battery thermal runaway. Hence, battery temperature control is taken into account along with aging prevention in most cases. To solve the optimal charging problem, it is necessary to formulate an effective charging strategy based on the battery model (Gao et al. 2020).

1.3 The Existing Charging Modes

In general, there are three charging approaches: conductive charging, inductive charging, and changing the depleted batteries mechanically. The conductive charging indicates that the onboard batteries are connected with a cable and plugged into an electricity provider. The cable can be fed from a standard electrical outlet or a charging station. The inductive method works through electromagnetic transmission without any contact between EV and the charging pile. There is the potential for 90–93% energy efficiency, but some energy loss could occur during the transfer. The third solution substitutes the discharged batteries with charged ones at the swapping station. However, this method requires the dimensions and standards of the equipped cells should be unified. At present, the conductive charging is preferred due to its low-cost and high-efficiency. The inductive charging method, also known as the wireless power transfer (WPT) technology, is becoming more popular with high-end EVs because it offers a better way to increase user comfort and could be a key feature for electric mobility.

In China, the current WPT technologies for EVs are still under exploration. Inter-operator charging has not been realized, and the application scenario, technical specifications as well as the business model, have not been made clear. Due to the Chinese Government's supporting policies, technical specifications, as well as the business model for WPT, are becoming more mature and the market share of wireless charging facilities is increasing (Yangbae 2016). For example, the Energy Industry Standard Committee sets up an interoperability test for WPT including five stages: test activity preparation, test platform preparation, product performance and safety test, interoperability bench test, and interoperability real vehicle test. Currently, they have entered the interoperability real vehicle test stage, and this test is expected to be completed in June 2020. Moreover, standards related to WPT are also in progress. At present, there are eight national standards for WPT systems of passenger cars involving general requirements, communication protocols, the electromagnetic environment, electromagnetic compatibility, and interoperability, etc.

The following two cases show the development and application of WPT for EVs in China.

1. ZTEV under ZTE Corporation has developed a series of WPT products with a high power ranging from 3 to 60 kW, which provide an automatic and free charging solution for intelligent connected vehicles (ICVs). In June 2012, ZTE developed the first-generation wireless charging prototype. In 2014, ZTEV launched the Smart Wireless Charging Program for urban buses and established wireless charging bus lines in more than 30 cities, including Chengdu, Xiangyang, Zhangjiakou, Dali, Lijiang, Guiyang, Changchun, Huidong, etc. Besides, ZTEV has also actively invested in the development of WPT systems for passenger vehicles. By far, it has developed 7/11 kW WPT systems for passenger vehicles and 30/60 kW for commercial vehicles.
2. In March 2016, Zonecharge demonstrated its wireless charging post of EVs for the first time. This 6.6 kW wireless charging post based on the magnetic coupling technology has a transmission distance of 20 cm and transmission efficiency of 90%. At the same time, another WPT system with a high power of 30 kW was displayed. Its transmission distance is up to 60 cm, and it enjoys a considerable degree of freedom in both the horizontal and vertical directions. At present, Zonecharge has developed a wireless charging solution with a maximum power of 30 kW and an efficiency of 85%.

In other countries, many companies and research institutes have also done a lot of research work in WPT for EVs. Some examples are listed below.

WiTricity has designed a 3.6–11 kW WPT system based on the widely used circular coil. It has provided a flexible system design for OEMs, Tier1, and EVSE after being tested by the U.S. Energy Information Administration (EIA). Besides, the DRIVE11 evaluation system is a set of reference designs developed by WiTricity for stop-and-charge wireless charging of EVs with a power of 11 kW and an efficiency of up to 94%. It also includes foreign object detection, living object detection, and position detection, which are applicable to vehicles with different chassis heights, including passenger cars, off-road vehicles, and light-duty trucks.

Bombardier's innovative PRIMOVE technology enables EVs to be charged in the dynamic state (dynamic charging) or static state (static charging), which can transfer energy efficiently and maintain the original performance of the vehicle. Its 22 kW PRIMOVE system has a maximum efficiency of 90%. From 2008 to 2009, a low floor tram equipped with PRIMOVE components and a test track were used to test the performance of the system under all typical operating conditions. In 2012, Bombardier cooperated with Brunswick and local transportation operators to transform part of the public transport network into an eco-friendly electric line, and put two buses equipped with the PRIMOVE system on a 12 km line for operation.

In 2014, Evatran cooperated with Toyota and General Motors to develop a WPT system for Evs, which became a third-party safety certified product in the United States. In 2016, Evatran developed the Plugless Power system with a power up to 7.2 kW to provide wireless charging for Tesla Model S. The efficiency of the 3.3 kW Plugless Power system is about 82.5%, the charging capacity per hour allows EVs to drive 20 km, and it only takes 10–12 h to fully charge Tesla Model S.

In May 2017, Qualcomm announced the dynamic WPT technology: Qualcomm Halo. The technology uses a multi-coil design, which enables efficient energy transmission even when the charging plate is not aligned with the vehicle. Qualcomm has a series of patented technologies and intellectual property rights and has actively participated in the formulation of various standards, including SAE, IEC, ISO, and GB standards. Qualcomm solutions have been integrated and tested on different vehicle platforms, including Renault Fluence, Nissan Leaf, BMW i3, BMW i8, and Honda Accord.

According to the abovementioned, it can be seen that WPT of EVs has been the research hotspot for experts and scholars at home and abroad in recent years. However, at present, the following problems remain to be solved (Qiu et al. 2013; Patil, et al. 2018; Lastname 2017).

(1) Short transmission distance: According to the relationship between the output power and the coupling coefficient, the increase of the coil distance leads to the decrease of the coupling coefficient, which causes the transmission efficiency to decrease and thus the output power decreases. The transmission distance limits the application scenario for EV wireless charging.
(2) Difficulty in control: For instance, a complex control system is needed in the adjustment of the operating state and circuit parameters of the power converter to achieve the effective control of the output power, which is difficult to design.
(3) High cost: To ensure a high transmission efficiency and reduce the magnetic flux leakage, it is necessary to use materials with high permeability and conduct the optimal design of the coil. This increases the complexity of circuit design. In addition, materials with high permeability are expensive, which is not conducive to commercialization.
(4) Concern for electromagnetic safety: According to ICNIRP 2010, exposure to an electromagnetic field beyond a certain limit may cause adverse health effects such as shocks and burns, elevated tissue temperatures and brain dysfunction which impairs the visual processing (a perception of faint flickering light in the periphery of the visual field) or motor coordination (stimulation of muscles). Besides, the central and peripheral nerves may be affected if the electromagnetic field exceeds the basic restrictions. More importantly, even lower than the limit, the electromagnetic field may interfere with medical electronic equipment and devices such as cardiac pacemakers.

(5) Sensitivity to change of system parameters: When the parameters of the system change, such as the self-inductance of the resonance coil, the equivalent capacitance, the distance between the resonance coils and the horizontal displacement of the coils, etc., the transmission characteristics of the system will greatly change. When displacement between coils occurs, the transmission efficiency will drop sharply and the coupling coefficient will decrease rapidly. A good impedance matching system is needed to ensure the transmission efficiency.

1.4 Prospects of Wireless Power Transfer Technologies

With the unceasingly thorough research to tackle the abovementioned problems one by one, the WPT technologies for EVs have a promising development prospect. By integrating WPT systems into the EV charging infrastructure, the future EVs can adopt the non-contact charging mode, which plays an important role in promoting the popularization and industrial application of new energy vehicles.

(1) The multi-layer wireless charging parking lot.

A patented conceptual EV charging station for urban development, called Charging Tower, was designed by Ennead Lab (Fernandes 2017; Shen et al. 2019). This urban roadside beacon for electric cars adopts both the stack-parking and wireless charging technologies to achieve a maximum number of charging cars and be pedestrian-friendly.

Compared with the traditional gas stations, EVs are environmentally friendly, but the charging time of EVs is usually longer than five hours, so the car owners are forced to wait. In this situation, the Charging Tower emerges. Charging Tower is divided into nine floors, which allows EVs to be parked in a multi-layer stacking way, which is much more efficient. The nine-story design of Charging Tower makes better use of the space in densely populated cities and can charge more EVs in a relatively small space at the same time. Therefore, the waiting time for car owners can also be greatly reduced.

Charging Tower not only allows more EVs to be charged at the same time but also provides the car owners with two charging options according to their own needs: standard charging and supercharging. Supercharging is much faster than standard charging, and its stations are housed at the ground level, while the standard charging stations are located at higher levels.

In the future, Charging Tower is expected to replace the existing gas stations and charging stations, so that the entire charging process can be completed when the EV is parked on the side of the road. Due to the two charging options provided by Charging Tower, for the car owners who choose standard charging, they can go to other places and have some fun during charging. For those who chose to supercharge their vehicles, Charging Tower is more like a traditional gas station, which enables them to go back on the road immediately after fast charging.

(2) Automatic parking and driving based on wireless charging.

At the current stage, if the EV needs to be charged, someone must plug the charging gun into the vehicle. But most of the charging posts are unattended now. If there is no one in the vehicle, how can the EV complete the charging process by itself? The answer to the problem of "unmanned vehicle and unattended post" is wireless charging. Wireless charging is the fundamental technology of intelligent EVs.

Undoubtedly, the WPT system simplifies the charging process of EVs. For static charging, simply park the car in the garage, and the car can automatically receive power from the primary side under the car; for dynamic charging, the car can be charged while running, and the endurance mileage is no longer a problem for EVs.

We are moving towards the future of automatic driving and intelligent cars and wireless charging of EVs is indispensable. With the combination of the WPT and automated driving technologies, we can easily realize the connection between the unmanned vehicle and the unattended post so that the car can be charged automatically for a long time. Furthermore, the combined technologies can automatically free up charging space for the other vehicles after one vehicle is fully charged, so that the charging port can be fully utilized, thus increasing efficiency and revenue.

The combination of automatic parking and driving with wireless charging can bring perfect experience to the owners of EVs. However, a completely driverless EV is a long-term goal and requires WPT technologies to pave the way for it.

(3) Automatic access to an energy Internet.

Due to the mobility of EVs, wireless charging will become the key to the automatic connection of EVs to the energy Internet. With the support of the bidirectional flow technology of electricity, we can realize the automatic bidirectional wireless charging and discharging for EVs, making the energy flow bi-directionally between the EV and the power grid, the EV becomes a highly autonomous mobile energy storage equipment and energy node. At the same time, under the new energy micro-grid system, wireless charging and discharging technology can achieve the safe and orderly charging and discharging of EVs and automatically participate in the micro-grid energy regulation. In this way, economic charging and grid coordination are promoted, EVs become an important part of the new energy interactive management system.

When the WPT technology of EVs is integrated with the new energy micro-network system, the EVs can autonomously store electricity during the off-peak period and sell electricity during the peak period of charging through the intelligent operation and maintenance of massive microsensors, helping the owners to pocket the difference. The whole process is automatically completed under the intelligent control of the system without human intervention, which is equivalent to the realization of "self-profit" by EVs.

Looking ahead, we can see that intelligent connected vehicles are trending. They can be traced back to the 1920s. In the 1940s, designer Norman Bel first proposed the systematic concept of autonomous vehicles. In the 1980s, autonomous vehicles entered substantial development at this stage, E. Dickmanns promoted the development of driverless vehicles and played a cornerstone role in the development of autonomous vehicles; in the 1990s, the Prometheus project laid the foundation for the development of autonomous vehicles in Europe. Until now, many autonomous projects are still based on this project. Also, in the 1990s, Carnegie Mellon University launched the Navlab series of autonomous vehicles based on its Navlab-1 prototype, and achieved tens of thousands of kilometers of road tests and completed crossing the United States. In this test, a total of 5000 km were traveled, of which 98% of the mileage was completed autonomously by autonomous vehicles.

In 2004, the Autonomous Driving Challenge Race launched by the U.S. Defense Advanced Research Projects Agency set off a new climax in the development of autonomous driving. The road segment was a 150-mile artificially designed road, but no team finished the competition. In the second competition in 2005, 5 of the 23 teams that successfully completed the competition. The previous two competition scenes were dominated by high-speed, desert, and other roads. The third autonomous driving challenge in 2007 was renamed the Unmanned City Challenge. The road environment is a typical urban road. A total of 6 teams completed the competition, verifying the feasibility of autonomous vehicles in the complex urban environment. Since then, the development of autonomous vehicles has entered the fast lane, and there has been a boom in driverless development worldwide (Buehler et al. 2009).

With the development of artificial intelligence technology and advanced sensing control technology, Intelligent Connected Vehicle (ICV) has achieved great development. The development of ICV has two technical paths: autonomous intelligence and networked connection, and finally, toward the integration of autonomous intelligence and networked connection. For example, the U.S. Department of Transportation in the "ITS Strategic Planning 2015–2019" clearly proposed that autonomous intelligent vehicles (Autonomous Vehicle) and connected vehicles (Connected Vehicle) are the two main paths to achieve autonomous driving, as shown in Fig. 1.16.

Internet of Vehicle (IOV) is based on the local area network (LAN), vehicle internetwork, and onboard internet (5G, etc.) according to the agreed communication protocol and data interaction standards. IOV communicates and exchanges information between V2X (X stands for the vehicle, road, pedestrian and Internet, etc.), realizing intelligent traffic management, intelligent dynamic information service, and intelligent vehicle control (Guerreroibanez et al. 2015).

Based on the information communication technology, the connected vehicles industry provides comprehensive information services through telematics, as well as V2V, V2I, V2P, and V2X connection and data interaction.

This new form of industry is a deep integration of the automotive, electronic information communication, and road transportation industries (Fig. 1.17).

Fig. 1.16 Three technical paths of the realization of Intelligent Vehicle

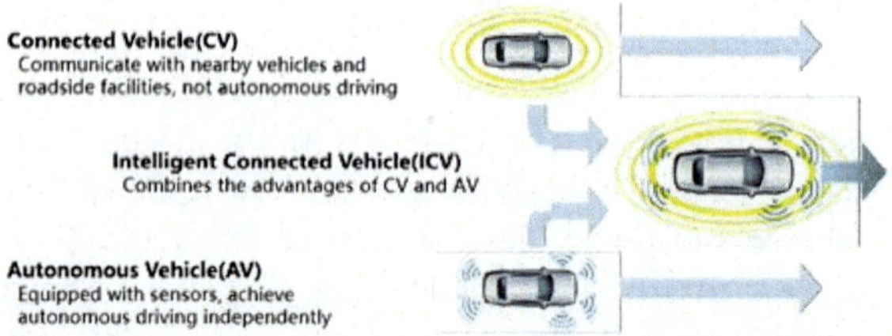

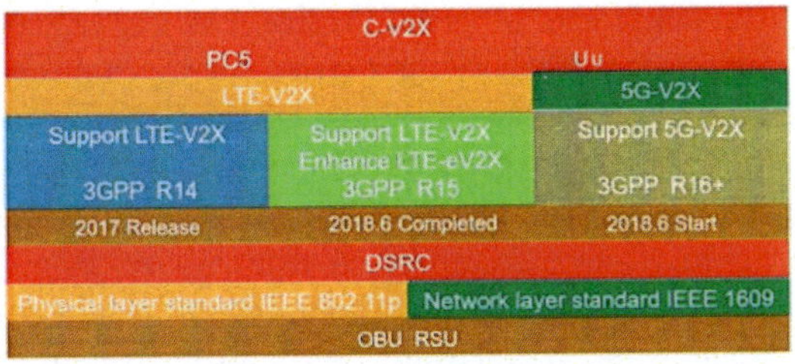

Fig. 1.17 Structure of V2X. *Source* IMT-2020(5G) Advancement Team

For autonomous vehicle (AV), that is equipped with advanced sensors, controllers, actuators, and integrates the theory of modern communication, information network, automatic control, and artificial intelligence.

It realizes the intelligent information exchange, solving problems of collaborative control and execution of intelligent decision-making in complex environment perception, and finally realizes safe, comfortable, and efficient driving, which can completely or partially replace the drivers' operations.

According to the degree of intelligence, SAE divides it into 6 levels. The higher the level, the higher the degree of automation. From low to high: no automation, driving assistance, partial automation, conditional automation, full automation (SAE International 2014) (Fig. 1.18).

From the technical point of view, autonomous vehicle (AV) consists of three main parts: perception and location, decision-making and planning, execution and control, and other functions in one integrated system (Paden et al. 2016).

Perception and location:

Lidar (Light Detection and Ranging): A sensor for accurately obtaining three-dimensional position information, which uses the laser's emission and reception, and

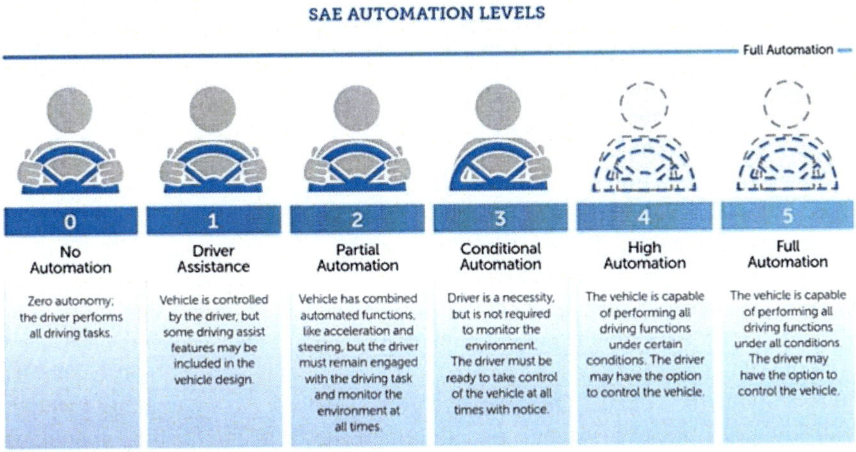

Fig. 1.18 SAE automation levels. *Source* SAE J3016-201,401, Taxonomy and definitions for terms related toon-road motor vehicle automated driving systems, New York, SAE International, 2014

can determine the position, size, exterior topography, and even the material of the object. It consists of three parts: the transmitting system, the receiving system, and information processing.

Laser radars' transmitters emit multiple laser beams. After the laser beam encounters the object, it is diffusely reflected and returned to the laser receiver to form a "point cloud." Commonly 4-wire, 8-wire, 16-wire, 32-wire, 64-line laser radar.

Advantages of Lidar: high angular resolution, high distance resolution, a wide range of speed measurement, and long measuring distance, ability to accurately track targets.

Camera: Compared with other sensors, the camera information is more abundant, and it is applied in the fields of scene reconstruction, pedestrian, and vehicle identification. Currently, a large number of startups carry out the research and application of intelligent connected vehicle techniques around the vision system.

According to the number of lenses, it can be divided into a monocular camera and binocular camera:

Monocular camera: small volume, low calculation requirement, but inaccurate ranging;

Binocular camera: large volume, high calculation requirement, but high ranging precision.

According to the camera function, in addition to the ordinary camera, there are fisheye camera and infrared camera:

Fisheye camera: The structure is modeled after fish eyes and has a wide imaging range;

Infrared camera: Infrared camera can image at night by collecting infrared light.

The car camera has been widely used in smart cars due to its low cost, rich imaging color, and small size. It has an irreplaceable position in applications such as lane recognition and traffic signal recognition.

Radar: Millimeter-wave (MMW) radar transmits FM continuous wave (FMCW) outward through the antenna, receives the reflected signal of the target, analyzes the difference in frequency, calculates the relative distance of obstacles, and uses Doppler mainly to calculate the relative speed of obstacles, to obtain the physical environment information around the car body quickly and accurately.

INS/GNSS (Global Navigation Satellite System): Refers to all satellite navigation systems, including global, regional, and enhanced, such as GPS (USA), Glonass (Russia), Galileo (Europe) and Beidou (China).

At present, most of the mass-produced passenger cars mainly focus on driving assistance, which is mainly Level1-Level2 in the classification of automatic driving levels. The driving assistance function is mainly aimed at an actuator of the vehicle and collects information through onboard sensors, and then generates instructions to drive the actuator to assist drivers to complete some actions, thereby ensuring the safety of the vehicle. Onboard driving assistance functions mainly include an adaptive cruise control system, automatic emergency braking, blind-spot monitoring system, lane departure warning system, and panoramic parking assistance system.

According to the aforementioned, we can see that the development of ICVs has a long way to go, and the WPT technology is indispensable for the complete realization of ICVs.

It is worth mentioning that there are many ways to realize WPT of electric energy, such as the inductive power transfer, the conductive power transfer, etc., but the conductive power transfer system has poor waterproof ability due to its inherent configuration, which has little practical application value. Therefore, at present, inductive power transfer is the most suitable technology for EV wireless charging. In this book, if not specified, WPT refers to the inductive power transfer technology.

References

Buehler M, Iagnemma K, Singh S (2009) The Darpa urban challenge: autonomous vehicles in city traffic

Fernandes M (2017) Charging tower

Gao Y, Zhang X, Guo B et al (2020) Health-aware multiobjective optimal charging strategy with coupled electrochemical-thermal-aging model for lithium-ion battery. IEEE Trans Industr Inf 5:3417–3429

Gao Y, Zhang X, Cheng Q et al (2019) Classification and review of the charging strategies for commercial lithium-ion batteries. IEEE Access 43511–43524

Goodenough JB, Mizuchima K (1981) Electrochemical cell with new fast ion conductors

Guerreroibanez JA, Zeadally S, Contrerascastillo J (2015) Integration challenges of intelligent transportation systems with connected vehicle, cloud computing, and internet of things technologies. IEEE Wirel Commun 6:122–128

Haering RR, Stiles JAR, Brandt K (1980) Lithium molybdenum disulphide battery cathode

Jian L (2018) Research status and development prospect of electric vehicles based on a hub motor

Li S, Zhao W, Cui X et al (2014) Lithium difluoro(sulfato)borate as a novel electrolyte salt for high-temperature lithium-ion batteries. Electrochimica Acta 327–333

Lin Q, Wang J, Xiong R et al (2019) Towards a smarter battery management system: a critical review on optimal charging methods of lithium ion batteries. Energy 220–234

Lu X, Wang P, Niyato D et al (2015) Wireless charging technologies: fundamentals, standards, and network applications. IEEE Commun Surv Tutorials 2:1413–1452

Mizushima K et al (1980) $LixCoO_2$ $(0<x<-1)$: A new cathode material for batteries of high energy density. Mater Res Bull

Nishi Y (2001) The development of lithium ion secondary batteries. Chem Rec 5:406–413

Paden B, Čáp M, Yong SZ et al (2016) A survey of motion planning and control techniques for self-driving urban vehicles. IEEE Trans Intelligent Vehicles 1:33–55

Patil D, Mcdonough et al (2018) Wireless power transfer for vehicular applications: overview and challenges. Curr Forestry Rep

Qiu C, Chau KT, Liu C et al (2013) Overview of wireless power transfer for electric vehicle charging

Rim CT, Mi C (2017) Wireless power transfer for electric vehicles and mobile devices. Introduction to static charging, John Wiley & Sons, Ltd

SAE International (2014) Taxonomy and definitions for terms related to on-road motor vehicle automated driving systems. SAE International. https://doi.org/10.4271/J3016_201401

Shen P, Liu Q, Liu G et al (2019) Robust reconfiguration of distribution network considering the access of photovoltaic electric vehicle charging tower

Yangbae C (2016) International standards for wpt ev. Auto J Korean Soc Autom Eng

Yin L, Liu J, Yang H et al (2011) The status and development trend of hydraulic hybrid vehicle

Yoshino A, Sanechika K, Nakajima T (1987) Secondary battery

Zou C, Hu X, Wei Z et al (2018) Electrochemical estimation and control for lithium-ion battery health-aware fast charging. IEEE Trans Industr Electron 8:6635–6645

Chapter 2
Basic Concepts of Static/Dynamic Wireless Power Transfer for Electric Vehicles

2.1 Basic Components of WPT System

In general, a WPT system for EVs is composed of two separate parts: primary and secondary sides, as shown in Fig. 2.1. The primary side is composed of an AC/DC power source, a primary power converter, primary compensators, a primary controller, and a transmitting device such as a coil (for inductive power transfer) or metal plate (for capacitive power transfer). The secondary side is composed of a receiving device (coil or metal plate), secondary compensators, a secondary controller, a secondary power converter, and a DC load (battery). A communication link is also included to transfer the control commands and system status between the primary and secondary sides. A protection module, including foreign object detection (FOD) and living objection detection (LOD), is also essential in the WPT system for safety.

The primary AC/DC power source is implemented to obtain the charging power from the utility grid. Due to the grid-interface regulations, the power-factor-correction (PFC) should be provided by the AC/DC power source to realize the unity power factor and low-harmonic distortion at the input side. The primary converter consists of a high-frequency (HF) inverter (for DC–AC power conversion) and an optional DC/DC converter for charging power regulation. Some WPT configurations regulate the charging current or voltage at the secondary side so that the DC/DC converter at the primary side is not always equipped. The HF inverter at the primary side is essential for the consideration of the size and efficiency of the transmitting/receiving device. The HF inverter operates at a soft-switching state, in which either the voltage or current of each switch becomes nearly zero when turns on or off, to reduce the power losses and to guarantee the reliability.

The secondary power converter consists of an HF rectifier and an optional DC/DC converter. The HF rectifier converts the AC voltage to DC voltage for supplying the DC load, e.g., the traction battery in EVs. The HF rectifier can either be passive or active-controlled, depending on whether the bidirectional power flow is needed. In

X. Zhang et al., *Wireless Power Transfer Technologies for Electric Vehicles*, Key Technologies on New Energy Vehicles,
https://doi.org/10.1007/978-981-16-8348-0_2

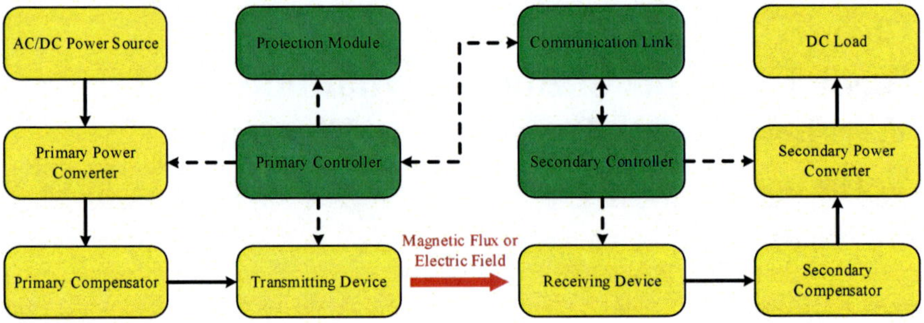

Fig. 2.1 Basic Components in the WPT system for EVs

some cases, the DC/DC converter at the secondary side can regulate the charging power by adjusting the DC voltage at the output side of the HF rectifier, such that the primary DC/DC converter can be avoided.

The primary and secondary compensators are usually resonant LC circuits for higher power efficiency. Since the distance between the transmitting and receiving devices for EV applications is usually 15–30 cm, the coupling effect between the two devices is relatively low. The primary and secondary compensators play important roles in eliminating the considerable reactive power of the HF inverter for the magnetic or electric field excitation between the transmitting and receiving side. Hence, the power factor of the primary HF inverter can be improved for higher transmitting efficiency. Meanwhile, by applying different combinations of resonant inductors and capacitors, either the constant current (CC) or the constant voltage (CV) output characteristic can be achieved in the WPT system, which greatly facilitates the battery charger control for EVs.

The wireless communication links between the primary and secondary sides exchange information such as output voltage, load current, operating status, and temperatures of components for the control of converters. The control commands, such as charging start/finish and cut-off for protection, are also transmitted via the communication links. The most common communication links are RF communication, including Bluetooth, 433M, ZigBee, and so on. Another approach is to use the magnetic or electric coupling communication by modulating the output or input voltage and current for communication.

The primary and secondary controllers are used to regulate the input and output power in WPT systems. The DC-link voltages and currents at the primary and secondary sides are monitored, whose information is transferred via wireless communication links between each side for controllers. Because the power components in WPT systems are separated at primary and secondary sides, the system status, such as no load, open circuit, short circuit, over voltage, under voltage, transient overshoot, in-rush current, and FOD/LOD signals, require highly reliable and low-latency communication links for real-time control. Despite difficulties and inefficiencies in removing the communication links, it would be ideal if all the charging power control and protection can be operated on one side. In the later chapters of this book, some

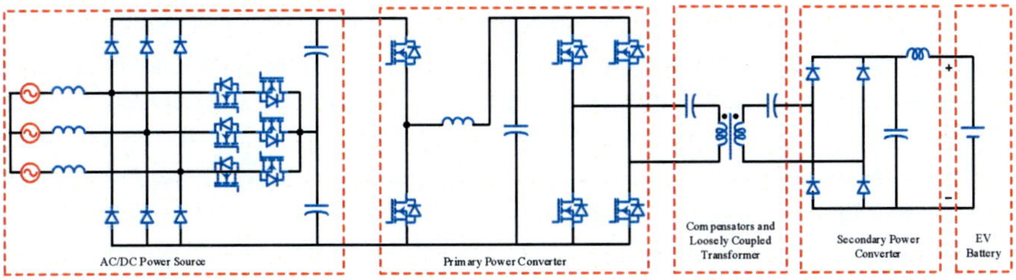

Fig. 2.2 Typical IPT system for EV applications

innovative design can realize a very simple power control strategy, in which the cumbersome communication is not needed during CV charging mode.

The transmitting and receiving devices are designed to transfer large power efficiently regardless of vertical and lateral misalignment between them. For the design of coils (for IPT) and metal plates (for CPT), mutual couplings of the magnetic flux and electric field, respectively, should be maximized for the given primary and secondary device dimensions. Another design consideration for the magnetic or electric couplers is that the mutual couplings can keep the same values when suffering lateral misalignment, which usually happens at parking spaces in EV applications. The structure and topology design for the magnetic couplers will be discussed in later chapters to alleviate the coupling variations with lateral misalignment of the primary and secondary coils.

The configuration of a typical IPT system for EV applications is shown in Fig. 2.2. The primary and secondary converters usually operate at 80.0–90.0 kHz according to the Standard SAE J2954 for light-duty EV/HEVs. An optional Buck converter is included in the primary power converter stage for regulating the charging power and realizing soft-start. The transmitting and receiving coils constitute a loosely coupled transformer, whose coupling coefficient depends on the vertical distance and horizontal misalignment between the coils. Ferrite cores are typically implemented in transmitting and receiving coils to increase the magnetic coupling and reduce the EMI radiation, which inevitably introduces iron losses and increases the system mass. The HF diode rectifier is the most common power stage to converter the AC voltage to DC voltage for the battery charging of EV/HEVs. In some cases (such as V2G), the diode rectifier can be replaced by the dual active bridge (DAB) topology to realize the bidirectional wireless power transfer between the vehicle and grid.

2.2 Basic Physical Principles of Wireless Power Transfer

In this section, noncontact electricity transmission methods are discussed. Based on the transmission principles, WPT systems could be classified as electromagnetic induction, magnetic resonance, electric coupling, electrical resonance, and microwaves. Based on the differences of the transmitter/receiver, WPT systems could be classified as inductive coupling, capacitive coupling, magneto-dynamic coupling, microwaves, and lasers. Based on the transmission distance, WPT systems could be classified as near-field techniques and far-field techniques. Near-field is also called non-radiative region, in this region, the energy flows back and forth periodically between the radiation source and the space around the radiation source. Far-field is also called the radiative region, energy in this region is separated from the radiator and emitted outward in the form of electromagnetic waves. Among them, inductive coupling and capacitive coupling are near-field non-radiative techniques, microwaves and lasers are far-field radiative techniques.

2.2.1 Electromagnetic Induction

The most common electromagnetic induction based device is transformer, where power is transferred between mutually isolated coils through the magnetic field. Structures of the electromagnetic induction based WPT system and the loosely coupled transformer are shown in Figs. 2.3 and 2.4. According to Ampere's theorem, the alternating current passing through L_1 generates an oscillating magnetic field in the silicon steel sheet or magnetic core. Then the changing magnetic field passes through the receiving coil and generates an AC potential according to Faraday's induction law, thereby generating an AC current in the receiving coil. The AC current generated can directly drive an AC load or be converted to a DC current by the rectifier to drive a DC load. The phenomenon of electromagnetic induction could be explained by Eqs. 3 and 4 of the MAXWELL equations,

$$\nabla \times H = J + \frac{\partial D}{\partial t} \tag{2.1}$$

$$\nabla \times E = -\frac{\partial B}{\partial t} \tag{2.2}$$

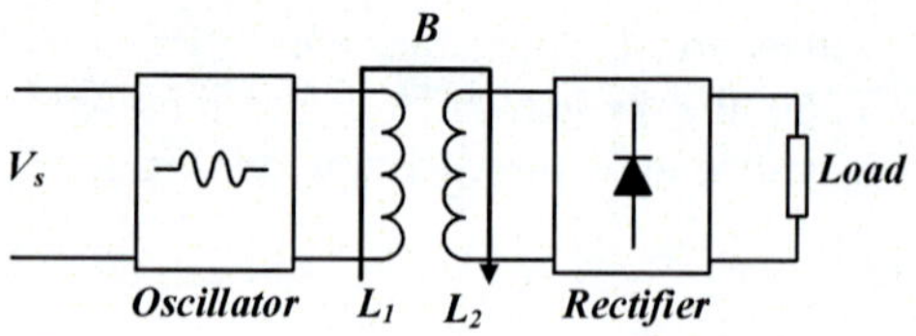

Fig. 2.3 The structure of the electromagnetic induction based WPT system

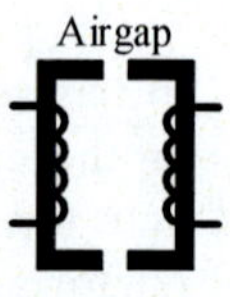

Fig. 2.4 The structure of the loosely coupled transformer

Equation (2.1) shows that the changing electric field generates the magnetic field, and Eq. (2.2) shows that the changing magnetic field generates the electric field.

Electromagnetic induction is the oldest and most widely used WPT technology. It has been widely used in commercial products such as electric toothbrushes and razors to reduce the risk of electric shock. Meanwhile, electromagnetic induction based WPT systems have broad prospects in biomedicine, electric vehicles, portable and handheld devices such as mobile phones and laptops.

The transmission power increases as the frequency and the mutual inductance of the coils increase. The concept of mutual inductance characterizes the magnetic flux generated by the transmitter coil L_1 passing through the receiving coil L_2. If L_1 and L_2 are aligned, and the distance is very close, the magnetic flux generated by L_1 completely passes through L_2, then the mutual inductance reaches its maximum value, and the wireless transmission loss is zero. The coupling coefficient $k = M/\sqrt{L_1 L_2}$ is usually used to represent the coupling of the coils. To achieve efficient and high power transmission, the coils must be very close. The value of the mutual inductance also depends on the geometry of the coils and the magnetic cores. Wide, flat coil shapes and bipolar structures are usually used to increase the coupling coefficient. The ferrite could effectively confine the magnetic field, reduce the divergence loss of the magnetic field lines, and improve the coupling coefficient.

In addition, the center of the inductive power transfer (IPT) system is the loosely coupled transformer. A large leakage inductance of the loosely coupled transformer may generate high reactive power, thus cause low efficiency and poor power transmission capability in the circuit parts of the system. Therefore, compensation circuits should be added to reduce the apparent power requirement of the power supply. Commonly used compensation circuits include SS, SP, PS, PP, LCL, LCC, etc. With compensation circuits, the input impedance of the system is resistive, which enables the system to operate at zero phase angle (ZPA) condition.

2.2.2 *Magnetic Resonance*

Traditional electromagnetic induction based WPT systems have a short transmission distance for high efficiency and stable power. In modern IPT systems, magnetic resonance is used to improve the transmission efficiency and distance. Magnetic resonance-based WPT system was first proposed by Tesla one century ago, and the experimental device of the wireless power supply is shown in Fig. 2.5 (Lee and Zhong 2012; Marincic 1982). The transmitter is composed of a high-frequency AC power supply, a transformer, a transmitting coil P, a spark gap switch S.G, and a capacitor C; the receiver is composed of a receiving coil S and a 40 w bulb. It was found that the usage of an L–C resonant circuit at the primary side could light the bulb at the secondary side. If an L–C resonant circuit with the same frequency is added at the secondary side, the distance to light the bulb could be increased, but the distance is still quite close (Wheeler 1943). In 2007, Prof. Marin Soljacic from MIT proposed the concept of magnetic field resonance, which makes the transmitter and receiver

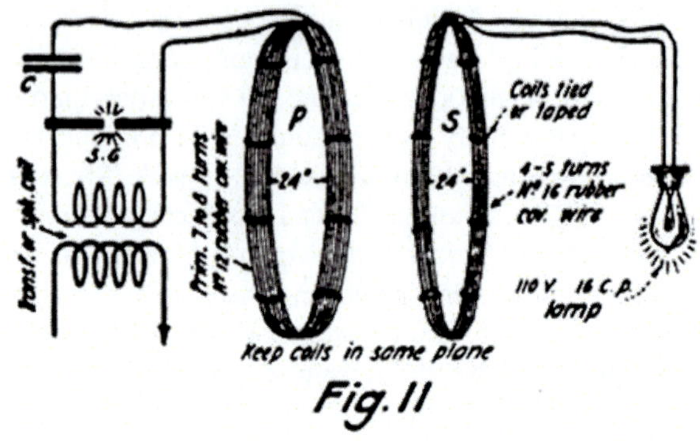

Fig. 2.5 The experimental device of wireless power supply proposed by Tesla (Lee and Zhong 2012)

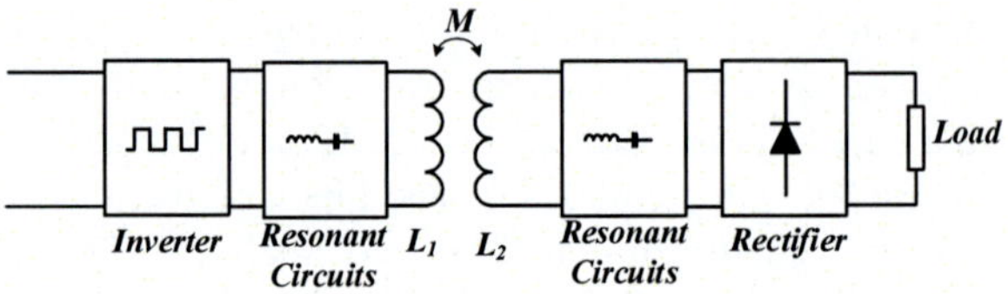

Fig. 2.6 The structure of the magnetic resonance-based WPT system

to resonate at the same frequency so that they can effectively exchange the energy at relatively long distance. The experiment done by his research team showed that two copper coils with the same design (coil radius is 25 cm) lighted a 60 W bulb at the distance of 2 m (8 times the coil radius) at 9.9 MHz, and the system efficiency is about 40% (Kurs et al. 2007). His results highly extend the transmission range of IPT systems (Fig. 2.6).

This phenomenon skillfully takes advantage of the "strong-coupled" regime in coupled-mode theory. At the evanescent (non-lossy) stationary near-field, high Q in the resonators leads to a much higher energy-transfer rate than intrinsic-loss due to internal damping. Thus, compared with the magnetic induction, magnetic resonance could deliver the same amount of the power with farther distance and weaker magnetic fields at fairly high efficiency. Here, the range up to 4–10 times of the coil diameter is called 'mid-range', and the range up to about the coil diameter is called 'short-range'. Magnetic resonance could utilize the sub-wavelength size of the resonant coils to effectively transfer energy, while magnetic induction could only utilize wavelength size of coils to transfer power. Therefore, magnetic resonance is a mid-range transfer method, while magnetic induction is a short-range transfer method (Karalis et al. 2008).

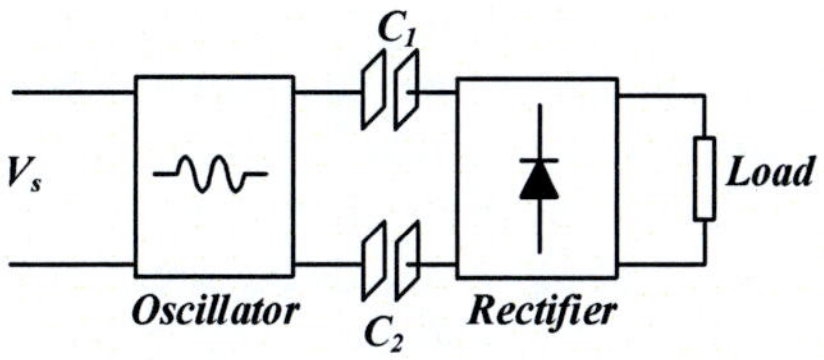

Fig. 2.7 The structure of the electric coupling based WPT system

2.2.3 *Electric Coupling*

Electric coupling is a method that directly uses the electric fields to transfer energy. As shown in Fig. 2.7, there are two transmitter plates and two receiver plates between the oscillator and the rectifier. Each transmitter plate is aligned with a receiver plate with the airgap. Therefore, two capacitors (C_1 and C_2) are formed, and the air is the dielectric. If the oscillator supplies high-frequency alternating voltage on the transmitter plate of C_1 and C_2, the same frequency alternating potential is induced through the oscillating electric field based on the electrostatic induction. It should be noted that the voltage phases applied on the transmitters of C_1 and C_2 should be inverted, and the phase of the induced alternating potentials in the receiver plate is also opposite with each transmitter plate. The transmitted power is proportional to the square of the voltage and the capacitance. The capacitance is relative to the area of the plate and the distance between the transmitter plate and the receiver plate.

Compared with magnetic coupling methods, the electric coupling could transfer lower power and may cause high voltage electric shock and noxious ozone production. In practice, the electric coupling has been applied in some low power applications such as portable devices, biomedical implants, and integrated circuits.

2.2.4 *Microwaves*

Microwave power transfer (MPT) is a method that uses microwaves to transfer energy. The microwave generator is utilized to transmit power to the free space and the power is received by the rectenna and converted to the DC voltage to supply the load. 2.45 GHz microwave is primarily used, and this method belongs to the far-field technique. The applications of this method include solar power satellite, drone aircraft, and some wireless devices, but this method is still limited by the low efficiency and power.

2.3 The Static Wireless Power Transfer System

Based on the state of the EVs during the charging process, the wireless charging system could be divided into two types: static wireless power transfer (SWPT) and dynamic wireless power transfer (DWPT). SWPT is suitable for parking lots, shopping malls, residential areas, and other applications. DWPT is to charge the vehicle while the car is running, so it allows electric vehicles to be equipped with smaller-capacity batteries. They all use the principle of magnetic field coupling, the circuit design of the system is the same, but the design of the magnetic circuit is different.

SWPT for EVs can be divided into the electric field based power transfer system and the magnetic field coupling based power transfer system according to the transmission mechanism, which is also known as the inductive power transfer (IPT) system and the capacitive power transfer (CPT) system, and IPT is widely applied on SWPT for EVs rather than CPT.

2.3.1 Working Principles of Inductive Power Transfer

An IPT system is composed of the primary side and the secondary side, as shown in Fig. 2.8. The primary side is placed on the ground, which is composed of an AC or DC power source, converters, the primary resonant circuit, and a primary coil (IPT) or metal plate (CPT). The secondary side is composed of a secondary coil (IPT) or metal plate (CPT), the secondary resonant circuit, secondary converters, and the DC or AC load, and the secondary side is placed in the electric vehicles (EVs). The two parts are magnetically coupled or electrical coupled through the coils or plates.

For an IPT system, the primary converters consist of a power factor correction (PFC), and an inverter operating normally at several kHz to several MHz for the high power and efficiency. PFC realizes the conversion of the low-voltage AC power from the power grid to the high-voltage DC power. Then this high-voltage DC power is converted into AC power of a prescribed frequency by the high-frequency inverter, and the AC power is injected into the primary compensation circuit and the coil; thus the power is transferred to the secondary sides by inductive coupling between two coils. Then the secondary high-frequency rectifier converts the AC power to the DC power, and a DC-DC converter can adjust the DC current or DC voltage to

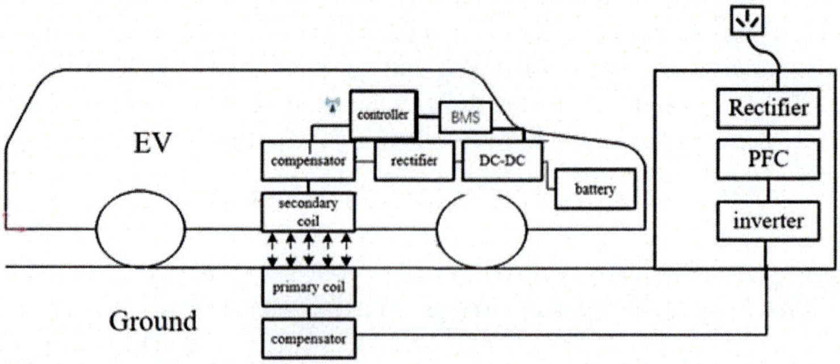

Fig. 2.8 Overall configuration of WPT

a manner suitable for charging according to the state of the battery. Above is the general working principle for an IPT system.

2.3.2 Compensation Circuits and Coils

From the working principle we can know that IPT transfers the power by inductive coupling of the coils having large air gap. So the efficiency of IPT will be lower when the air gap between primary and secondary coils is increased, and a compensation circuit is introduced to compensate the leakage flux due to the air gap.

The primary and secondary compensation circuit are mostly lossless *L–C* circuits in the WPT system. They play the roles of reducing the reactive power caused by the loosely coupling of the coils between the ground and the vehicle, and reducing the harmonic current and voltage. By applying compensation circuits, the power factor of the charging system and the transmission efficiency can be improved dramatically. Currently, various compensation circuits varying from the simplest basic compensation circuits to some derived compensation circuits like LCC and SPS.

The simplest compensation circuits use just two capacitors, connected in series or parallel with the primary and secondary coils. Thus four different topologies can be obtained, such as series-series (SS), series-parallel (SP), parallel-series (PS), and parallel-parallel (PP) compensation circuits as shown in Fig. 2.9. By designing capacitance parameters under different compensation methods, all topologies can behave as a current source or voltage source. Among all, the S–S topologies are preferred for EV applications due to the variable coupling coefficient in practical application (Sallan et al. 2009). More analysis is shown in specialized chapters, so they are not further studied in this chapter.

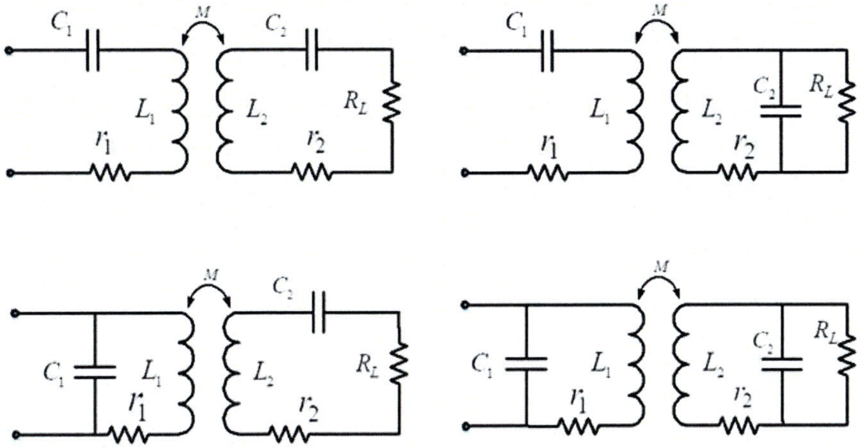

Fig. 2.9 Four compensation circuits with two capacitors

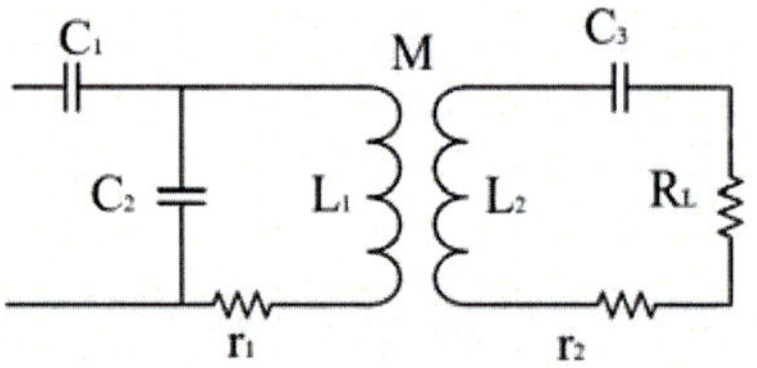

Fig. 2.10 SPS compensation topology

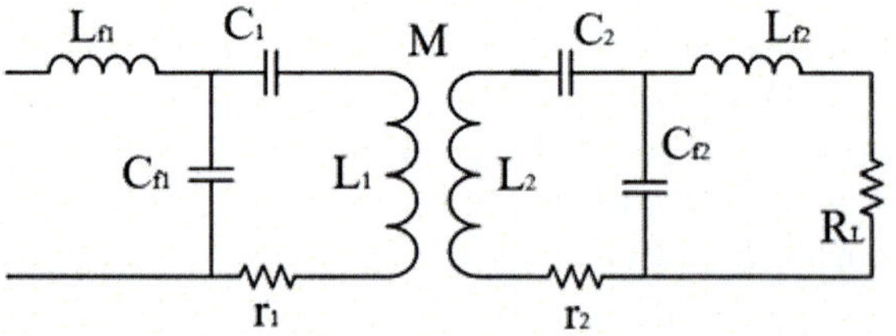

Fig. 2.11 LCC compensation topology

Apart from the compensation topologies above, some derived compensation circuits have also been proposed. One example is the SPS compensation circuit, as shown in Fig. 2.10. In SPS compensation topology, another capacitor is applied, and compared with the conventional topologies. This can perform better in case of the misalignment between the primary and secondary coils. Another category of compensation topology is LCC, with an inductor together with two capacitors, as shown in Fig. 2.11. These topologies also have different compensation methods, leading to quite different characteristics. They are described in detail in later chapters.

As to coils for SWPT, circular and square coils are often used for the transmitting and receiving coils in the magnetic circuit structure of the static wireless charging system. Currently, double-D (DD) and double-D-quadrature (DDQ) are the mainstream coil design shapes. The electric vehicle is stationary during charging, so the coupling coefficient is constant during the charging process.

2.3.3 Configuration to Capacitive Power Transfer

The configuration of a CPT system is similar to IPT except for the primary and secondary plates which are used to form capacitors to transfer power. The CPT system utilizes electric fields, instead of magnetic fields in IPT to transfer power. However, CPT systems mainly focus on short-distance (around 1 mm) applications, the transfer distance is far less than the ground clearance of EVs, limiting its application on SWPT for EVs. Therefore, CPT is not analyzed in detail in the book.

2.3.4 Research Status on SWPT

Many institutions, companies, and universities have committed to the surveying of SWPT. By now, prototypes of SWPT have been built: a team of KAIST realizes the transmission power of 6.6 kW at a distance of 20 cm in 2014, and the overall efficiency can reach 95.57% (Moon et al. 2014); a team of Qualcomm realizes the design of a 3.3 to 20 kW system with the efficiency over 90% (Qualcomm 2020); a team of Chongqing University realizes a 10 kW wireless power system by using D–D coils (Hu 2015); Evatran develops the PLUGLESS wireless power transfer system, which can transfer the power of 3.6 kW and 7.2 kW (EVATRAN 2020); and the world's first car equipped with wireless charging function-BMW 530e iPerformance has also been launched in 2018 (Witricity 2020). SWPT is developing quickly and of a promising future.

2.3.5 Discussions for SWPT

By now, IPT is the mostly analyzed method for SWPT, which can transfer large power with high efficiency. Compared with the traditional plug-in charging method, SWPT has many advantages:

1. Stronger environment adaptability: traditional plug-in charging method needs to plug and unplug cables frequently. The devices may age faster, and in thunderstorms and humid weather, it may not be convenient to operate. While for SWPT, the system is sealed in the shell, so it is of higher reliability.
2. Higher space utilization: since the SWPT system is buried or half-buried under the ground, so more devices can be installed under the same construction area, which is suitable for large-scale cluster construction.
3. Safer SWPT system has no exposed conductive points, and there is no need for people to touch the charging equipment. So it can eliminate the safety problems caused by leakage of electricity and protect personal safety.

However, there are still some problems we need to pay attention to and study further:

1. The coil misalignment: The survey indicates that if the primary and secondary coils offset, the transfer power will decrease significantly.
2. Lack of interoperability: Since the batteries on the market are different, and the ground clearance of different cars is also different, these parameters will affect the design of the system. So the design of the general SWPT system is essential.
3. Foreign object detection (FOD) design: When the WPT system is working, a high-frequency alternating magnetic field is excited in the area around the transmission coils. If objects containing metals fall into the magnetic field, eddy current is generated to heat the metals, resulting in fire hazards. Meanwhile, the electromagnetic radiation may also have potential harm on animals staying

close to the system. Thus a WPT must be equipped with foreign object detection devices, including metal object detection and living object detection.

2.4 The Dynamic Wireless Power Transfer System

The power supply rail buried under the ground transmits electric energy in the form of a high-frequency alternating magnetic field to the electric power pick-up mechanism of the vehicle receiving end running on a certain range on the ground, and then supplies power to the on-board energy storage device so that the electric vehicle can be equipped with a small number of batteries group to extend its cruising range. At the same time, the power supply becomes safer and more convenient.

The receiving coil of the magnetic circuit mechanism of the dynamic wireless power transfer is the same as the receiving coil used in the static wireless charging system, while the structure of the transmitting coil is different. According to the shape of the transmitting coil, the transmitting end can be divided into three types: array type, segmented rail type, and long rail type. The array-type transmitting coils are arranged in a static charging coil array; while the structure of segmented rail coils and long rail coils is characterized by their lengths being much larger than their widths, this magnetic circuit mechanism also exists in unipolar and bipolar. Figure 2.12 is a schematic diagram of the structure of three kinds of electric vehicle dynamic wireless charging systems.

2.4.1 Three Types of Electric Vehicle Dynamic Charging Systems

Coil array rail electric vehicle dynamic wireless charging system: the coil array dynamic wireless charging is achieved by laying coil arrays under the predetermined driving route of an electric vehicle. When the electric vehicle is driving on this road, electric energy is wirelessly transmitted to the on-board energy storage battery in real-time based on the principle of electromagnetic coupling through the coil array buried under the road surface, so that electric vehicles are less equipped or even need not be equipped with energy storage battery packs. This method extends EVs' cruising range, and improves the convenience and safety of power supply.

Segmented rail and long rail electric vehicle dynamic wireless charging systems: the currently known power transmission rails include centralized long rails and segmented rails. The centralized long rails have a serious loss and low power transmission efficiency. Therefore, to improve the efficiency of power transmission, a more practical solution is the wireless power supply technology of multi-segment rail switching. Only the corresponding rail under the vehicle is turned on. When the vehicle leaves, the power supply to the rail is cut off, and the other section of the rail at the corresponding position of the vehicle is turned on. Therefore, the wireless power

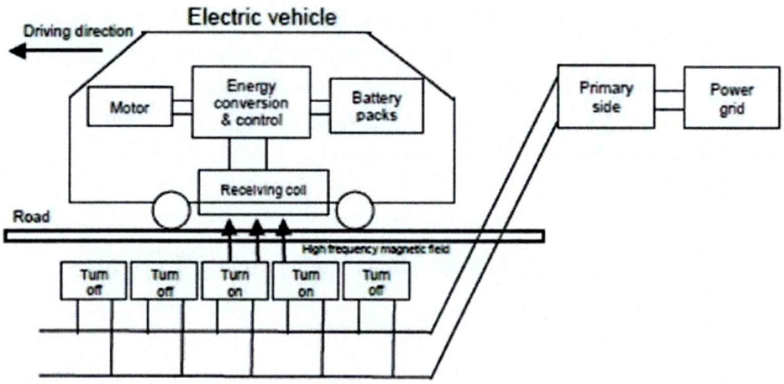

(a) Coil array electric vehicle dynamic wireless charging system;

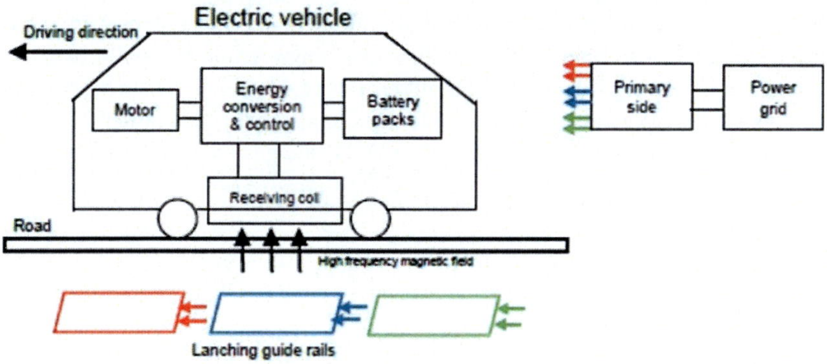

(b) Segmented rail electric vehicle dynamic wireless charging system;

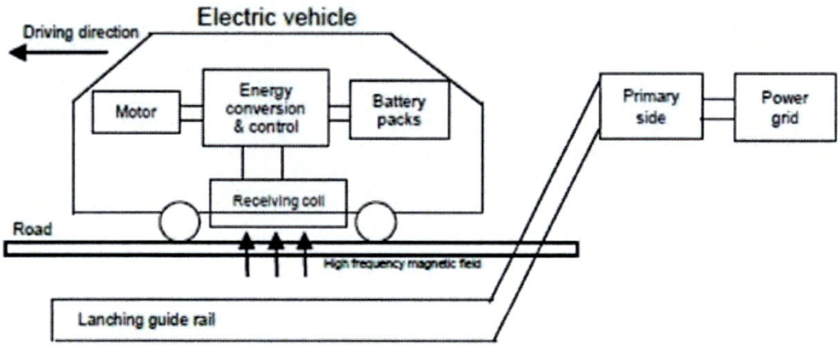

(c) Long rail electric vehicle dynamic wireless charging system;

Fig. 2.12 Electric vehicle dynamic wireless charging system supply technology of the segmented guide rail needs to detect the position of the vehicle to switch on and off the power transmission guide rail. At present, it mainly uses the method of installing a sensor on the guide rail or adding additional signal transmission coils on the vehicle and the guide rail to detect relevant parameters to perceive the location of electric vehicles.

2.4.2 Comparison of Three Kinds of Dynamic Charging Systems

The characteristics of the above three dynamic systems are shown in Table 2.1.

Comparing the three types of electric vehicle dynamic wireless charging systems, it is found that the long rail type has only a single-stage rail. When the rail breaks down, it will cause the problem that the whole line cannot work, but it has the advantage of less output voltage fluctuation. In the coil array type rail, the coils are independent of each other, and the failure of a single guide rail will not affect the coils of other guide rails, and the reliability is strong. However, due to a large number of coil switching positions, the output voltage is unstable. The segmented guide rail is a structure between the two, which has the advantages of both while reducing the impact of the two shortcomings on the system.

Table 2.1 Features of three kinds of electric vehicle dynamic charging systems

Classification	Coil numbers and charging efficiency	Voltage fluctuation	Construction cycle and cost
Coil array	Primary-side transmitting coils number is very large, complex control, high system efficiency	The voltage at the switching position fluctuates, and the number of switching points is large	Long construction period and high cost
Segmented rail	Primary launching guide rails number is large, complicated control, high system efficiency	The voltage at the switching position fluctuates, and the number of switching points is large	Moderate difficulty in construction and high cost
Long rail	Single launching guide rail, simple control, and low system efficiency	The output voltage fluctuation is small during the driving of electric vehicles	Short construction period and low cost

2.4.3 Research Status of Dynamic Wireless Charging Technology for Electric Vehicles

At present, foreign research teams such as the University of Auckland in New Zealand, the University of Tokyo in Japan, the Oak Ridge National Laboratory in the United States, and the Korea Advanced Institute of Science and Technology (KAIST) have launched a series of studies on the technical difficulties and key issues related to the electric vehicle dynamic wireless power supply. It focuses on system modeling methods, electrical energy conversion topology, electromagnetic coupling mechanism optimization design, and electromagnetic shielding technology. The University of Auckland in New Zealand and the German company Kang Wen developed the world's first wireless charging bus with a power of 30 kW, and also developed a prototype of a 100 kW wirelessly powered train with a train track length of 400 m (Chen et al. 2015a).

KAIST refers to electric vehicles using dynamic wireless charging technology as online electric vehicles (OLEVs). In 2013, two OLV electric bus lines in Gumi City were put into operation. The total length of the line is 24 km, the transmission power is 100 kW, and the efficiency is 85% (Choi et al. 2015).

The Oak Ridge National Laboratory of the United States researched the coupling mechanism, transmission characteristics, dielectric loss, and electromagnetic radiation of dynamic wireless charging of electric vehicles. Its ground launch device uses a full-bridge inverter and two primary windings in series. The experimental results show that the transmission power and efficiency are greatly affected by the location of electric vehicles (Miller et al. 2015; Onar et al. 2013).

The University of Tokyo in Japan proposes a secondary side maximum efficiency control method based on a DC/DC converter. The coupling coefficient is estimated online in real-time through the primary-side equivalent impedance, and the feedforward controller is used to change the input duty cycle of the DC/DC converter to achieve maximum efficiency control (Kobayashi et al. 2015).

In terms of wireless power supply technology for rail trains, the Korea Railway Research Institute (KRRI) designed and studied the entire rail train wireless power supply system, and made an experimental device with a power of 1 MW and a track length of 128 m. The coupling mechanism adopts a long straight guide rail at the transmitting end, and the coupling performance is enhanced by two small U-shaped magnetic cores. Due to the long track and large inductance, in order to reduce the voltage stress of the capacitor, the capacitor is dispersed in the transmitting coil (Kim et al. 2015; Lee et al. 2016, 2015).

In addition, the German Bombardier is also at a leading level in the field of wireless power supply for electric vehicles and trams. Due to commercialization, its corresponding technical information is less (Winter et al. 2013).

Southeast University conducted an in-depth study on the effect of the size of the primary and secondary coils on dynamic wireless energy transmission on transmission efficiency and lateral shift and proposed a method based on frequency control to achieve the optimal energy transmission efficiency of the system (Chen et al.

2015b; Qiang and Tan 2012). Based on the coupled-model theory, Tianjin Polytechnic University analyzed the influence of the change of the natural resonance frequency of the transmitting coil and receiving coil of the high-speed train wireless power supply system on the system transmission efficiency under motion, and proposed a frequency tracking with adjustable power factor at the transmitting end (Zhang et al. 2015), and in 2013 proposed the idea of applying dynamic wireless energy transmission technology to the charging of high-speed railway trains and established a high-speed rail charging sand table model, which attracted widespread attention. In addition, the literature (Tian 2012) proposed a nonlinear programming mathematical model for the segmented power rail system, analyzed the length of the segmented rails and the rapid switching of the rails, and obtained better planning and control results.

2.4.4 Problems and Technical Difficulties of Electric Vehicle Dynamic Wireless Charging Technology

The current problem of dynamic wireless charging is the stability, the efficiency, and the economy of its system power supply. There is relative movement between the receiving coil of the vehicle and the ground transmitting coil, the stability of the power transmission of the dynamic wireless charging system needs to be improved, and the transmission efficiency is lower than that of the static wireless charging. In terms of the initial cost of operation and construction, the dynamic wireless charging system is high and requires road and road reconstruction while using more inverters and coils.

(1) The design problem of the high-performance coupling mechanism

Compared with the unipolar long coil guide rail, the bipolar power supply guide rail has the characteristics of high power density, compact size, strong lateral shift adaptability, low exposure level to the magnetic field on both sides of the track, and low ground construction difficulty. With low volume and construction costs, bipolar rail is suitable for large-scale engineering applications. However, its magnetic field distribution is uneven, resulting in discontinuous energy transmission, which not only affects system stability but also reduces the energy transmission power and efficiency. The structure needs to be further optimized to improve the average transmission efficiency and average transmission power of the dynamic wireless power supply.

(2) Robust control of energy transmission

In the dynamic wireless power supply system of the bipolar power supply rail, due to the influence of multi-parameter disturbances such as the relative position change of the coupling mechanism, the uneven distribution of the magnetic field between the segmented rails, and the different subgrade media, the energy transmission is in a rapid nonlinear change process. Stability and improving system response speed have become the research goals of dynamic WPT control strategies.

(3) Electromagnetic compatibility issues

The problem of electromagnetic compatibility is closely related to the quality of energy transmission, the electromagnetic interference caused to the system, and the impact on the human body. Only by effectively solving the problem of electromagnetic compatibility can the system be safe, reliable, and stable. It can be seen that how to ensure the electromagnetic compatibility of the system efficiently and reliably becomes the main content of the research under the condition of minimally affecting the system efficiency.

2.4.5 Discussions for DWPT

The research on dynamic wireless charging technology for electric vehicles at home and abroad is still in the stage of continuous development and improvement. Although certain theoretical research and experimental results have been achieved, there are still many problems to be solved in the engineering and commercialization of this technology.

References

Chen L, Nagendra GR, Boys JT et al (2015a) Double-coupled systems for IPT roadway applications. IEEE J Emerging Selected Topics Power Electronics 3(1):37–49

Chen Z, Jing W, Huang X et al (2015b) A promoted design for primary coil in roadway-powered system. IEEE Trans Magnetics 51(11):1–6

Choi SY, Gu BW, Jeong SY et al (2015) Advances in wireless power transfer systems for road way-powered electric vehicles. IEEE J Emerging Selected Topics Power Electronics 3(1):18–36

EVATRAN. PLUGLESS. https://www.pluglesspower.com/shop/. Accessed 26 May 2020

Hu C (2015) Research on output performance and parameters optimization of magnetic coupler for EVs' wireless power supply. College of Automation of Chongqing University, Chongqing, China

Karalis A, Joannopoulos JD, Soljačić M (2008) Efficient wireless non-radiative mid-range energy transfer. Ann Phys 323:34–48

Kim JH, Lee BS, Lee JH et al (2015) Development of 1-MW inductive power transfer system for a high-speed train. IEEE Trans Ind Electronics 62(10):6242–6250

Kobayashi D, Imura T, Hori Y (2015) Real-time coupling coefficient estimation and maximum efficiency control on dynamic wireless power transfer for electric vehicles. In: Conference of the IEEE industrial electronics society, Yokohama, Japan: 13–18

Kurs A, Karalis A, Moffatt R et al (2007) Wireless power transfer via strongly coupled magnetic resonances. Science 317:83–86

Lee CK, Zhong WX (2012) Ron Hui SY (2012) Recent progress in mid-range wireless power transfer. IEEE Energy Convers Congress Exposition (ECCE) 2012:3819–3824

Lee SH, Lee BS, Lee JH (2016) A new design methodology for a 300 kW, low flux density, large air-gap, on-line wireless power transfer system. IEEE Trans Ind Appl 52(5):4234–4242

Lee JH, Lee BS, Lee SH et al (2015) A study on the control of output current to respond to the load changes of the wireless power transfer. In: Transportation electrification conference and expo, Long Beach, CA, USA, pp 3073–3080

Marincic AS (1982) Nikola Tesla and the wireless transmission of energy. IEEE Trans Power Appar Syst 10:4064–4068

Miller J, Onar C, Chinthavali M (2015) Primary-side power flow control of wireless power transfer for electric vehicle charging. IEEE J Emerging Selected Topics Power Electronics 3(1):147–162

Moon S, Kim B, Cho S et al (2014) Analysis and design of a wireless power transfer system with an intermediate coil for high efficiency. IEEE Trans Ind Electron 61:5861–5870

Onar OC, Miller JM, Campbell SL et al (2013) A novel wireless power transfer for in motion EV/PHEV charging. In: 28th annual IEEE applied power electronics conference and exposition, Long Beach, CA, USA, pp 3037–3038

Qiang H, Tan L et al (2012) Achieving maximum power transfer of inductively coupled wireless power transfer system based on dynamic tuning control. Sci China (series e) 42(7):830–837 (in Chinese)

Qualcomm. Download documents about Qualcomm Halo TM technology. https://www.qualcomm.com/products/halo/resources. Accessed 8 May 2020

Sallan J, Villa J, Llombart A et al (2009) Optional design of ICPT systems applied to electric vehicle battery charge. IEEE Trans Ind Electron 56:2140–2149

Tian Y (2012) Research on key issues of sectional track-based wireless power supply technology for electric vehicles. College of Automation of Chongqing University, Chongqing, China (in Chinese)

Wheeler L (1943) II—Tesla's contribution to high frequency. Electr Eng 62:355–357

Winter J, Mayer S, Kaimer S et al (2013) Inductive power supply for heavy rail vehicles. In: Electric drives production conference, Nuremberg, Germany, 9p

Witricity. BMW drives the future of electric vehicle wireless charging with WiTricity technology. http://witricity.com/bmw-drives-future-electric-vehicle-wireless-charging-witricity-technology/. Accessed 26 May 2020

Zhang X, Su H, Yang Q et al (2015) Research on maximum transmission efficiency of resonance coupling wireless transmission in high-speed train system. Trans China Electrotechnical Soc 30(Supplement):308–312

Chapter 3
Resonant Circuit Analysis Theories

3.1 Introduction to Coupled Coils

For the analysis of a generalized IPT system, the modeling of the two coupled coils is essential. Let us first consider the situation shown in Fig. 3.1. Note that the polarity of the coils is shown by the dot in the figures, and the definition of currents at the primary and secondary coils is symmetrical. There are two different conditions for the coupled coils, with the secondary side dot at the upper and lower side. Thus the secondary current direction is different, but the direction of magnetic flux by the secondary current is unchanged, which is always opposite of the primary magnetic flux direction according to Lenz's law. Furthermore, when explaining internal electromagnetic processes throughout this chapter, it is assumed that all coils are free from any parasitic parameters by applying a lumped parameter method. The operating frequency of the IPT is low enough that we do not have to consider electromagnetic radiation and transmission delay. The coils can be coreless or with a core as long as it is not nonlinear.

According to Ampere's law, magnetic flux is generated from the current i_1 as shown in Fig. 3.1. The primary magnetic flux (ϕ_1) can be divided into two parts, magnetic flux intersecting the secondary coil (ϕ_{11}) and leakage flux (ϕ_{12}), for the secondary coil, the intersected magnetic flux is ϕ_{21} and the leakage flux is ϕ_{22}. Here the effect of transmission delay is not counted in the coupled coil model. Then we can get the primary and secondary coils' voltage equation according to the physical model, the voltage of the coil for Fig. 3.1 can be determined by the time derivation of magnetic flux according to Faraday's law as follows:

$$u_1 = N_1\left(\dot{\phi}_{11} + \dot{\phi}_{12} - \dot{\phi}_{21}\right) \tag{3.1a}$$

$$u_2 = -N_2\left(\dot{\phi}_{21} + \dot{\phi}_{22} - \dot{\phi}_{12}\right) \tag{3.1b}$$

X. Zhang et al., *Wireless Power Transfer Technologies for Electric Vehicles*, Key Technologies on New Energy Vehicles,
https://doi.org/10.1007/978-981-16-8348-0_3

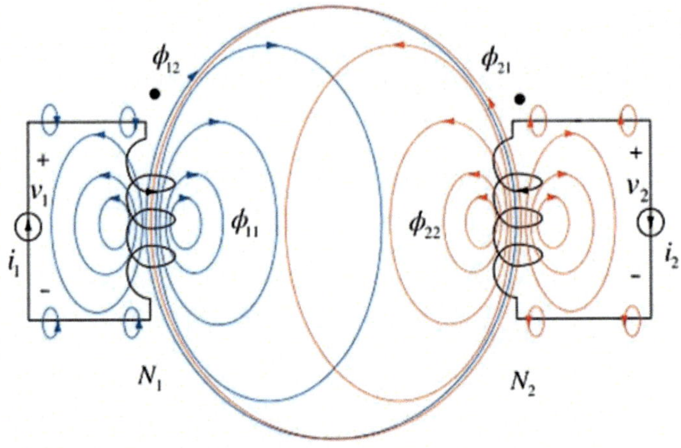

Fig. 3.1 Two coupled coils

For another case of Fig. 3.1, since the direction of the secondary current is opposite, the polarity of the voltage u_2 is reversed, but the direction of the magnetic flux keeps the same. It is necessary to be cautious with the polarity of these parameters like current, voltage, and magnetic flux.

From Ampere's Law, it is known that magnetic flux is proportional to its corresponding current. Moreover, the inductance can be canonically defined as follows:

$$v = N\dot{\phi} = L\dot{i} \Rightarrow L = \frac{N\phi}{i} \tag{3.2}$$

By applying (3.2) to (3.1a, b), we can get equations as follows:

$$v_1 = N_1(\dot{\phi}_{11} + \dot{\phi}_{12} - \dot{\phi}_{21}) = (L_{l1} + L_{12})\dot{i}_1 - L_{21}\dot{i}_2/n \tag{3.3a}$$

$$v_2 = -N_2(\dot{\phi}_{21} + \dot{\phi}_{22} - \dot{\phi}_{12}) = -(L_{l2} + L_{21})\dot{i}_2 + nL_{12}\dot{i}_1 \tag{3.3b}$$

where the turn radio is defined as follows:

$$n = \frac{N_1}{N_2} \tag{3.4}$$

From (3.3a, b), the physical magnetic model is transformed into a lumped-circuit-element model, as shown in Fig. 3.2. The primary and secondary coils are divided into the leakage inductance (L_{l1}) corresponding to the leakage magnetic flux (ϕ_{11}) and the mutual inductance corresponding to the intersecting magnetic flux ($\phi_m = \phi_{12} - \phi_{21}$).

It can be seen that the leakage magnetic flux can be replaced by the leakage inductance, and in the subsequent sections, we will introduce some circuit models

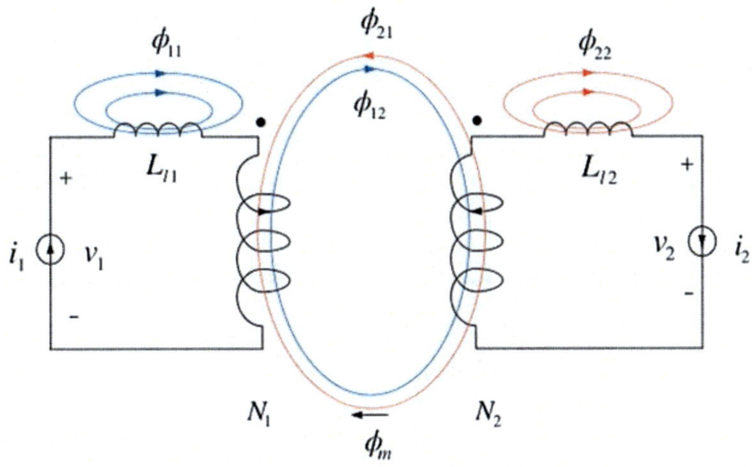

Fig. 3.2 Magnetic coupling coil model without leakage inductances

for the magnetic model of Fig. 3.2, which is not analyzable by circuit theories due to the variables of magnetic flux.

3.1.1 Equivalent Transformer Model

Since the magnetic coupling part of Fig. 3.2 must be replaced by an electric circuit to get an equivalent circuit of the IPT. Hence, let us find the relationship between L_{12} and L_{21} in Fig. 3.2. By applying (3.2), L_{12}, and L_{21} can be calculated as follows:

$$L_{12} = \frac{N_1\phi_{12}}{i_1}, L_{21} = \frac{N_2\phi_{21}}{i_2} \tag{3.5}$$

According to KVL law of magnetic circuits, we can know that:

$$\phi R_C = N_1 i_1 - N_2 i_2 \tag{3.6}$$

where R_c is defined as the magnetic resistance of the magnetic coupling coil. Assuming that the secondary current or the primary current is zero, then Eq. (3.5) can be rewritten as follows:

$$L_{12} = \frac{N_1\phi_m}{i_1} = \frac{N_1 N_1 i_1}{i_1 R_C} = \frac{N_1^2}{R_C}, L_{21} = \frac{N_2\phi_m}{-i_2} = \frac{N_2^2}{R_C} \tag{3.7}$$

Thus, the inductance has the following relationship:

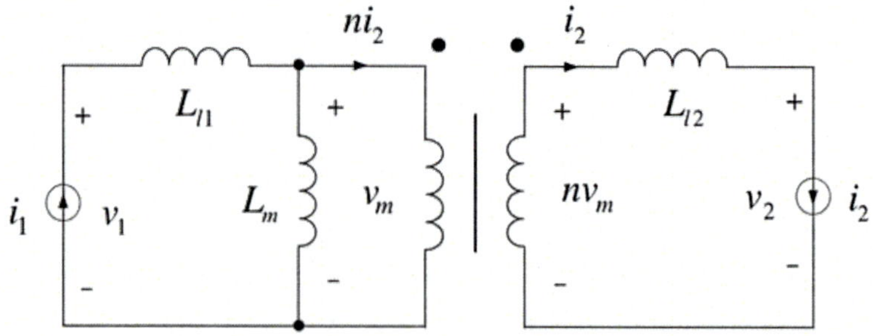

Fig. 3.3 A transformer model for coupled coils

$$L_{21} = n^2 L_{12} = n^2 L_m,\ for L_{12} \equiv L_m \tag{3.8}$$

Applying (3.8) to (3.3a, b) results in the following:

$$v_1 = (L_{l1} + L_{12})\dot{i}_1 - \frac{L_{21}\dot{i}_2}{n} = L_{l1}\dot{i}_1 + L_1\left(\dot{i}_1 - n\dot{i}_2\right) = L_{l1}\dot{i}_1 + v_m \tag{3.9a}$$

$$v_2 = -(L_{l2} + L_{21})\dot{i}_2 + nL_{12}\dot{i}_1 = -L_{l2}\dot{i}_2 + nv_m \tag{3.9b}$$

$$for v_m = L_m\left(\dot{i}_1 - n\dot{i}_2\right), i_m = i_1 - ni_2 \tag{3.9c}$$

The equivalent circuit for (3.9a, b, c) is shown in Fig. 3.3. The transformer is an ideal transformer with a turn ration of n. The magnetic inductance is represented by L_m, which is defined as:

$$L_m \equiv \frac{N_1 \phi_m}{i_m} \tag{3.10}$$

As shown in Fig. 3.3, the equivalent circuit of IPT is very similar to a conventional transformer. But since the large air gap of IPT, the values of leakage inductances are too large to be neglected, which are quite small compared with L_m in the conventional transformer. The leakage inductances in the IPT are larger than the magnetizing inductance, which also means the coupling coefficient k is relatively small (0.2–0.4) (Ahmad et al. 2018). Moreover, the voltages and currents of the primary and secondary coils may not be in the same phase anymore.

The advantages of the equivalent transformer model shown in Fig. 3.3, is that it can apply to an arbitrary waveform of currents. Since the model is similar to a transformer model, it is convenient and useful for designing IPT coils.

3.1.2 M-Model

Mutually coupled model, which is also called the "M-model", is another useful model for IPT. The two coupled coils are described by another method, as shown in Fig. 3.4, with:

$$v_1 = L_1 \dot{i}_1 - M \dot{i}_2 \tag{3.11a}$$

$$v_2 = -L_2 \dot{i}_2 + M \dot{i}_1 \tag{3.11b}$$

(3.11a, b) can be deduced from (3.9a, b, c) of the transformer model as follows:

$$v_1 = L_{l1} \dot{i}_1 + L_m \left(\dot{i}_1 - n \dot{i}_2 \right) = L_1 \dot{i}_1 - M \dot{i}_2 \tag{3.12a}$$

$$v_2 = -(L_{l2} + L_{21}) \dot{i}_2 + n L_{12} \dot{i}_1 = -L_{l2} \dot{i}_2 + M \dot{i}_1 \tag{3.12b}$$

$$for M \equiv n L_m, L_1 = L_{l1} + L_m, L_2 = L_{l2} + n^2 L_m \tag{3.12c}$$

From (3.12a, b, c), we can know that the M-model removes the transformer, turn ratio, and magnetic inductance. At the same time, dependent voltage sources are introduced in the model, as shown in Fig. 3.4b, and the polarity of voltage sources can also be reversed if the polarity of the current is defined reversely. The M-model applies the mutual inductance M which is dependent on the turn ratio and the magnetic inductance L_m to represent the relationship between primary and secondary coils. This M-model simplifies the equivalent transformer model and can be convenient for circuit formula derivation.

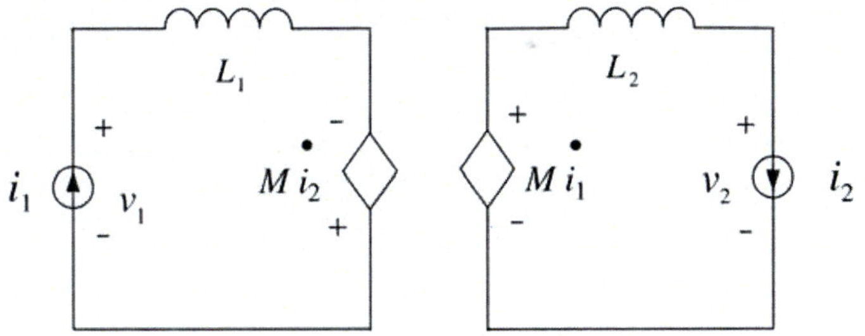

Fig. 3.4 M-model for the two coupled coils

3.1.3 T-Model

Another category of the model for the coupled coils is T-model, which can be derived from the M-model above, as shown in Fig. 3.5, which resembles the letter "T".

The T-model circuit equations are as follows:

$$v_s = L_1\dot{i}_1 - M\dot{i}_2 = (L_1 - M)\dot{i}_1 + M\left(\dot{i}_1 - \dot{i}_2\right) \tag{3.13a}$$

$$v_L = -L_{l2}\dot{i}_2 + M\dot{i}_1 = -(L_2 - M)\dot{i}_2 + M\left(\dot{i}_1 - \dot{i}_2\right) \tag{3.13b}$$

This T-model is not an exact equivalent circuit of the coupled coils because there is no galvanic isolation compared with the transformer model and M-model, but this model is exactly equivalent to other models when it comes to circuit-based analyses. The advantage of the T-model is its simplicity and can be used for circuit-based analyses except for its physical meaning.

The equivalent transformer model can also be transformed into the corresponding T-model. According to the impedance transformation characteristics of ideal transformers, we can convert the secondary parameters like voltage, current, and impedance to the primary.

3.2 Gyrator Circuit Model

3.3 Introduction

Compensation circuits are crucial for an IPT system to improve the transfer power and efficiency by canceling out the reactive powers. By now, there are many types of compensation circuits reported in the literature (Ahmad et al. 2018; Li and Mi 2015,

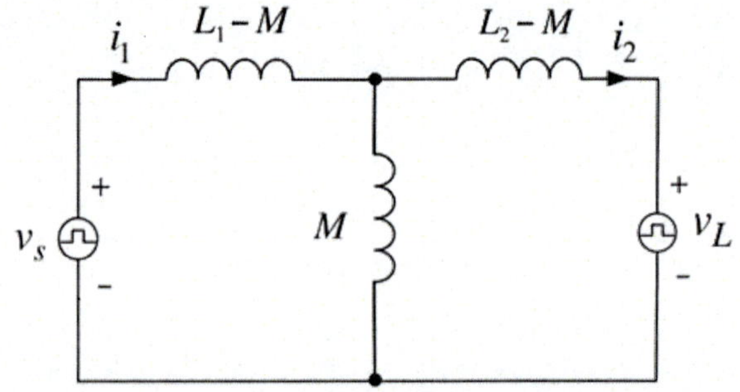

Fig. 3.5 T-model for the two coupled coils

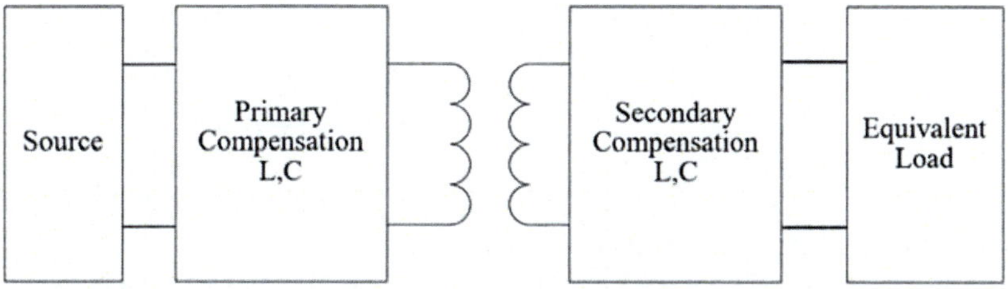

Fig. 3.6 Configuration of compensation circuits

2016; Zhang and Mi 2016), and each compensation circuits have its own characteristics such as source-to-load voltage/current gain, load-independent output characteristics, etc. We must select the most appropriate compensation circuits according to the application requirements.

A typical configuration for compensation circuits is shown in Fig. 3.2. The primary and secondary compensation circuits are usually combinations of capacitors and inductances (Li and Mi 2015, 2016; Zhang and Mi 2016). Analyses of these compensation circuits have been mostly performed by applying various models in Sect. 3.1 and solving the equations derived from the models. But when it comes to several complicated compensation topologies, solving the equations is time-consuming. As an alternative to the problems, a graphical approach for modeling IPTs is proposed by Sohn et al. (2016), they have found that all IPTs have the nature of a gyrator. A gyrator is one of the basic circuit elements, which was introduced by Tellegen (1948). By utilizing the gyrator approach, analyses of compensation circuits can be simplified (Fig. 3.6).

3.3.1 Representation of Compensation Circuits with Gyrators

In Fig. 3.7a, based on the Kirchhoff's voltage law, the relationships between U_1, I_1, U_2, I_2 are as follows,

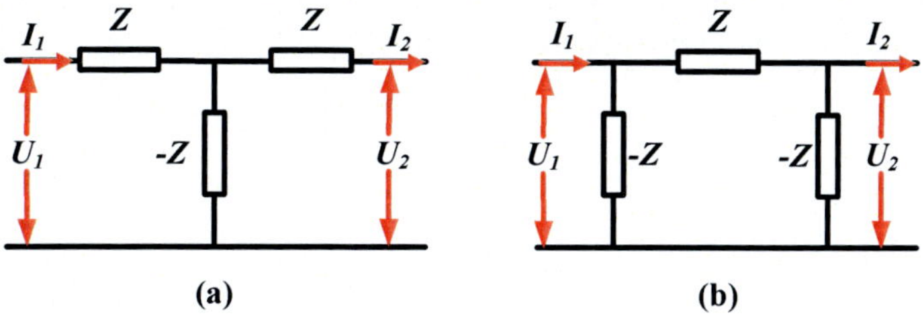

Fig. 3.7 **a** The structure of the T-network. **b** The structure of Π-network

$$U_1 = I_1 \times Z + (I_1 - I_2) \times (-Z) = I_2 \times Z \tag{3.14}$$

$$U_2 = -I_2 \times Z + (I_1 - I_2) \times (-Z) = -I_1 \times Z \tag{3.15}$$

where $\boldsymbol{U_1}$, $\boldsymbol{I_1}$, $\boldsymbol{U_2}$, and $\boldsymbol{I_2}$ are the phasors of the input current and voltage and the output current and voltage, respectively.

In Fig. 3.7b, based on the Kirchhoff's current law, the relationships between U_1, I_1, U_2, I_2 are as follows,

$$I_1 = \frac{U_1}{-Z} + \frac{U_1 - U_2}{Z} = -\frac{U_2}{Z} \tag{3.16}$$

$$I_2 = -\frac{U_2}{-Z} + \frac{U_1 - U_2}{Z} = \frac{U_1}{Z} \tag{3.17}$$

It is obvious that (1), (2), and (3), (4) are the same, which means T-network and Π-network are equivalent. And it should be noticed that these equations are suitable for the whole angular frequency domain. The equivalent series resistances (ESRs) of the inductively coupled coils, compensation inductors, and the compensation capacitors are assumed to be negligible for the simplicity of theoretical analysis. Then the impedance Z is pure imaginary in the resonant networks because it is composed of the compensation inductors and capacitors. Thus, the equations as follows are obtained,

$$\left(\frac{1}{Z}\right)^* = -\frac{1}{Z} \tag{3.18}$$

$$I_2 = G^* \times U_1 \tag{3.19}$$

$$U_2 = G^* \times I_1 \tag{3.20}$$

where $G = \frac{1}{Z}$. Based on the definition of the gyrator, the T-network and Π-network in Fig. 3.7 could be substituted with a gyrator, as shown in Fig. 3.8. Based on (5), if U_1 is the constant voltage input, I_2 is the constant current output. A similar situation occurs when I_1 is the constant current input.

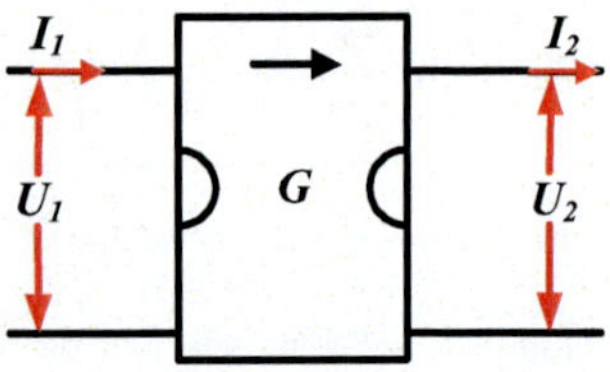

Fig. 3.8 The representation of the T-network and the Π-network with the gyrator

3.4 Gyrators in the Resonant Circuits

Figure 3.9 shows the implementation of a gyrator with reactive components. As shown in Fig. 3.9a, b, if the Eq. (3.21) exists, they could be implemented with Fig. 3.9c.

$$\omega^2 = \frac{1}{LC} \tag{3.21}$$

Figure 3.10 shows the representation of a series–parallel resonant circuit with a gyrator. If the inductor L resonates with the capacitor C at the angular frequency ω, the series-connected capacitor and inductor could be added in the series–parallel structure (Fig. 3.10a) as shown in Fig. 3.10b. The added inductor and capacitor construct a short circuit which does not affect the previous circuit. Then a gyrator could be constructed, as shown in Fig. 3.10c.

Figure 3.11 shows the representation of coupled inductors with a gyrator. Figure 3.11a shows coupled inductor L_1 and L_2 with the mutual inductance of M. Figure 3.11b shows the mutual inductance based equivalent circuit of Fig. 3.11a. The blue section in Fig. 3.11c is the T-circuit, which could be converted to a gyrator shown in Fig. 3.11d.

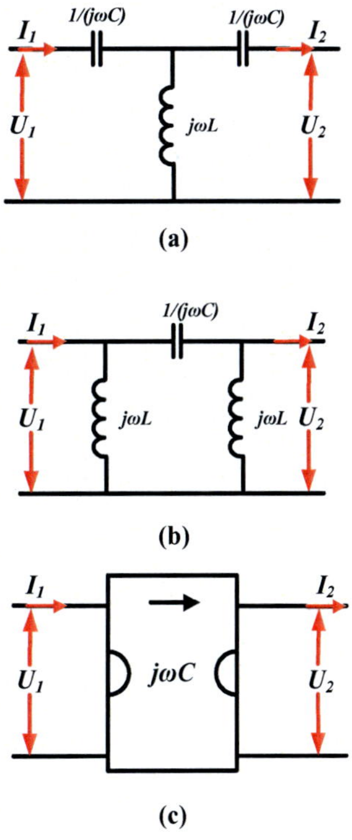

Fig. 3.9 Implementation of a gyrator with reactive components

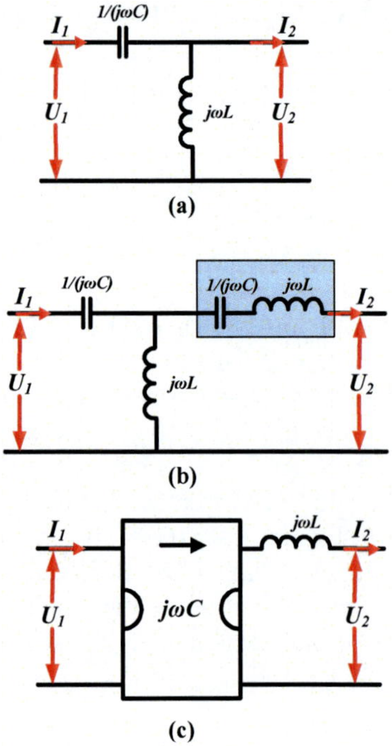

Fig. 3.10 Representation of a series–parallel resonant circuit with a gyrator

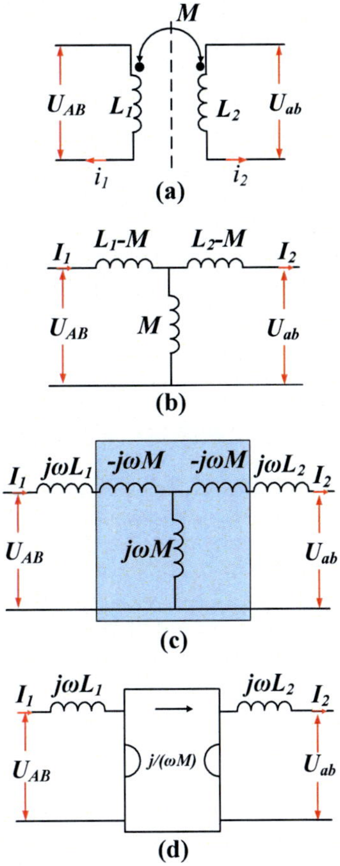

Fig. 3.11 Representation of coupled inductors with a gyrator

References

Ahmad A, Alam MS, Chabaan R (2018) A comprehensive review of wireless charging technologies for electric vehicles. IEEE Trans Transp Electrification 4(1):38–63

Li S, Mi CC (2015) A double-sided LCC compensation network and its tuning method for wireless power transfer. IEEE Trans Veh Technol 64(6):2261–2273

Li W, Mi CC (2016) Comparison study on SS and double-sided LCC compensation topologies for EVPHEV Wireless Chargers. IEEE Trans Veh Technol 65(5):4429–4439

Sohn YH, Choi BH, Rim CT (2016) Gyrator-based analysis of resonant circuits in inductive power transfer systems. IEEE Trans Power Electron 31(10):6824–6843

Tellegen BDH (1948) The gyrator, a new electric network element. Philips Res Rep 3:81–101

Zhang W, Mi CC (2016) Compensation topologies of high-power wireless power transfer systems. IEEE Trans Veh Technol 65(6):4768–4778

Chapter 4
Resonant Compensation Topologies

4.1 Circuit Principle of Compensation Topologies

The resonant compensation topology is the most important part of the wireless charging system. To analyze the effect of the compensation network, the circuit principle and characteristics should be first analyzed (Li and Mi 2014). Figure 4.1 gives the structure and the equivalent circuit of the coupling coils. L_1 and L_2 are the self-inductance of the primary coil and secondary coil, M is the mutual inductance of the primary coil and secondary coil, I_1 and I_2 are the current phasors flowing through the primary coil and secondary coil.

Figure 4.1 shows the equivalent circuit model of the coupled inductance of the coupled coil. It can be seen that the primary current induces a voltage in the secondary loop, and the secondary current also induces a voltage in the primary loop. The induced voltage is the energy transmission medium between the primary circuit and secondary circuit. If the energy transmission is bidirectional, the induced voltage can either absorb energy or release energy. Suppose $\dot{S}_{12}$ is the released or absorbed apparent power phasor from L_1 to L_2, and $\dot{S}_{21}$ is the absorbed or released apparent power phasor from L_2 to L_1. Ideally, they can be calculated as

$$\begin{aligned}\dot{S}_{12} &= \dot{U}_{12}\dot{I}_2^* = j\omega M\dot{I}_1\dot{I}_2^* = j\omega MI_1I_2(\cos\theta_{12} - j\sin\theta_{12}) \\ &= \omega MI_1I_2\sin\theta_{12} + j\omega MI_1I_2\cos\theta_{12}\end{aligned} \tag{4.1}$$

$$\begin{aligned}\dot{S}_{21} &= \dot{U}_{21}\dot{I}_1^* = j\omega M\dot{I}_2\dot{I}_1^* = j\omega MI_1I_2(\cos\theta_{21} - j\sin\theta_{21}) \\ &= \omega MI_1I_2\sin\theta_{21} + j\omega MI_1I_2\cos\theta_{21}\end{aligned} \tag{4.2}$$

where θ_{12} is the phase angle that the primary coil current leads the secondary coil current, θ_{21} is the phase angle that the secondary coil current leads the primary coil current, so $\theta_{12} = -\theta_{21}$. Then Eq. (4.2) can be converted to

X. Zhang et al., *Wireless Power Transfer Technologies for Electric Vehicles*,
Key Technologies on New Energy Vehicles,
https://doi.org/10.1007/978-981-16-8348-0_4

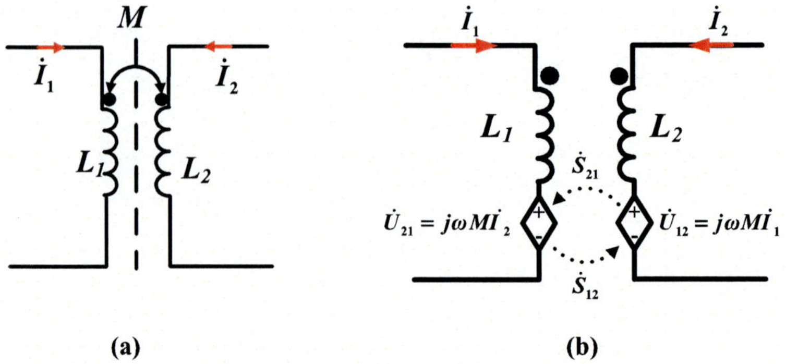

Fig. 4.1 The structure and the equivalent circuit of the coupling coils

$$\dot{S}_{21} = -\omega M I_1 I_2 \sin\theta_{12} + j\omega M I_1 I_2 \cos\theta_{12} \tag{4.3}$$

As can be seen from Eqs. (4.1) and (4.3), the apparent power exchanged between the coils is divided into active power term (real term) and reactive power term (imaginary term). The positive or negative of the active power term determines the direction of power transmission. When θ_{12} is positive, namely, the primary coil current leads the secondary coil current, the active power of $\dot{S}_{12}$ is positive and the active power of $\dot{S}_{21}$ is negative. The primary coil transmits power to the secondary coil. On the contrary, when θ_{12} is negative, the primary coil receives power from the secondary coil.

In addition, when $\theta_{12} = 90°$, namely, the angle that the primary coil current leads the secondary coil current is 90°, the reactive power term in Eqs. (4.1) and (4.3) is zero, and only active power is transmitted by the coils. In the case of the same output power, the larger the reactive power transmitted by the coil, the larger its internal resistance loss and magnetic loss. Therefore, to maximize the transmission efficiency of the coil, it is necessary to minimize the reactive power transmitted by the coil, that is, the leading angle is close to 90°. According to $\dot{U}_{12} = j\omega M\dot{I}_1$, it can be concluded that the angle that the secondary induced voltage leads the secondary coil current is 180°. The reference direction of the secondary coil current is the direction that the current flows into the secondary coil, so the secondary induced voltage should be in phase with the current flows out of the secondary coil, that is, for the secondary side induced voltage, the secondary side impedance is resistive. The power factor of the secondary side circuit is 1. In summary, to eliminate the reactive power of the coil and reduce the apparent power level of the coil, thereby maximizing the transmission efficiency of the coil, it is necessary to make the secondary impedance resistive.

Figure 4.2 is the impedance distribution diagram of the AC–AC power conversion circuit at different levels, where Z_L is the load impedance, Z_s is the secondary impedance, Z_r is the impedance reflected from the secondary side to the primary side, and Z_{in} is the input impedance. The secondary impedance Z_s can be expressed as

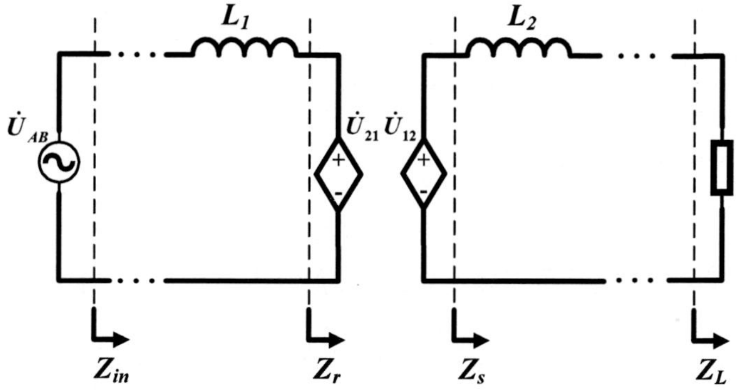

Fig. 4.2 Circuit impedance of the AC–AC power converter

$$Z_s = j\omega L_2 + Z_L \tag{4.4}$$

If the load impedance Z_L is resistive, the secondary impedance is inductive at this time, and the secondary impedance needs to be capacitively compensated to make it resistive, namely,

$$Z_y = -j\omega L_2 \tag{4.5}$$

It can be seen that by compensating the inductive impedance of the self-inductance of the secondary coil, the reactive power transmitted by the coils is eliminated and the coil loss is reduced. Similarly, for inverters that provide input voltage U_{AB}, reactive power will increase the input apparent power level, increase the output current of the inverter, thus increase the switching loss and conduction loss of the inverter, and also raises the inverter switch selection criteria and the system cost. Therefore, to eliminate the reactive power of the inverter and reduce the inverter losses, it is necessary to make the system input impedance Z_{in} resistive.

The primary reflected impedance Z_r can be calculated as,

$$Z_r = \frac{j\omega M\dot{I}_2}{\dot{I}_1} = \frac{j\omega M}{\dot{I}_1} * \left(-\frac{j\omega M\dot{I}_1}{Z_s}\right) = \frac{\omega^2 M^2}{Z_s} \tag{4.6}$$

Obviously, if the secondary side is compensated so that Z_s is resistive, then Z_r is also resistive. Similarly, if the primary side is compensated so that the compensation impedance is Z_x, the input impedance Z_{in} can be calculated as

$$Z_{in} = Z_x + j\omega L_1 + Z_r \tag{4.7}$$

To make Z_{in} resistive, let $Z_x = -j\omega L_1$. It can be seen that if the secondary side is compensated first to make the reflected impedance Z_r resistive, then the reactive power of the inverter can be eliminated by only compensating the inductive impedance generated by the self-inductance of the primary coil.

The circuit characteristics of several basic and composite compensation networks and their application prospect in high-power electric vehicle wireless charging systems are discussed as follows.

4.2 Basic Compensation Topologies

The basic compensation network uses the simplest capacitor compensation method for the primary and secondary sides, respectively. According to the connection of the compensation capacitor to the coil (series or parallel), four basic compensation networks can be classified as shown in Fig. 4.3, which are SS, SP, PS, and PP compensation networks (Wang et al. 2004, 2005). Where U_{AB} is the input voltage, R is the system load, L_1 is the self-inductance of the primary coil, C_1 is the primary compensation capacitor, L_2 is the self-inductance of the secondary coil, C_2 is the secondary compensation capacitor, and M is the mutual inductance between the coils. In the following, the circuit characteristics of the four basic compensation networks will be analyzed according to the parameter design.

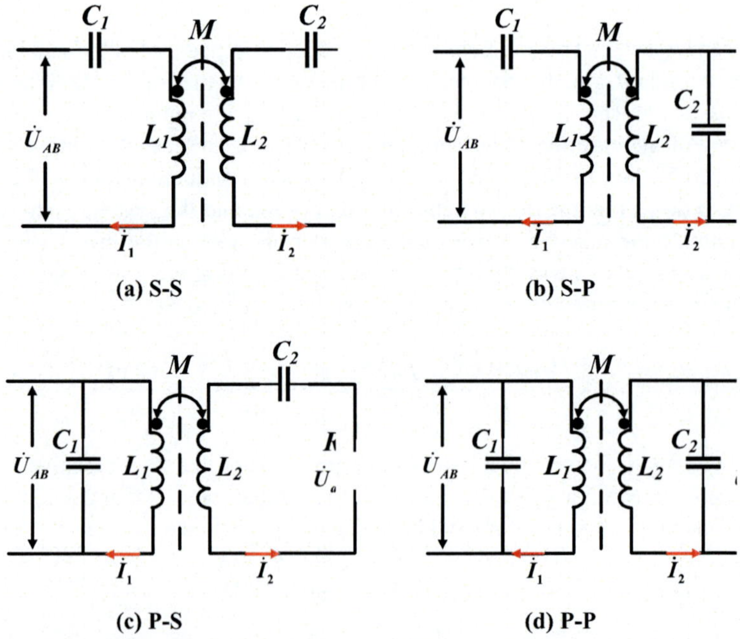

Fig. 4.3 Four basic compensation networks

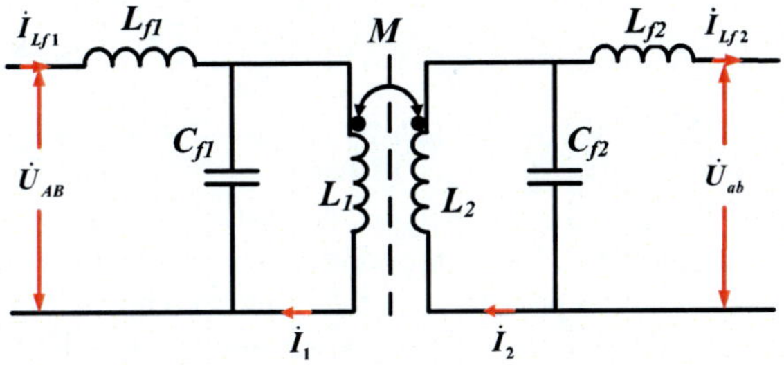

Fig. 4.4 Double-sided LCL compensation network

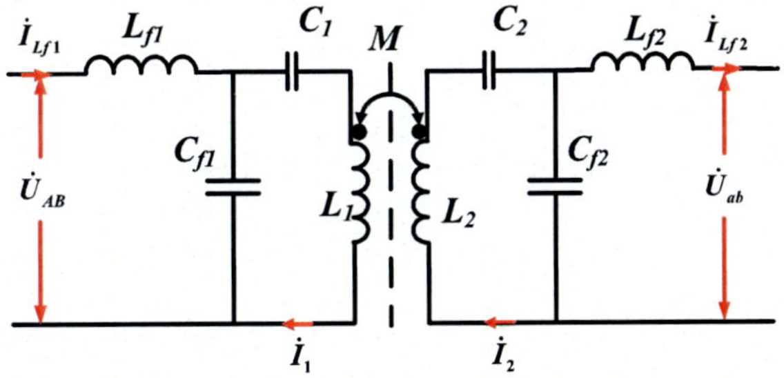

Fig. 4.5 Double-sided LCC compensation network

For the secondary serial compensation network (Fig. 4.3a, c), the secondary impedance Z_{s_s} can be calculated as (Fig. 4.4)

$$Z_{s\_s} = j\omega L_2 + \frac{1}{j\omega C_2} + R \tag{4.8}$$

For the secondary parallel compensation network (Fig. 4.5b, d), the secondary impedance Z_{s_p} can be calculated as

$$Z_{s\_p} = j\omega L_2 + \frac{1}{j\omega C_2 + \frac{1}{R}} \tag{4.9}$$

Assuming that the system operation angular frequency is the secondary resonant angular frequency ω_o, then the value of the secondary compensation capacitance can be calculated as

$$C_2 = \frac{1}{\omega_o^2 L_2} \tag{4.10}$$

Based on Eq. (4.10), Eqs. (4.8) and (4.9) can be simplified as

$$Z_{s\_s} = R \tag{4.11}$$

$$Z_{s\_p} = \frac{j\omega_o L_2}{1 + j\omega_o C_2 R} \tag{4.12}$$

When the secondary side is series compensated, the secondary impedance is purely resistive, the secondary power factor is 1, and the power transferred by the coils is all active power, which can reduce the apparent power level and eliminate the reactive power loss of the coils. When the secondary side is parallel compensated, the secondary impedance is in the form of complex impedance, which cannot eliminate the system reactive power transmitted by the coils.

Substituting Eqs. (4.11) and (4.12) into Eq. (4.6), the primary impedances Z_{r_s} and Z_{r_p} reflected from the secondary side can be obtained as follows

$$Z_{r\_s} = \frac{\omega_o^2 M^2}{R} \tag{4.13}$$

$$Z_{r\_p} = \frac{M^2 R}{L_2^2} - j\frac{\omega_o M^2}{L_2} \tag{4.14}$$

Based on Eqs. (4.7) and (4.13), the input impedance Z_{in_ss} of the SS compensation network can be calculated as

$$Z_{in\_ss} = j\omega_o L_1 + \frac{1}{j\omega_o C_1} + \frac{\omega_o^2 M^2}{R} \tag{4.15}$$

To make Z_{in_ss} purely resistive, let the imaginary part in Eq. (4.15) be zero, so the input impedance Z_{in_ss} and the configured primary compensation capacitor C_1 are obtained as

$$Z_{in\_ss} = \frac{\omega_o^2 M^2}{R}, C_1 = \frac{1}{\omega_o^2 L_1} \tag{4.16}$$

Similarly, for the SP compensation network, the input impedance Z_{in_sp} and the configured primary compensation capacitor C_1 can be calculated as

$$Z_{in\_sp} = \frac{M^2 R}{L_2^2}, C_1 = \frac{L_2}{(L_1 L_2 - M^2)\omega_o^2} \tag{4.17}$$

For the PS compensation network, the input impedance Z_{in_ps} and the configured primary compensation capacitor C_1 can be calculated as

$$Z_{in\_ps} = \frac{M^2 R \omega_o^2}{C_1^2 M^4 \omega_o^6 + R^2\left(-1 + C_1 L_1 \omega_o^2\right)^2}, C_1 = \frac{L_1 R^2}{\omega_o^2\left(L_1^2 R^2 + M^4 \omega_o^2\right)} \tag{4.18}$$

For the PP compensation network, the input impedance Z_{in_pp} and the configured primary compensation capacitor C_1 can be calculated as

$$
\begin{aligned}
Z_{in\_pp} &= \frac{M^2 R}{L_2\left(L_2 - C_1 L_1 L_2 \omega_o^2 + C_1 M^2 \omega_o^2\right)} \\
C_1 &= \frac{L_1 L_2^4 - L_2^3 M^2}{M^4 R^2 + L_1^2 L_2^4 \omega_o^2 - 2 L_1 L_2^3 M^2 \omega_o^2 + L_2^2 M^4 \omega_o^2}
\end{aligned}
\tag{4.19}
$$

According to the compensation principle of resonant circuits, Eqs. (4.16)–(4.19) give the calculation of the primary compensation capacitance in SS, SP, PS, and PP compensation networks, respectively. If the SP, PS, and PP compensation networks are used, the value of the primary compensation capacitance is affected by the coil coupling inductance M in addition to the coil self-inductance and operation frequency. In particular, in the PS and PP compensation networks, the value of the primary-side compensation capacitor is also related to the load size. During the actual use of the EV wireless charging system, the relative positions of the primary and secondary coils are highly susceptible to change, and the coupling inductance changes with the misalignment, while the primary compensation capacitance is difficult to be adjusted with the coupling inductance. In addition, the equivalent load of electric vehicle power battery changes with the SOC value, which means that the primary compensation capacitance of PS and PP compensation networks also needs to be adjusted with SOC value. Therefore, the SP, PS, and PP compensation networks cannot meet the actual demand of wireless charging for electric vehicles.

Compared with the other three compensation networks, the SS compensation network not only has the characteristic of unity power factor, which minimizes the system apparent power and reactive power loss but also has the characteristic that the system parameters are independent of the coupled inductor and load. According to Kirchhoff's law, the circuit parameters of the SS compensation network can be calculated as

$$
\begin{cases}
\dot{I}_1 = \dfrac{\dot{U}_{AB}}{Z_{in\_ss}} = \dfrac{\dot{U}_{AB} R}{\omega_o^2 M^2} = -\dfrac{\dot{U}_{ab}}{j\omega_o M} \\
\dot{I}_2 = -\dfrac{j\omega_o M \dot{I}_1}{Z_{s\_s}} = \dfrac{\dot{U}_{AB}}{j\omega_o M} \\
\dot{U}_{ab} = \dot{I}_2 R = \dfrac{\dot{U}_{AB} R}{j\omega_o M}
\end{cases}
\tag{4.20}
$$

As can be seen, the secondary current of the SS compensation network is only related to the input voltage and mutual inductance and is not related to the output voltage or load. According to Eq. (4.20), the output power of the SS compensation network can be calculated

$$
P_{ab} = \mathrm{Re}\left(\dot{U}_{ab} \dot{I}_2^*\right) = \frac{U_{ab} V_{AB}}{\omega_o M} = \frac{U_{ab} U_{AB}}{\omega_o k \sqrt{L_1 L_2}}
\tag{4.21}
$$

where k is the coupling coefficient of L_1 and L_2, $k = \frac{M}{\sqrt{L_1 L_2}}$

4.3 LCL and LCC Compensation Topologies

As shown in Figs. 4.4 and 4.5, LCL and LCC compensation networks are two familiar composite compensation networks (Keeling et al. 2009; Li et al. 2014). They have similar T-shaped structures and are composed of the compensation inductor L_{f1} or L_{f2}, the parallel compensation capacitor C_{f1} or C_{f2}, and self-inductance L_1 or L_2 of the coils. Compared to the LCL compensation networks, the LCC compensation network has a compensation capacitor C_1 or C_2 in series with the self-inductance of the coils.

LCL and LCC also have similar parameter design methods, let each circuit loop is in a resonant state, which can be expressed as

$$LCL:\begin{cases} j\omega_o L_{f1}+\dfrac{1}{j\omega_o C_{f1}}=0,\ j\omega_o L_1+\dfrac{1}{j\omega_o C_{f1}}=0 \\ j\omega_o L_{f2}+\dfrac{1}{j\omega_o C_{f2}}=0,\ j\omega_o L_2+\dfrac{1}{j\omega_o C_{f2}}=0 \end{cases} \tag{4.22}$$

$$LCC:\begin{cases} j\omega_o L_{f1}+\dfrac{1}{j\omega_o C_{f1}}=0,\ j\omega_o L_1+\dfrac{1}{j\omega_o C_1}+\dfrac{1}{j\omega_o C_{f1}}=0 \\ j\omega_o L_{f2}+\dfrac{1}{j\omega_o C_{f2}}=0,\ j\omega_o L_2+\dfrac{1}{j\omega_o C_2}+\dfrac{1}{j\omega_o C_{f2}}=0 \end{cases} \tag{4.23}$$

Simplify Eqs. (4.22) and (4.23), corresponding parameter relations can be obtained as

$$\begin{aligned} &LCL: L_{f1}=L_1, L_{f2}=L_2 \\ &LCC: L_{f1}=L_1-\frac{1}{\omega_o^2 C_1}<L_1, L_{f2}=L_2-\frac{1}{\omega_o^2 C_2}<L_2 \end{aligned} \tag{4.24}$$

From Eq. (4.24), in the LCL compensation network, the value of the compensation inductance L_{f1} or L_{f2} is the same as the self-inductance L_1 or L_2; and in the LCC compensation network, because the series compensation capacitor C_1 or C_2, the compensation inductance L_{f1} or L_{f2} can be designed according to actual needs, and the value can be much smaller than the self-inductance of the coils. Reducing the value of the compensation inductance helps reduce the loss of the compensation circuit and reduces the size of the system.

In the following, according to Kirchhoff's law, the electrical parameters of the LCL and LCC compensation networks can be obtained separately,

$$LCL:\begin{cases}\dot{U}_{AB}=j\omega_oL_{f1}\dot{I}_{Lf1}+\dfrac{\dot{I}_{Lf1}-\dot{I}_1}{j\omega_oC_{f1}}\\ \dot{U}_{AB}=j\omega_oL_{f1}\dot{I}_{Lf1}+j\omega_oL_1\dot{I}_1-j\omega_oM\dot{I}_2\\ j\omega_oM\dot{I}_1=j\omega_oL_2\dot{I}_2+\dfrac{\dot{I}_2-\dot{I}_{Lf2}}{j\omega_oC_{f2}}\\ j\omega_oM\dot{I}_1=j\omega_oL_{f2}\dot{I}_{Lf2}+j\omega_oL_2\dot{I}_2+\dot{U}_{ab}\end{cases}\tag{4.25}$$

Substituting Eq. (4.24) into Eq. (4.25),

$$\begin{cases}\dot{I}_{Lf1}=\dfrac{jM\dot{U}_{ab}}{\omega_oL_{f1}L_{f2}},\ \dot{I}_{Lf2}=-\dfrac{jM\dot{U}_{AB}}{\omega_oL_{f1}L_{f2}}\\ \dot{I}_1=-\dfrac{j\dot{U}_{AB}}{\omega_oL_{f1}},\ \dot{I}_2=\dfrac{j\dot{U}_{ab}}{\omega_oL_{f2}}\end{cases}\tag{4.26}$$

$$LCC:\begin{cases}\dot{U}_{AB}=j\omega_oL_{f1}\dot{I}_{Lf1}+\dfrac{\dot{I}_{Lf1}-\dot{I}_1}{j\omega_oC_{f1}}\\ \dot{U}_{AB}=j\omega_oL_{f1}\dot{I}_{Lf1}+\dfrac{\dot{I}_1}{j\omega_oC_1}+j\omega_oL_1\dot{I}_1-j\omega_oM\dot{I}_2\\ j\omega_oM\dot{I}_1=j\omega_oL_2\dot{I}_2+\dfrac{\dot{I}_2}{j\omega_oC_2}+\dfrac{\dot{I}_2-\dot{I}_{Lf2}}{j\omega_oC_{f2}}\\ j\omega_oM\dot{I}_1=j\omega_oL_2\dot{I}_2+\dfrac{\dot{I}_2}{j\omega_oC_2}+j\omega_oL_{f2}\dot{I}_{Lf2}+\dot{U}_{ab}\end{cases}\tag{4.27}$$

Substituting Eq. (4.23) into Eq. (4.27)

$$\begin{cases}\dot{I}_{Lf1}=\dfrac{jM\dot{U}_{ab}}{\omega_oL_{f1}L_{f2}},\ \dot{I}_{Lf2}=-\dfrac{jM\dot{U}_{AB}}{\omega_oL_{f1}L_{f2}}\\ \dot{I}_1=-\dfrac{j\dot{U}_{AB}}{\omega_oL_{f1}},\ \dot{I}_2=\dfrac{j\dot{U}_{ab}}{\omega_oL_{f2}}\end{cases}\tag{4.28}$$

Equations (4.26) and (4.28) are exactly the same. When the input voltage and operating frequency remain unchanged, the output current remains the same independent of load changes.

The output power of the LCL and LCC compensation network can be calculated as

$$P_{ab}=\mathrm{Re}\big(\dot{U}_{ab}\dot{I}_{Lf2}^{*}\big)=\frac{MU_{AB}U_{ab}}{\omega_oL_{f1}L_{f2}}=\frac{k\sqrt{L_1L_2}U_{AB}U_{ab}}{\omega_oL_{f1}L_{f2}}\tag{4.29}$$

Although the output power expressions of the LCL and LCC compensation networks are the same, the LCL compensation network can transmit less system power than the LCC compensation network. Because in the LCL compensation network, the compensation inductance value must be the same as the self-inductance, and in the LCC compensation network due to the existence of the series compensation capacitor, the value of L_{f1} and L_{f2} can be much smaller than the self-inductance, which can greatly improve the power transfer capability of the LCC compensation network.

4.4 The Comparison of Compensation Topologies

Compared with other compensation networks, SS and LCC compensation networks have three main advantages, which make them more suitable for electric vehicle wireless charging systems. Firstly, both have unit power factor, which can minimize the apparent power level of the system and reduce the reactive power loss of the system; secondly, both the parameter design and resonant frequency of the compensation circuit is independent of the coupled inductance and the load, which reduces the sensitivity of the system to the natural environment and usage conditions, and enhances the robustness of the system; finally, both have constant current output characteristics, which makes the system power decrease to zero when the load is short-circuited, which has good output short-circuit protection characteristics.

However, SS and LCC compensation networks have different structures and circuit characteristics. The application prospects of SS- and LCC compensation networks in the high-power wireless charging system of electric vehicles will be analyzed in the following aspects: the number of components, transmission power, system anti-misalignment ability, and compensation capacitor voltage stress.

(1) **System structure**

The SS compensation network has a simple structure, and the number of components in the SS compensation network is significantly less than that of the double-sided LCC compensation network. Additional components will increase system losses. The loss of compensation capacitance and compensation inductance can be represented by equivalent internal resistances in the circuit. The internal resistance R_C of the compensation capacitor C and the internal resistance R_L of the compensation inductor L can be calculated as

$$\begin{aligned} R_C &= \frac{D_C}{\omega C} \\ R_L &= \frac{\omega L}{Q} \end{aligned} \tag{4.30}$$

where DC represents the loss factor of the capacitor, Q represents the quality factor of the inductance, and ω represents the operation angular frequency of the resonant circuit. Therefore, ideally, the highest efficiency of the SS compensation network is higher than that of the double-sided LCC compensation network.

(2) **Transmission power**

According to the Eq. (4.21), the output power of the SS compensation network is related to the input/output voltage of the system, the size of the coupling inductance, and the operation frequency of the system. According to Eq. (4.29), the output power of the double-sided LCC compensation network is not only related to the input/output voltage of the system, the size of the coupling inductance, and the operation frequency of the system but also related to the added compensation inductance. Although the compensation inductance of the double-sided LCC topology needs to satisfy the Eq. (4.23), due to the existence of the compensation capacitors C_1 and C_2, the value of the compensation inductance is more flexible. In the premise of known input/output voltage, coil parameters (size, structure, and relative position), and system operating frequency, there is still a larger choice in terms of designing the output power in the double-sided LCC topology. In contrast, in the SS topology, the output power has been determined by the existing coil parameters, and cannot be flexibly adjusted according to actual needs.

Assuming that the input/output voltage and operation frequency of the two topologies are the same, the output power ratio δ can be obtained

$$\delta = \frac{P_{ab\_LCC}}{P_{ab\_SS}} = \frac{M^2}{L_{f1}L_{f2}} = \frac{k^2 L_1 L_2}{L_{f1}L_{f2}} \tag{4.31}$$

Depending on the value of the compensation inductance, the double-sided LCC topology can transmit more power than the SS topology (δ > 1), and it can also meet the low-power charging requirements (δ < 1). It is worth noting that the maximum values of the compensation inductances L_{f1} and L_{f2} are L_1 and L_2, respectively. At this time, the double-sided LCC topology is converted to the LCL topology, and the minimum transmission power of the double-sided LCC topology can be calculated as

$$P_{ab\_LCC\_\min} = \frac{kU_{AB}U_{ab}}{\omega\sqrt{L_1 L_2}} = k^2 P_{ab\_SS} \tag{4.32}$$

The minimum transmission power of the double-sided LCC topology is the same as the transmission power of the LCL topology, and is k^2 times that of the SS topology. The wireless charging system of electric vehicles usually adopts loosely coupled coils, and k < 0.3 at this time. Then, the minimum transmission power of the double-sided LCC topology is less than 10% of the SS transmission power.

(3) **Coil misalignment**

In practical applications, because the secondary side is mounted on an electric vehicle, the misalignment of coils often occurs. The coupling coefficient is the parameter that is most prone to change in the wireless charging system. The coupling coefficient when the misalignment occurs is usually smaller than that when the coils are aligned, and in extreme cases, the coupling coefficient is zero when the coils are completely misaligned. According to Eq. (4.21), the output power of the SS topology is inversely proportional to the coupling inductance, that is, when the misalignment is greater, the coupling coefficient is smaller, and the output power of the system is greater. According to the Eq. (4.20), as the coupling inductance decreases, the primary coil current increases. In extreme cases, the mutual inductance drops to zero, and a short circuit occurs on the primary side, which may damage the inverter. Obviously, the misalignment characteristics of the SS compensation network do not satisfy the actual situation. Adding a coil position detection system or a system power detection system can avoid extreme situations and ensure the safe operation of the system. However, this will increase cost and system complexity.

On the contrary, according to Eq. (4.29), the output power of the double-sided LCC topology is proportional to the coupling inductance, that is, when the misalignment is larger, the coupling coefficient is smaller, and the output power of the system is smaller. According to Eq. (4.28), as the coupling inductance decreases, the primary coil current remains constant, and the inverter output current decreases. In extreme cases, the coupling inductance drops to zero and the inverter output current also drops to zero. It can be seen that the coil misalignment will only affect the output power of the double-sided LCC topology, and will not pose a safety threat to the system operation, and no additional protection system is required.

(4) **Current and Voltage stress of the components**

To compare the component stress of the SS compensation network and the LCC compensation network, it is first necessary to control the input/output voltage, self-inductance of coils, coupling coefficient, and output power to be the same. At this time, $\delta = 1$. Normally, the primary/secondary side parameters of the LCC compensation network are symmetrical, and the Eq. (4.28) can be transformed into

$$\begin{aligned} \left|\dot{I}_{1\_LCC}\right| &= \left|-\frac{j\dot{U}_{AB}}{\omega_o L_{f1}}\right| = \left|-\frac{j\dot{U}_{AB}}{\omega_o M}\right| = \frac{U_{ab}}{U_{AB}}\left|\dot{I}_{1\_SS}\right| \\ \left|\dot{I}_{2\_LCC}\right| &= \left|\frac{j\dot{U}_{ab}}{\omega_o L_{f2}}\right| = \left|\frac{j\dot{U}_{ab}}{\omega_o M}\right| = \frac{U_{AB}}{U_{ab}}\left|\dot{I}_{2\_SS}\right| \end{aligned} \tag{4.33}$$

The current ratio of the two topologies is related to the input/output voltage ratio, and the ratio of the primary side is opposite to the secondary side. If $U_{AB} = U_{ab}$, the coil currents are the same. The total loss of the coil can be calculated as

$$P_{coil\_LCC} = \left(I_{1\_LCC}^2 + I_{2\_LCC}^2\right)R_L = \frac{U_{AB}^2 + U_{ab}^2}{\omega_o^2 M^2}R_L = P_{coil\_SS} \tag{4.34}$$

If the same output power is guaranteed, the coil losses of the two topologies are the same, and there is no significant difference in transmission efficiency.

On the other hand, the compensation capacitor has high voltage stress, which puts forward high requirements on the selection and quantity of the capacitor and the design of the capacitor circuit. For example, to ensure that the voltage stress does not exceed the withstand voltage of the capacitor, the number of parallel capacitors is required to increase to reduce the peak voltage of each capacitor; to ensure circuit safety, it is necessary to increase the distance between capacitor pins and other circuits, which increase the system volume. According to Eqs. (4.20) and (4.28), assuming that the input voltage is close to the output voltage, that is, $U_{AB} \approx U_{ab}$, the capacitor voltages of the two compensation networks can be derived as:

$$\begin{aligned}
\dot{U}_{C1_SS} &= \frac{\dot{I}_{1_SS}}{j\omega_o C_1} = \frac{1}{k}\dot{U}_{ab} = \frac{1}{k}\dot{U}_{AB} \\
\dot{U}_{C1_LCC} &= \frac{\dot{I}_{1_LCC}}{j\omega_o C_1} = -\frac{\dot{U}_{AB}}{\omega_o^2 C_1 L_{f1}} = \left(1 - \frac{1}{k}\right)\dot{U}_{AB} \\
\dot{U}_{Cf1_LCC} &= \frac{\dot{I}_{Lf1_LCC} - \dot{I}_{1_LCC}}{j\omega_o C_{f1}} = \dot{U}_{ab} - \dot{U}_{AB} < 2\dot{U}_{AB}
\end{aligned} \tag{4.35}$$

Since the coupling coefficient of the coils is generally less than 0.3, so $\frac{1}{k} > \frac{1}{k} - 1 > 2$. Furthermore, $U_{C1\_SS} > U_{C1\_LCC} > U_{Cf1\_LCC}$ can be obtained. It can be seen that under the same conditions, the compensation capacitor of the LCC topology bears smaller voltage stress than the SS topology.

(5) **Transmission efficiency**

Although under the same conditions, the SS topology has higher operation efficiency due to its fewer components. However, in actual use, since coil misalignment is prone to occur, the coupling coefficient of the coils is usually smaller than the coupling coefficient under alignment. Both the primary and secondary coil currents of the SS topology increase with the decrease of the coupling coefficient. Due to the low quality factor (usually 200–400), the coil loss is the largest. The increase of the coil current will bring about a rapid increase in the overall loss of the system, and the system efficiency will decrease rapidly with the decrease of the coupling coefficient. In contrast, the primary and secondary coil currents of the LCC topology have nothing to do with the coupling coefficient. This means that when the coupling coefficient changes, the working efficiency of the LCC topology will remain relatively stable. In short, although the SS topology can provide higher operating efficiency points, the LCC topology can provide more stable and overall higher system operating efficiency within the allowable misalignment range.

In summary, compared to the SS compensation network, the LCC compensation network has a higher power transmission capacity, and the transmission power can be flexibly adjusted according to the actual situation; when the coil misalignment occurs, it can still work normally without additional protection system; when the same power is transmitted, the stress of the compensation capacitor is smaller, which is conducive to capacitor selection and circuit design; overall more efficient and stable. Therefore, the LCC topology is more suitable for electric vehicle applications.

References

Keeling NA, Covic GA, Boys JT (2009) A unity-power-factor IPT pickup for high-power applications. IEEE Trans Industr Electron 57(2):744–751
Li S, Mi CC (2014) Wireless power transfer for electric vehicle applications. IEEE J Emerging Selected Topics Power Electronics 3(1):4–17
Li S, Li W, Deng J, Nguyen TD, Mi CC (2014) A double-sided LCC compensation network and its tuning method for wireless power transfer. IEEE Trans Veh Technol 64(6):2261–2273
Wang CS, Covic GA, Stielau OH (2004) Power transfer capability and bifurcation phenomena of loosely coupled inductive power transfer systems. IEEE Trans Industr Electron 51(1):148–157
Wang CS, Stielau OH, Covic GA (2005) Design considerations for a contactless electric vehicle battery charger. IEEE Trans Industr Electron 52(5):1308–1314

Chapter 5
Magnetic Couplers

The magnetic couplers are regards as the essential part of the inductive power transfer system. Electric energy is transmitted between the transmitter and the receiver coils through a certain distance (the air gap). The purpose of the research on magnetic couplers is to achieve high power and high efficiency in power transfer while using less material and taking up a relatively small area. Meanwhile, the designed magnetic couplers should be able to maintain robust when the relative position of the transmitter and the receiver coils changes. This chapter introduces basic magnetic coupling theory based on the principle of electromagnetic induction, the basic structures of magnetic coupling coils and their performances, and the performance of the prototype of DDQ coils.

5.1 Design Principle of Magnetic Couplers

The main difference between wireless charging and plug-in charging is that the magnetic coupling coils in wireless charging can be regards as a loosely coupled transformer. The compensation circuit and the magnetic coupling coils can be analyzed separately to analyze the basic coupling theory of the wireless charging magnetic coupling coils. The simplified model is shown in Fig. 5.1. L_1 and L_2 are the self-inductance of the transmitter coil and the receiver coil. R_1 and R_2 are the parasitic resistance of the transmitter coil and the receiver coil. $\dot{I}_1$ and $\dot{I}_2$ are the currents that flow into the transmitter coil and the receiver coil. $\dot{U}_{21}$ is the induced voltage of the receiver coil in the transmitter coil and $\dot{U}_{12}$ is the induced voltage of the transmitter coil in the receiver coil. $\dot{S}_{12}$ is the apparent power transferred from the transmitter coil to the receiver coil and $\dot{S}_{21}$ is the apparent power transferred from the receiver coil to the transmitter coil. S_1 is the apparent power injected into the transmitter coil. S_2 is the apparent power injected into the receiver coil. S_3 is the apparent power

X. Zhang et al., *Wireless Power Transfer Technologies for Electric Vehicles*,
Key Technologies on New Energy Vehicles,
https://doi.org/10.1007/978-981-16-8348-0_5

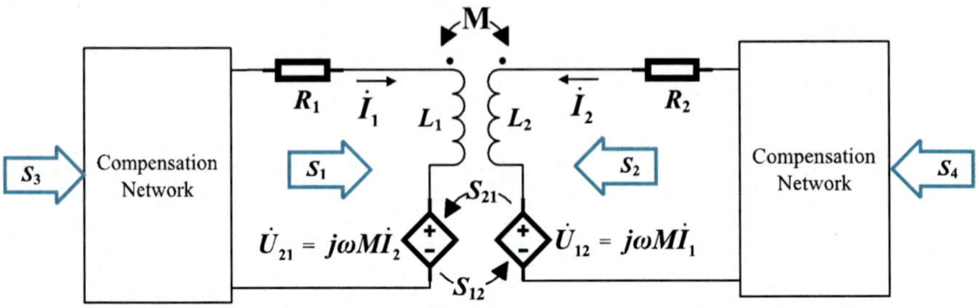

Fig. 5.1 Two-direction power transfer of wireless charging

injected into the primary side. S_4 is the apparent power injected into the secondary side.

As mentioned above, the magnetic coupling coils for wireless charging can be regarded as a loosely coupled transformer. The coupling coefficient of the coils is less than 0.5 (Li and Mi 2015). In order to maximize the transmission power and the power transmission efficiency when the coil is excited by a certain current, capacitance is utilized in the compensation circuit to resonate with the self-inductance or leakage inductance of the magnetic coupling coils. Moreover, the phase of the induced current in the receiver coil ($\dot{I}_2$) should lag the excitation current of the transmitter coil ($\dot{I}_1$) by 90°. The induced voltage of the receiver coil in the transmitter coil ($\dot{U}_{21}$) and the current in the receiver coil should be at the same phase. From the $\dot{U}_{12}$ side, the secondary side should show pure resistance and in this case, the apparent power injected into the primary side (S_3) should be minimum. Then from Fig. 5.1, we have:

$$U_{12}= I_2(R_2+R_{Le})= \omega M I_1= \omega k\sqrt{L_1 L_2}I_1 \tag{5.1}$$

where R_{Le} is the equivalent load resistance, k is the mutual inductance coefficient between the transmitter coil and the receiver coil. The transmission efficiency of coils can be:

$$\eta = \frac{\dot{I}_2 R_{Le}}{\dot{I}_1 R_1 + \dot{I}_2 R_2 + \dot{I}_2 R_{Le}} \tag{5.2}$$

The quality factors of two sides can be defined by $Q_1 = \omega L_1/R_1$ and $Q_2 = \omega L_2/R_2$, respectively. ω is the resonant angular frequency. Then, Eq. (5.1) can be rewritten as:

$$\eta = \frac{R_{Le}}{\frac{(R_2+R_{Le})^2}{k^2 Q_1 Q_2 R_2}+R_2+R_{Le}} \tag{5.3}$$

Define $a = R_{Le}/R_2$, then Eq. (5.3) can be:

$$\eta = \frac{1}{\frac{a+\frac{1}{a}+2}{k^2 Q_1 Q_2} + \frac{1}{a} + 1} \tag{5.4}$$

To maximize η, we need to solve these equations:

$$\frac{\partial}{\partial a}\eta(a) = 0,\ \frac{\partial^2}{\partial^2 a}\eta(a) < 0 \tag{5.5}$$

Considering $a = \sqrt{1 + k^2 Q_1 Q_2}$, η has its maximum value and it can be:

$$\eta_{\max} = \frac{k^2 Q_1 Q_2}{\left(1+\sqrt{1+k^2 Q_1 Q_2}\right)^2} \tag{5.6}$$

From Eq. (5.6), it can be found that the maximum transmission efficiency of the magnetic coupling coils is affected by the mutual inductance coefficient between the transmitter coil and the receiver coil, and their quality factors. Figure 5.2 shows the relation between the maximum transmission efficiency of the magnetic coupling coils and the quality factors of coils at different mutual inductance coefficients. From the Figure, we can find that the maximum transmission efficiency of the coils is proportional to both the mutual inductance coefficient and the quality factor. Moreover, when the mutual inductance coefficient and the quality factor reach a certain value, the maximum transmission efficiency of the coils can be an appreciable value. For example. in the case that $k > 0.2$ and $Q > 200$, the maximum transmission efficiency of the coils can theoretically be more than 95%, which meets the charging demand for electric vehicles. Therefore, the transmission efficiency of the magnetic coupling coils can be high enough to provide theoretical feasibility for the wireless

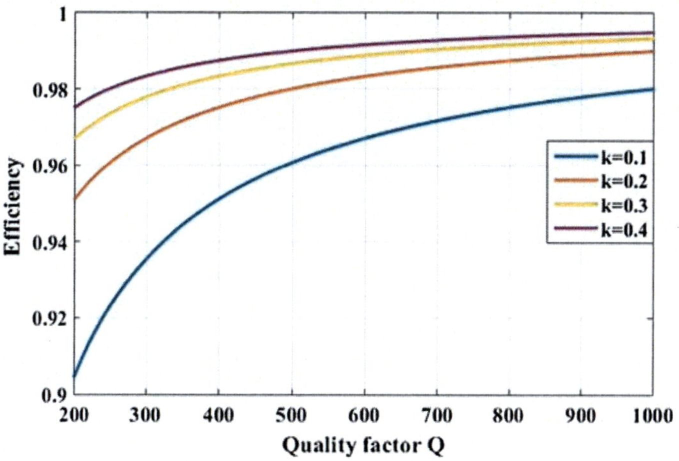

Fig. 5.2 Maximum transfer efficiency versus coupling coefficient and quality factor of coils

charging. In the actual case, this requirement can be met via optimizing the magnetic coupling coils. As mentioned above, the magnetic coupling coils, which have higher coupling coefficient k and higher quality factor Q, have better performance. This is the basic principle of optimization analysis of the magnetic coupling coils.

5.2 Structure of the Magnetic Coupler

The transmission power and the efficiency of the magnetic coupling coils can affect the performance of the entire wireless charging system, which makes it the most critical part of the whole system. The magnetic couplers used currently can be divided into two types: single-side flux magnetic coupler and doubled-side flux magnetic coupler. They are distinguished by whether the magnetic field is distributed on both sides of the coil or one side. For the doubled-side flux magnetic coupler, it mainly refers to the solenoid coil (Fig. 5.3). It can be seen from Fig. 5.3 that the magnetic field is distributed to both sides of the coil. However, in a wireless charging system, only one side of the magnetic field can interact with the magnetic field generated by the other coils. Hence, the magnetic field distributed to the other side is regarded as the leakage magnetic field. In addition, a shielding layer, like the aluminum pad, is usually set behind the coil and the ferrite layer on the side that does not interact with the magnetic field of the other coil to reduce the radiation of the magnetic field to the surrounding environment. This means that there is a large energy loss for doubled-side flux magnetic couplers and results in a decrease in the transmission power and efficiency of the coil. Hence, the performance of doubled-side flux magnetic couplers is relatively poor, and this type of coils has been gradually eliminated in recent years (Budhia et al. 2010). Figure 5.4 shows three typical structures of single-side flux magnetic couplers: circular coils, rectangular coils, and bipolar coils. Differed from the doubled-side flux magnetic coupler, the ferrite layer is set on the side that does not interact with the magnetic field of the other coil for a single-side flux magnetic coupler. The ferrite layer is usually formed by power ferrite strips, which is composed of soft magnetic materials, and the way that power ferrite strips are arranged is determined by different design requirements. The ferrite layer can reduce the leakage magnetic field of the coil, thereby improving the transmission

Fig. 5.3 Double-sided coil

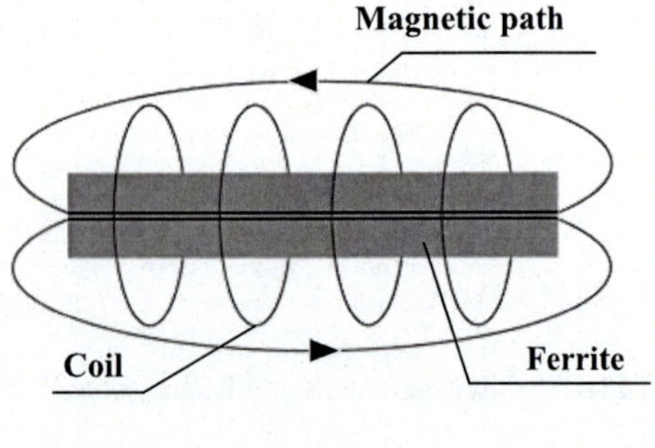

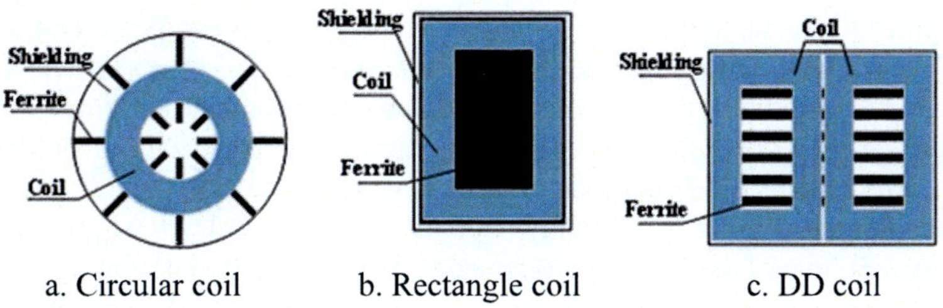

Fig. 5.4 Structure of single-sided coils

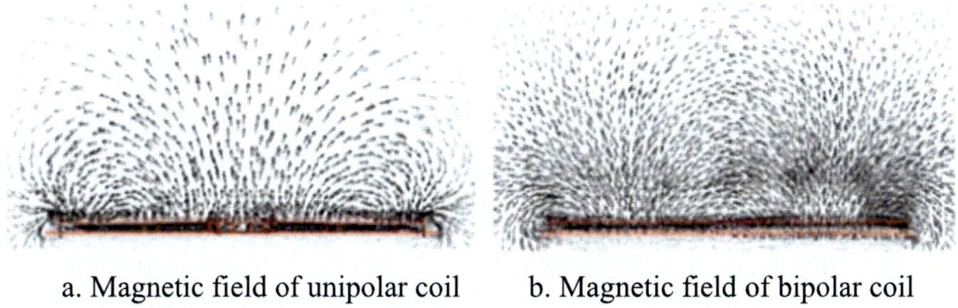

Fig. 5.5 Magnetic distribution of single-sided coils

efficiency of the coil. Moreover, the ferrite layer and the shielding layer can reduce the radiation of the magnetic field to the surrounding. Unipolar coils and bipolar coils are distinguished by the number of magnetic poles generated by the coil. For circular coils and rectangular coils, the magnetic field generated by them has a magnetic pole, and therefore they are unipolar coils. Figure 5.5a shows typical magnetic field distribution of unipolar coils, and the magnetic field on the central axis is vertically arranged. For bipolar coils, like the DD coil, its typical magnetic field distribution is shown in Fig. 5.5b. The magnetic field of the bipolar coil on the central axis is arranged horizontally.

The following mainly compares the performance of the unilateral coil. Among the single-sided coils, the researches on circular coils and DD coils are the most, so the performance of circular coils and DD coils are mainly compared below.

There are many parameters that should be taken into consideration when designing a magnetic coupling coil. For example, the size of the coil, the distance between the transmitter coil and the receiver coil (the air gap), the self-inductances of the transmitter coil and the receiver coil, the mutual inductance of the transmitter coil and the receiver coil, the current injected into the coils and the voltage set of the coils. To compare the transmission performance of each type of coils (mainly transmission power and transmission efficiency), some design properties should be consistent. In the actual design process, to confirm design parameters, at first, we need to follow the guideline for wireless electric vehicle charging (WEVC), which is published by SAE (Society of Automotive Engineers). Then, based on the size of the chassis of the electric vehicle, the size of the magnetic coupling coil can be determined. The distance between the transmitter coil and the receiver coil can be decided by the height of the chassis of the electric vehicle. The voltage and the current injected into the coils can be determined by the voltage level of the battery. In this chapter, the size of the primary side (including the magnetic coupling coil, the ferrite layer, and the shielding) is limited to 500 mm × 500 mm, and the height of the primary side is less than 50 mm. The size of the secondary side is limited to 450 mm × 350 mm, and the height of the secondary side is less than 50 mm. The distance between the coils is 150 mm. The coil is wound with 0.1 mm × 1500 strands of Litz wire, and the diameter of the Litz wire is about 5.5 mm.

It is known from the previous section that the maximum transmission efficiency of the magnetic coupling coils is proportional to the coupling coefficient of the coils and the quality factor of each coil. In the structure of the single-sided flux magnetic coil, the ferrite layer is set on the side of the coil that does not participate in the magnetic field coupling, which can reduce the leakage of the magnetic field. Hence, the quality factor of the coil is promoted. Generally, the quality factor of a single-sided flux magnetic coil is higher than 200. Therefore, the transmission performance of the coils is mainly affected by the coupling coefficient between the transmitter coil and the receiver coil. This means the first target of coil optimization is to promote the coupling coefficient. We utilize the magnetic field finite element simulation software, ANSYS Maxwell, to simulate and optimize magnetic coupling coils, design the structure of coils, and compare the performance of different types of coils.

5.2.1 Double D Coils

The 3D model of the DD structure coils is shown in Fig. 5.6. The whole structure can be divided into the primary side and secondary side. Each side has the same structure, which is composed of the DD coil, ferrite layer, and aluminum plate (the

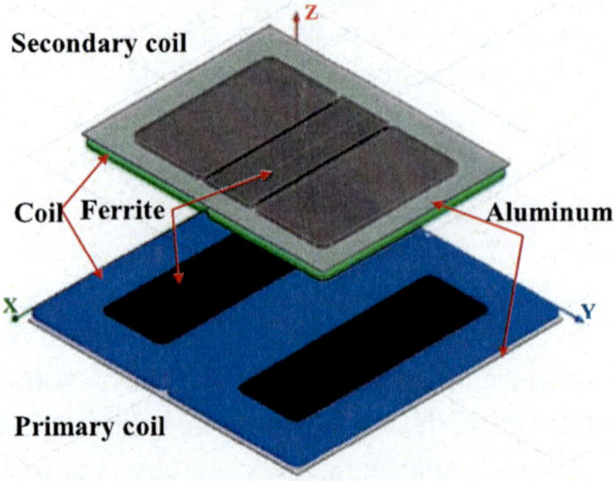

Fig. 5.6 3D model of DD structure coils

shielding). The DD coil is formed by two rectangular coils connected in series and wound as "□□". This winding method makes the magnetic fields generated by the two rectangle coils connected in series and generates two magnetic poles. The utilized ferrite material is PC40 from TDK Corporation, and its shape is a concave plate. The ferrite in the middle of the two rectangular coils of the DD coil is convex, and the surrounding is concave. In other words, the thickness of the ferrite layer decreases from the center of the magnetic coupler to the side in the *Y*-direction (as shown in Fig. 5.6). This structure of the ferrite layer can promote the coupling coefficient between the coils (Mohammad and Choi 2018). The shielding layer used to shield the magnetic field and reduce the radiation of the magnetic field to the surrounding is an aluminum plate. As the main target of coil optimization is to promote the coupling coefficient of the coils, the first step is to set the turn number of the primary coil N and the turn number of the secondary coil n of the DD coil as variable, and the size of the ferrite layer changes as the turn number of coils changes. Other design parameters are fixed. The simulation result is shown in Fig. 5.7. It can be found that the coupling coefficient increases first and then decreases as the turn number of coils increases. When the turn number of the primarY-side coil is 12 and the turn number of the secondarY-side coil is 8, the coupling coefficient between the coils is maximum, which is 0.2764. Therefore, the turn number of the primary coil and the secondary coil can be determined, which are 12 and 8, respectively. After determining the turn number of coils, the structural parameters of the DD coil are shown in Fig. 5.8 and some of the simulation output values are listed in Table 5.1.

When an electric vehicle is charged during parking, misalignment should be taken into concern. The vertical distance between the coils will be different due to the different load of the car, which is vertical misalignment. Horizontal misalignment can be divided into two types based on the position of the vehicle: left-to-right (or door-to-door) and front-to-rear. The misalignment research focuses on the performance of the magnetic coupler (including changes in coupling coefficient and self-inductance) when the relative position of the transmitter coil and the receiver coil changes. The

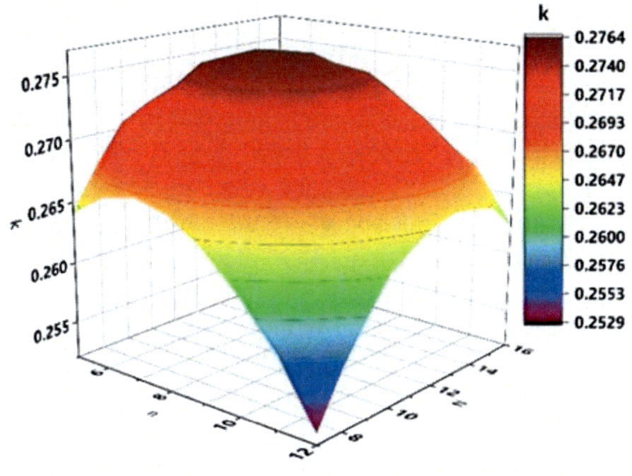

Fig. 5.7 Coupling coefficient versus turn number of DD coils

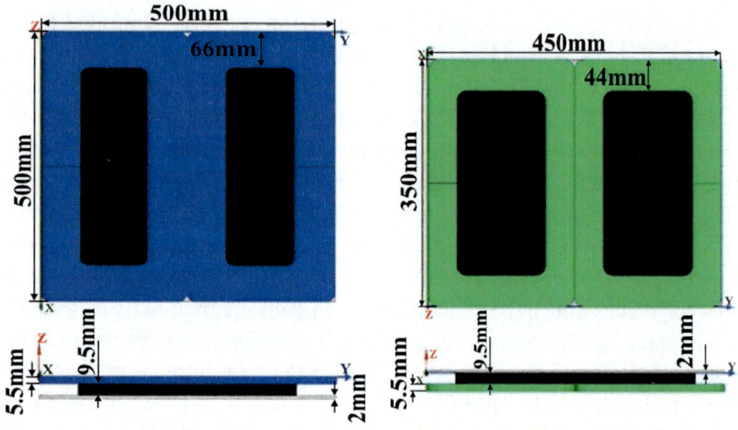

a. Primary coil b. Secondary coil

Fig. 5.8 Structure parameters of DD coils

Table 5.1 Parameters of DD coils

Parameter	Self-inductance of primary coil L_1	Self-inductance of secondary coil L_2	Coupling coefficient k
Value	323.80 μH	127.83 μH	0.2764

vertical misalignment can be realized by changing the relative positions of the coils in the Z-direction and horizontal misalignment can be achieved by changing in the X-direction and Y-direction (as shown in Fig. 5.6).

Based on the previously optimized parameters, the simulation results on misalignment research can be obtained by ANSYS Maxwell software. Figure 5.9 shows how the coupling coefficient of the DD coils changes as horizontal and vertical misalignment happens and Fig. 5.10 illustrates how the self-inductances of the primary coil and the secondary coil changes as the relative position of DD coils changes.

From Fig. 5.9, it can be found that the change of the coupling coefficient is symmetrical in the positive and negative offset directions as far as X and *Y-direction* is concerned, but as the misalignment in Y-direction increases, the coupling coefficient decreases dramatically while it is relatively steady in the X-direction. When the misalignment is 100 mm, the coupling coefficient decreases by approximately 42% in the Y-direction while the coupling coefficient decreases by approximately 10% in the X-direction. Therefore, the coupling coefficient changes with the relative position in the Y-direction is more sensitive than that in the X-direction. Once an EV vehicle is parked, it is more convenient to adjust the position of the vehicle in its heading direction (front-to-rear misalignment) with the help of some detection tools. It is more difficult to adjust the left-to-right misalignment compared to the former one. Therefore, when setting the DD coils, the X-direction should be in door-to-door direction, and the Y-direction should be in front-to-rear direction as DD coils

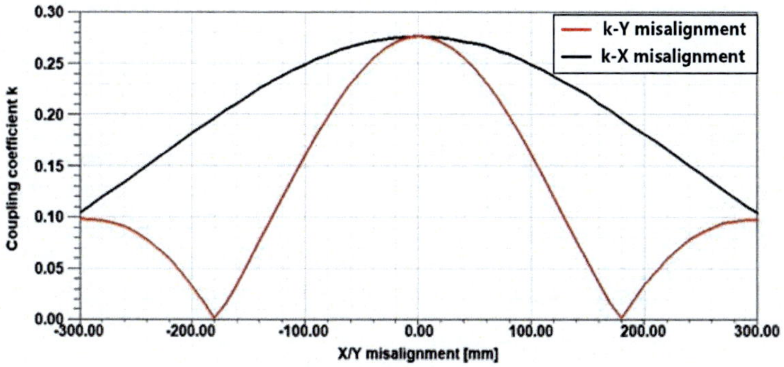

a. Coupling coefficient v.s. *X*- and *Y*-misalignment

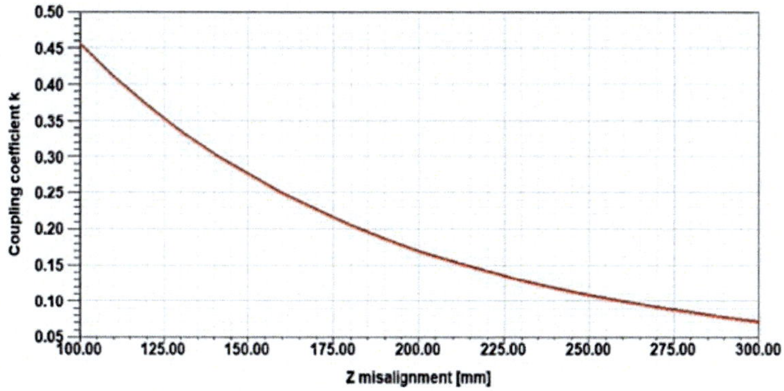

b. Coupling coefficient v.s. *Z*-misalignment

Fig. 5.9 Coupling coefficient versus relative position between two-sided coils of DD coils

show a better misalignment tolerance in the *X*-direction. As for the *Z*-direction, the coupling coefficient decreases as the vertical distance between the coils increases. When the rated distance of coils increases from 150 to 200 mm, the coupling coefficient decreases by about 38.2%. When the rated distance of coils decreases from 150 to 100 mm, the coupling coefficient increases by about 65%.

As it is shown in Fig. 5.10, the self-inductances of the primary coil and the secondary coil change as the relative position of the coils changes in the *X*/*Y*/*Z*-direction, but the amplitude of change is relatively small compared to that of the coupling coefficient. The change of the self-inductances is symmetrical in the positive and negative offset directions when misalignment happens in the *X*-direction or *Y*-direction, which is similar to that of the coupling coefficient. When the misalignment is 300 mm, the changes in the self-inductance of the coils are within 7% for both *X*-misalignment and *Y*-misalignment. For misalignment happened in the *Z*-direction, the self-inductance of the coils decreases as the distance between the coils increases. When the rated distance of coils increases from 150 to 200 mm or decreases from 150

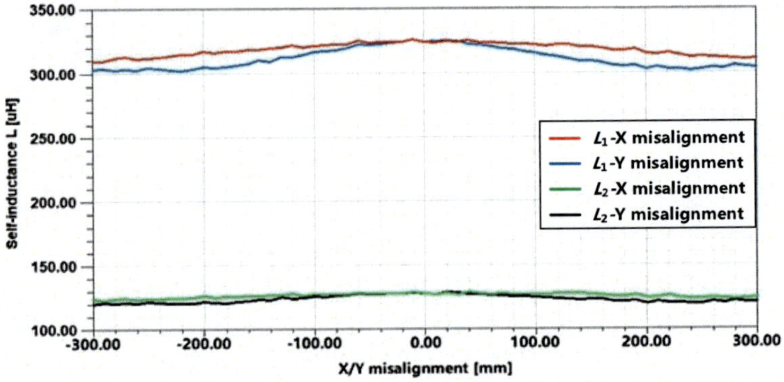

a. Coil self-inductance v.s. X and Y-misalignment

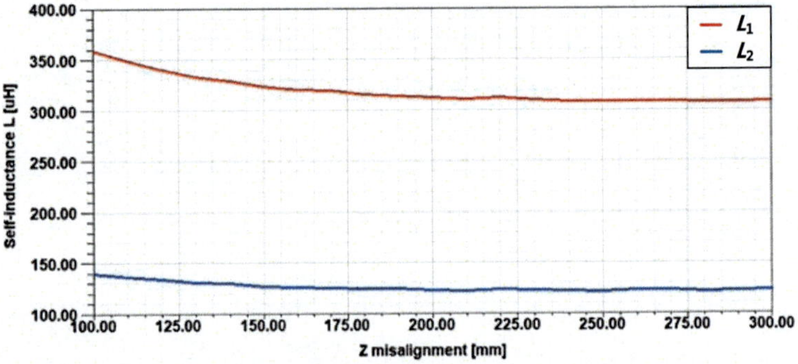

b. Coil self-inductance v.s. Z-misalignment

Fig. 5.10 Coil self-inductance versus relative position between two-sided coils of DD coils

to 100 mm, the change is within 10%. Therefore, the change of the self-inductance of the coils can be neglected when there is a slight change in the relative position of the coils.

5.2.2 *Circular Coil*

The 3D model of the circular coils is shown in Fig. 5.11. Like DD coils, the primary side and the secondary side of circular coils have the same structure, which is composed of the circular coil, ferrite layer, and aluminum plate (the shielding). The structure of the ferrite layer is in the shape of the wheel, which can promote the coupling coefficient (Budhia et al. 2011).

The optimization method of circular coils is the same as that of DD coils, which is to set the turn number of the primary coil N and the turn number of the secondary coil

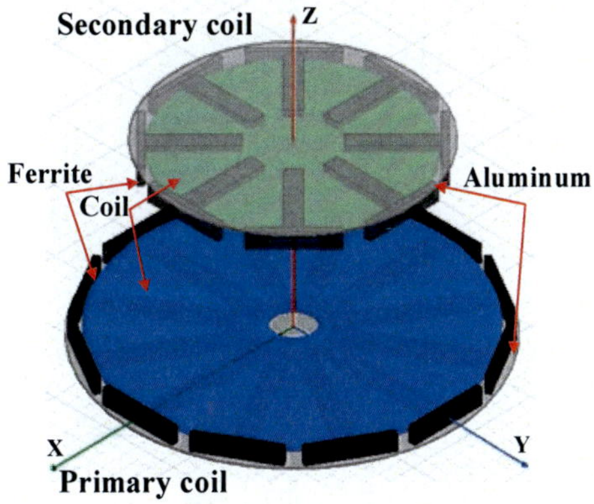

Fig. 5.11 3D model of circular coils

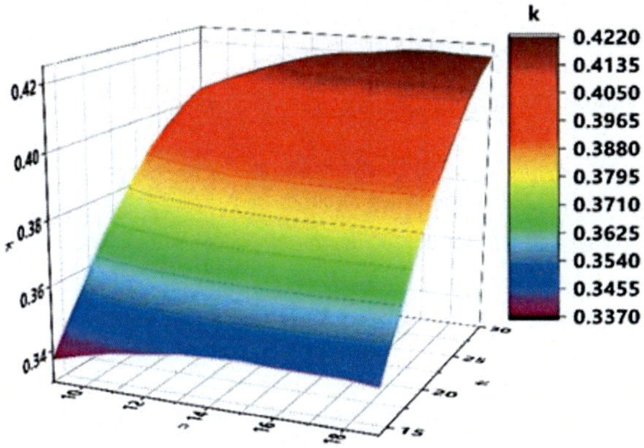

Fig. 5.12 Coupling coefficient versus turn number of circular coils

n of the circular coils as variables, and the arrangement of the ferrite strips changes as the turn number of coils changes, and keep other design parameters fixed. The simulation result is shown in Fig. 5.12.

The coupling coefficient increases as the turn number of the coils increase, and the coupling coefficient is significantly higher than that of DD coils. In order to compare different types of coils, the fixed parameters of the different coils should be the same, which is to keep the self-inductances of the coils as consistent as possible. As mentioned in the previous wireless charging coupling theory analysis, when $k > 0.2$ and $Q > 200$, the maximum transmission efficiency of the coils can theoretically be more than 95%, which meets the charging demand for electric vehicles. Therefore, the coupling coefficient can be decided based on a specific situation in the optimization process. In this section, the turn number of the primary coil N is 20, and the turn number of the secondary coil n is 15, which gives a coupling coefficient of 0.3836. Since the turn number of coils are determined, the structural parameters of the circular coil are shown in Fig. 5.13, and some of the simulation results are listed in Table 5.2.

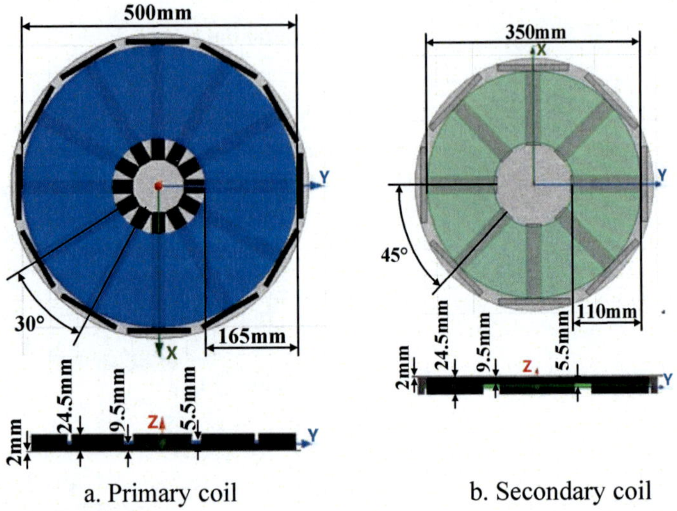

Fig. 5.13 Structure parameters of circular coils

Table 5.2 Parameters of circular coils

Parameter	Self-inductance of primary coil L_1	Self-inductance of secondary coil L_2	Coupling coefficient k
Value	339.55 μH	125.21 μH	0.3836

After design parameters are determined, the misalignment research of circular coils can be conducted. Figure 5.14 demonstrates how the coupling coefficient of the circular coils changes as horizontal and vertical misalignment happens. Figure 5.15 illustrates how the self-inductances of the primary coil and the secondary coil change as the relative position of circular coils changes.

As can be seen from Fig. 5.14, since the circular coil is symmetrical in both *X*- and *Y*-direction, the value and the trend of the coupling coefficient are consistent in these two directions, and the change of coupling coefficient is symmetrical in the positive and negative offset directions.

When the misalignment is 100 mm, the coupling coefficient decreases by about 24.9% in the *X*- and *Y*-direction. The coupling coefficient decreases as the vertical distance between the coils increases in the *Z*-direction. When the vertical distance of coils increases from 150 to 200 mm, the coupling coefficient decreases by about 38.1%. When the vertical distance of coils decreases from 150 to 100 mm, the coupling coefficient increases by about 63.8%. As it's shown in Fig. 5.15, the self-inductances of the primary coil and the secondary coil change as the relative position of the coils changes in the *X*/*Y*/*Z*-direction, but the amplitude of change is relatively small compared to that of the coupling coefficient. The change of the self-inductances is also symmetrical in the positive and negative offset directions when misalignment

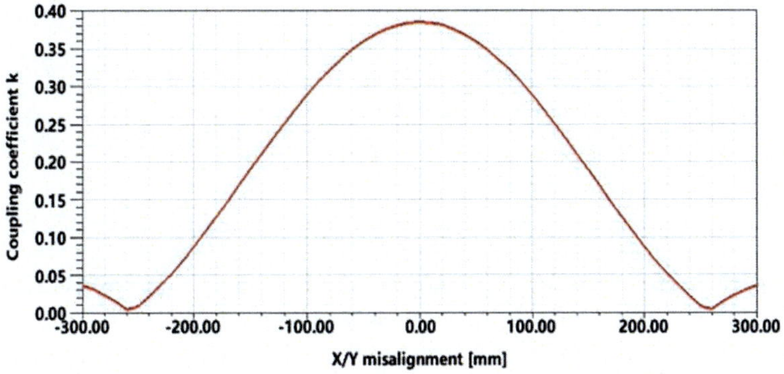

a. Coupling coefficient v.s. X/Y-misalignment

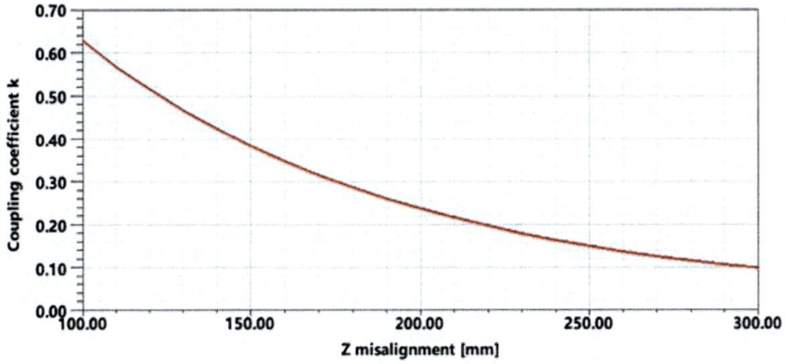

b. Coupling coefficient v.s. *Z*-misalignment

Fig. 5.14 Coupling coefficient versus relative position between two-sided coils of circular coils

happens in the *X*-direction or *Y*-direction. When the misalignment is 300 mm, the changes in the self-inductance of the coils are within 2% for both *X*-misalignment, and *Y*-misalignment. For misalignment happened in the *Z*-direction, the self-inductance of the coils decreases as the distance between the coils increases. When the rated distance of coils increases from 150 to 200 mm or decreases from 150 to 100 mm, the change is within 2%. Therefore, the change of the self-inductance of the coils can be neglected when there is a slight change in the relative position of the coils.

5.2.3 Double D Quadrature Coils (DDQ)

From the previous sections, it can be found that circular coils and DD coils have their own advantages and disadvantages. Although the coupling coefficient of the circular coils is higher than that of DD coils, it decreases dramatically when *X*-misalignment or *Y*-misalignment happens. This means that circular coils show a poor misalignment

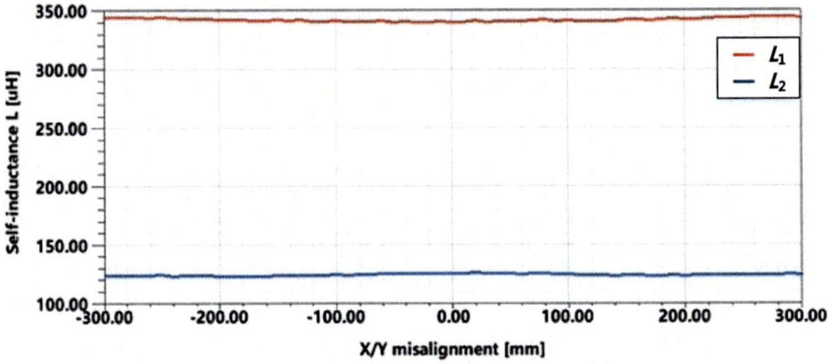

a. Coil self-inductance v.s. X/Y-misalignment

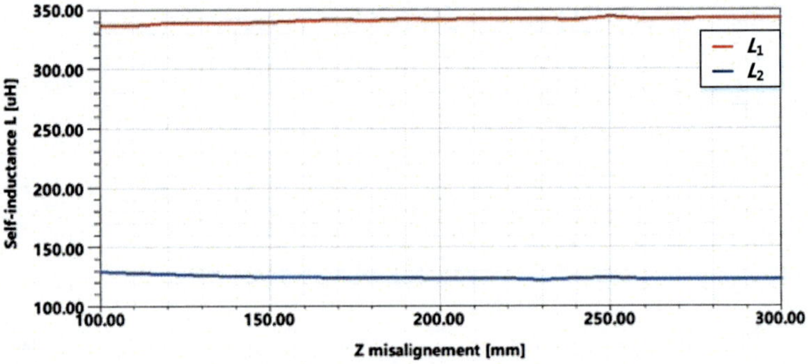

b. Coil self-inductance v.s. Z-misalignment

Fig. 5.15 Coil inductance versus relative position between two-sided coils of circular coils

tolerance. As for DD coils, this type of coils shows DD coils show an appropriate misalignment tolerance in the X-direction while its misalignment tolerance in the Y-direction is quite poor. In order to promote the misalignment tolerance of magnetic couplers, a design that combines the unipolar coil and the bipolar coil is introduced and researched in this section. This type of magnetic coupler utilizes a DD coil as the primary side and combines a unipolar coil (rectangle coil) and a bipolar coil (DD coil) as its secondary side. The secondary side coil is called the Double D Quadrature coil (DDQ coil), and the 3D model of this type of magnetic coupler is shown in Fig. 5.16.

The primary side of the DD-DDQ magnetic coupler shows the same structure as the DD coil, while the structure of the secondary side is composed of a DD coil and a rectangle coil integrated in parallel. The circuit of the DD-DDQ magnetic coupler and its compensation circuit is shown in Fig. 5.17. The output side of the DD coil and the rectangle coil of the secondary side are respectively connected with an LCC compensation circuit and a rectifier to supply power to the battery together. Due to the different structures of the primary side and the secondary side, the structure of the ferrite layer of the primary side is a concave plate, while the ferrite layer of the secondary side adopts a combination of plate structure and strip structure. The

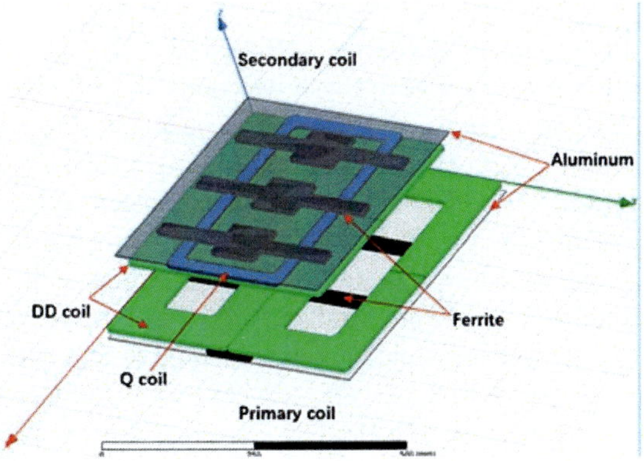

Fig. 5.16 3D model of DDO structure coils

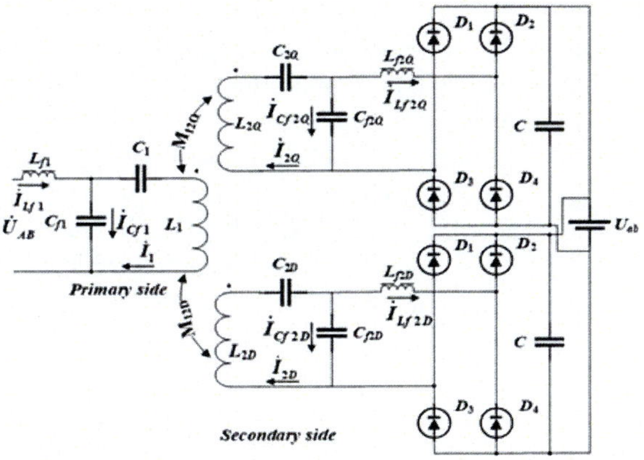

Fig. 5.17 Equivalent circuit of double-sided LCC compensation topology based on DDQ coils

aluminum plates are both utilized on two sides for shielding. The Q coil is set in the center of the secondary side DD coil, where the magnetic field generated by the secondary side DD coil is orthogonal to the Q coil, as shown in Fig. 5.18a. The magnetic flux generated by the secondary side DD coil passes through the Q coil

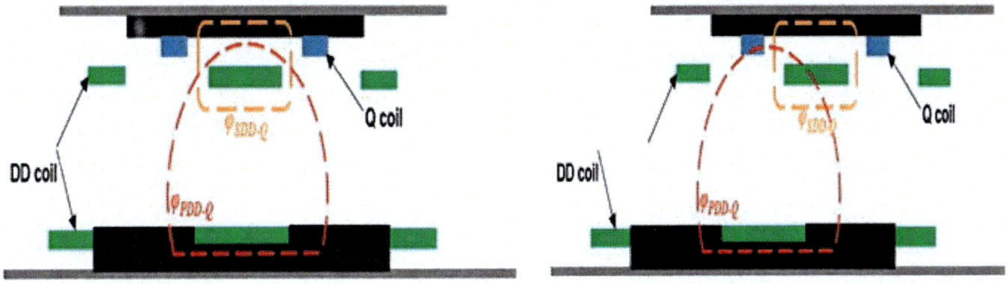

a. No misalignment b. *Y*-misalignment

Fig. 5.18 Magnetic field distribution versus relative position between two-sided coils of DDQ coils

($\varphi_{SDD\text{-}Q}$), and the flux path is shown in Fig. 5.18, which suggests that the magnetic field between the secondary side DD coil and the Q coil is decoupled and the mutual inductance between these two coils is zero. When the misalignment is zero, The magnetic flux generated by the primary side DD coil passes through the Q coil ($\varphi_{PDD\text{-}Q}$), and the flux path in Fig. 5.18a shows that the magnetic field between the primary side DD coil and the Q coil is decoupled. Therefore, the mutual inductance of the primary side DD coil and the Q coil is zero. When misalignment in *Y*-direction happens, shown in Fig. 5.18b, the mutual inductance between the secondary DD coil and the Q coil is still zero. However, in this case, the quantity of magnetic flux, which goes into the Q coil, is inequal to that goes out of the Q coil. And the mutual inductance of these two coils cannot maintain zero.

Compared with the DD-DD coils, the coupling between the DD coils on both sides of the DD-DDQ magnetic coupler is completely the same as that of DD-DD coupler. Hence, the DD-DDQ coils can show the performance similar to DD-DD coils. Meanwhile, due to the existence of the secondary side Q coil, when misalignment in *Y*-direction happens, the primary side DD coil and the Q coil can gradually generate mutual inductance, which makes the parallel secondary side DD coil and Q coil simultaneously induced. Then, the output energy is transmitted to the back-end circuit. Therefore, DD-DDQ magnetic coupler increases the utilization rate of the magnetic field generated by the primary side coil and promotes the transmission power and efficiency when misalignment in the *Y*-direction taken place. When there is a misalignment in the *X*-direction, the primary side DD coil and the Q coil are still decoupled, and the performance of DD coils on two sides is the same as that of DD-DD magnetic coupler.

Based on characteristics of DDQ coils, in order to be comparable to circular coils and DD coils, the value of the overall coupling coefficient of the DD-DDQ coil is the sum of coupling coefficient between primary side DD coil and secondary side DD coil, and coupling coefficient between primary side DD coil and secondary side Q coil. The self-inductance of secondary DDQ coil is the sum of self-inductances of secondary side DD coil and the Q coil.

To optimize DD-DDQ coils, we take the same methods used in other types of magnetic couplers. The turn number of secondary DD coil and the Q coil are the same. The simulation result is shown in Fig. 5.19. As the turn number of coils increases, the overall coupling coefficient increases first and then decreases. When the turn number of the primary-side coil $N = 13$ and the turn number of the secondary-side coil $n = 7$, the coupling coefficient between the coils is maximum, which is 0.2490. Therefore, the turn number of the primary coil and the secondary coil can be determined, which are 13 and 7, respectively. After determining the turn number of coils, the structural parameters of the DD coil are shown in Fig. 5.20, and some of the simulation output values are listed in Table 5.3.

Since the design parameters of DD-DDQ magnetic coupler are determined, the misalignment research can be conducted. Figure 5.21 demonstrated how the coupling coefficient changes as horizontal and vertical misalignment happens, where k is the overall coupling coefficient of two sides, k_{1D} is the coupling coefficient of primary and secondary DD coils, and k_{1Q} stands for the coupling coefficient of primary DD

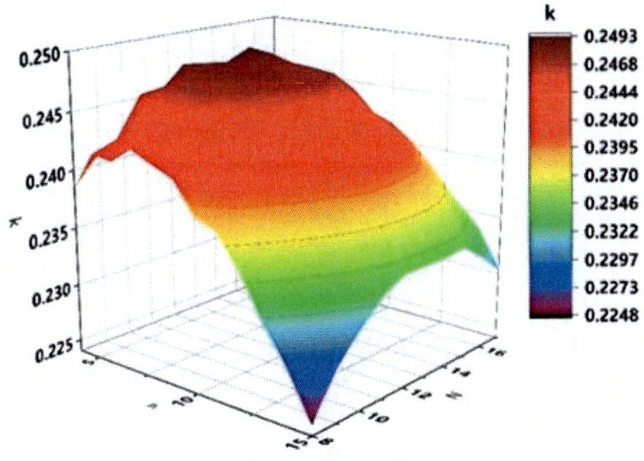

Fig. 5.19 Coupling coefficient versus number of turn number of DDQ coils

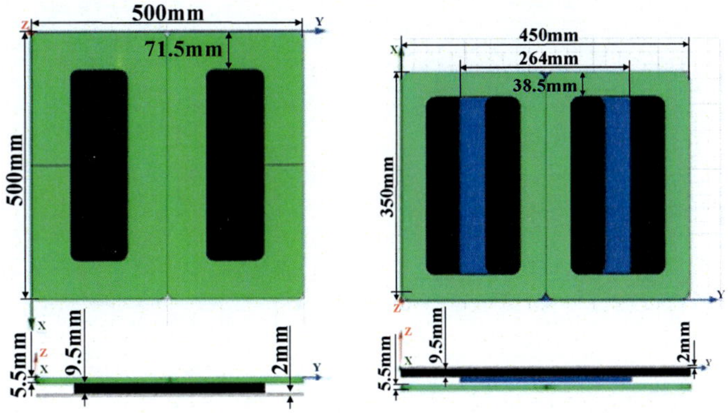

a. Primary coil b. Secondary coil

Fig. 5.20 Structure parameters of DDQ coils

Table 5.3 Parameters of circular coils

Parameter	Self-inductance of primary coil L_1	Self-inductance of secondary coil L_2	Coupling coefficient k
Value	350.33 μH	125.18 μH	0.2490

coil and the Q coil. Figure 5.22 illustrates how the self-inductances of the primary coil and the secondary coil change as the relative position of coils changes. From Fig. 5.21, it can be found that the change of the coupling coefficient is symmetrical in the positive and negative offset directions as far as the *X*/*Y*-direction is concerned. The overall coupling coefficient decreases as the *X*-misalignment increases. And it increases at first and then decreases as the *Y*-misalignment increases. This is a combination of how secondary DD coil and Q coil react as the misalignment happens. Compared to the circular coil and DD coil, DDQ coil shows a promotion

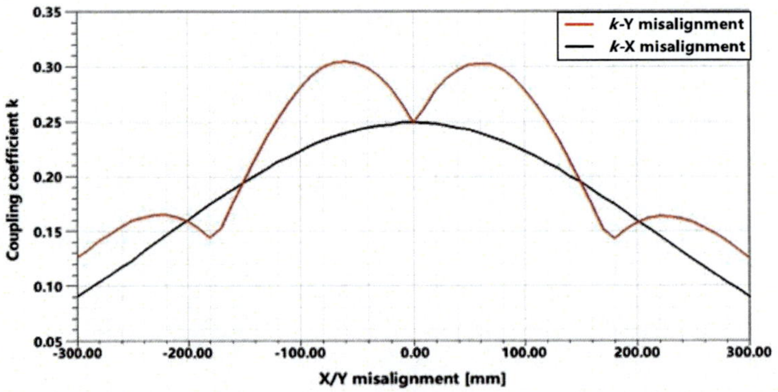

a. Overall coupling coefficient v.s. *X*/*Y*-misalignment

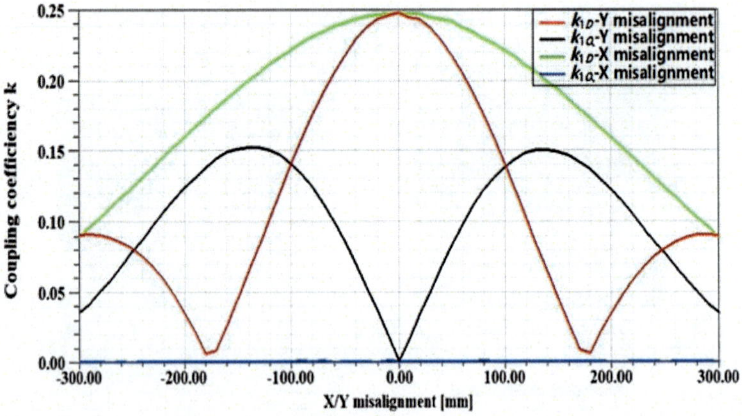

b. Coupling coefficient v.s. *X*/*Y*-misalignment

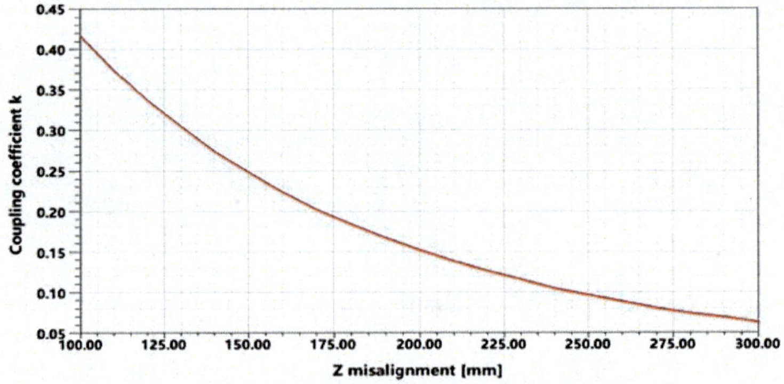

c. Overall coupling coefficient v.s. *Z*-misalignment

Fig. 5.21 Coupling coefficient versus relative position between two-sided coils of DDQ coils

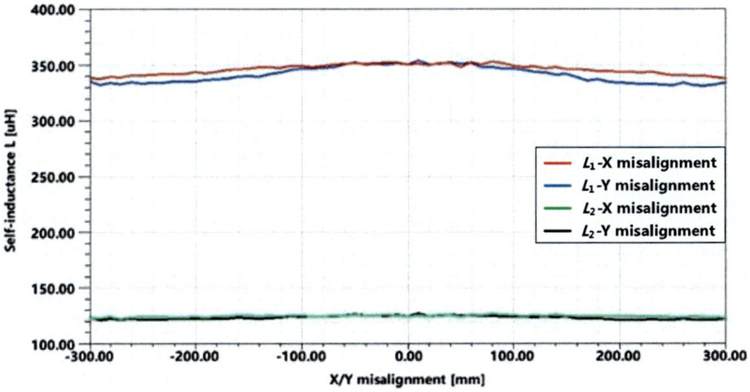

a. Coil inductance v.s. *X*/*Y*-misalignment

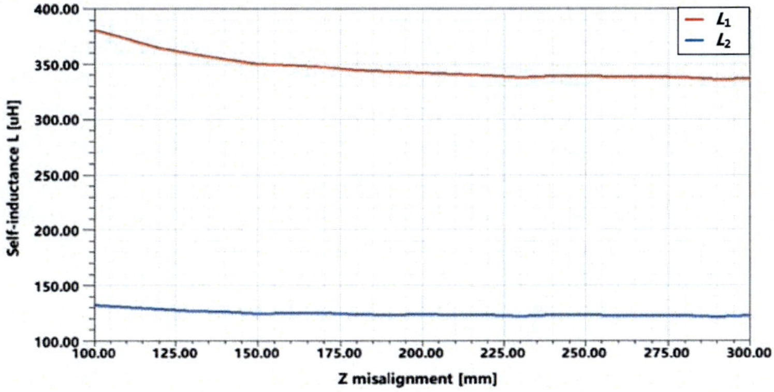

b. Coil inductance v.s. *Z*-misalignment

Fig. 5.22 Coil inductance versus relative location between two-sided coils of DDQ coils

in *Y*-misalignment tolerance. When the misalignment is 100 mm in the *Y*-direction, the coupling coefficient increases by approximately 12%. The coupling coefficient decreases by approximately 10.5% when the same misalignment in the *X*-direction. Like setting DD coils, the *X*-direction should be in door-to-door direction, and the *Y*-direction should be in front-to-rear direction when setting DDQ coils to the EV. For misalignment happened in the *Z*-direction, the overall coupling coefficient decreases as the misalignment increases. When the air gap increases from 150 to 200 mm, the coupling coefficient decreases by about 38.8%. When the rated distance of coils decreases from 150 to 100 mm, the coupling coefficient increases by about 67.2%.

Like circular coils and DD coils, the self-inductances of DD-DDQ coils show little change as misalignment happens. Form Fig. 5.22, it can be found that when the misalignment is 300 mm in *X*- or *Y*-direction, the change in the self-inductance of the coils is within 6%. Moreover, this change is less than 9% when the air gap increases from 150 to 200 mm or decreases from 150 to 100 mm.

5.2.4 Comparison of the Different Structure of Coils

Table 5.4 show a comparison of the three mentioned coils. It can be found that the self-inductances of these coils show little change when misalignment happens. In the term of coupling coefficient, these three coils show similar performance in Z-misalignment. However, DDQ coils show better performance when there is a horizontal misalignment. Therefore, from the simulation results, the DDQ coils show better performance. However, as mentioned above, the circuit of the DD-DDQ magnetic coupler is different from the other two types of coils. Its secondary side adopts a two LCC compensation circuit that connected in parallel. This topology may affect the performance of the front-end inverter and the performance of the back-end rectifier. Therefore, the compensation circuit should be taken into consideration when comparing the performance of the system of different coils.

To compare the effect of the compensation circuit, the parameters of the system should be confirmed first. Both circular coils and DD coils utilize double side LCC topology. The primary side of DDQ coils uses one LCC compensation circuit, and the secondary side uses two compensation circuits. There are some variables that can affect the output voltage and power of coils in actual cases, such as the misalignment of coils and the voltage change of battery during the charging process. To compare the performance of different coils, some variables are fixed. The rated voltage at the input of the inverter is set to 356 V, the rated voltage at the output of the rectifier is 300 V. The rated output power of the system is 3 kW. After the coil parameters are confirmed, we can optimize the parameter of the compensation circuit using the method mentioned in Chap. 4. Table 5.5 shows the parameters of the compensation circuit of different magnetic couplers. Then, the simulation is conducted in Pspice. As the DC-DC output power and efficiency of the three coils are symmetrical about

Table 5.4 Performance comparison of the three coils

	Offset (mm)	DD			Circular			DDQ		
		k	L_1(μH)	L_2(μH)	k	L_1(μH)	L_2(μH)	k	L_1(μH)	L_2(μH)
x	0	0.2764	323.80	127.83	0.3836	339.55	125.21	0.2490	350.33	125.18
	± 100	−10.0%	−0.8%	−0.2%	−24.9%	+ 0.3%	−0.5%	−10.5%	−0.3%	+ 0.4%
	± 150	−20.9%	−1.3%	−0.1%	−50.8%	+ 0.6%	−1.1%	−21.8%	−1.3%	−0.7%
	± 200	−34.3%	−3.0%	−2.4%	−77.4%	+ 0.7%	−1.5%	−35.9%	−1.8%	−0.3%
y	0	0.2764	323.80	127.83	0.3836	339.55	125.21	0.2490	350.33	125.18
	± 100	−42%	−2.8%	−2.0%	−24.9%	+ 0.1%	−0.1%	−12%	−1.1%	−0.5%
	± 150	−79.8%	−4.6%	−3.2%	−50.8%	+ 0.3%	−0.5%	−22%	−2.5%	−0.8%
	± 200	−87.6%	−6.7%	−6.1%	−77.4%	+ 0.6%	−1.1%	−36.7%	−4.6%	−2.9%
z	100	+ 64.9%	+ 10.8%	+ 9.9%	+ 63.8%	+ 0.7%	−1.5%	+ 67.2%	+ 8.8%	+ 6.1%
	150	0.2764	323.80	127.83	0.3836	339.55	125.21	0.2490	350.33	125.18
	200	−38.2%	−3.5%	−3.2%	−38.1%	+ 0.5%	−1.2%	−38.8%	−2.3%	−0.7%

Table 5.5 Parameters of the three coils

Primary side	Circular		DD	DDQ	Secondary side	Circular	DD	DDQ	
								DD	Q
	C_1(nF)	13.995	14.04	12.515	C_2(nF)	54	45.5	82	162
	R_{C1}(Ω)	0.027	0.027	0.030	R_{C2}(Ω)	0.008	0.010	0.006	0.003
	C_{f1}(nF)	39.377	47.318	49.947	C_{f2}(nF)	73.04	85.51	106.24	206.23
	R_{Cf1}(Ω)	0.095	0.0791	0.075	R_{Cf2}(Ω)	0.051	0.044	0.035	0.018
	L_{f1}(μH)	89.03	74.092	70.192	L_{f2}(μH)	48	41	33	17
	R_{f1}(Ω)	0.118	0.098	0.093	R_{f2}(Ω)	0.063	0.054	0.043	0.023

the zero point when misalignment happens in *X*- or *Y*-direction, the simulation result only shows the positive axis.

Figure 5.23 demonstrates how the *X*-misalignment affects the DC-DC output power and efficiency. As the *X*-misalignment increases, the output power of DD and DDQ coils show a similar tendency of decreasing, which is much slower than that of circular coils. When the *X*-misalignment is around 200 mm, the output power of the circular coil decreases dramatically and then reaches the null point. The three different coils show similar value, and the tendency of changes in efficiency when the misalignment in the X-direction is less than 200 mm. Although their curves of efficiency show no obvious regularity, the value is around 96–99%, which meets the requirement of wireless charging.

The relationship between *Y*-misalignment and the DC-DC output power and efficiency can be found in Fig. 5.24. From the power curve of the DD coil, it can be observed that the output power of the DD coil decreases rapidly at first until reaching a null point, and then it increases as the *Y*-misalignment increases. However, the null

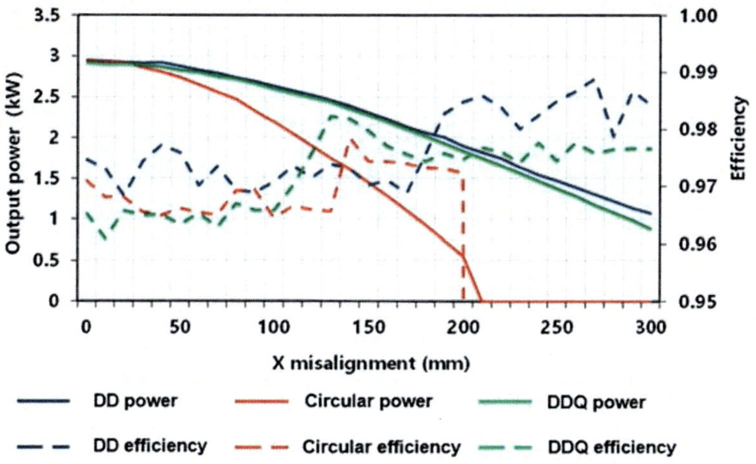

Fig. 5.23 DC-DC output power and efficiency versus *X*-misalignment of the three coils

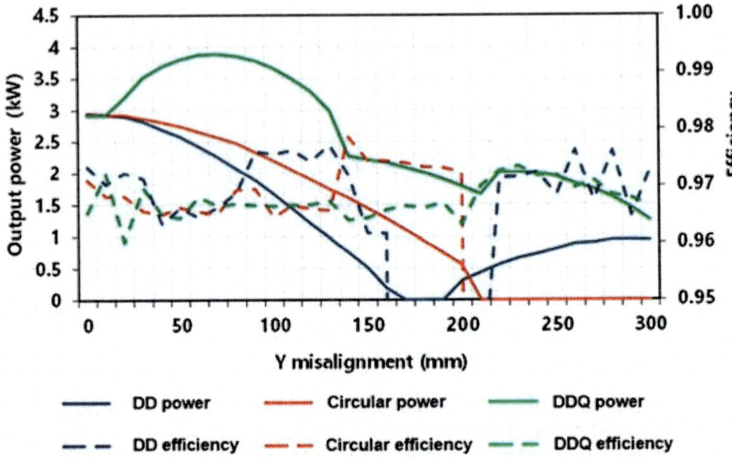

Fig. 5.24 DC-DC output power and efficiency versus *Y*-misalignment of the three coils

point shows up in a large misalignment that is not allowed in actual application. The curve of the circular coil is the same as that in the X-misalignment case. DDQ coil shows a tendency that the output power increases first and then decreases as the offset becomes larger, and the trend of decreasing is slower than the other two kinds of coils. The efficiency of the three coils fluctuates between 96 and 98%.

Figure 5.25 shows the curve of DC-DC output power and efficiency when the air gap changes. The curves of the output power of the three coils are similar, which decreases as the air gap increases. Meanwhile, the efficiency curves of these coils indicate that the efficiency can reach more than 95% as the air gap is less than

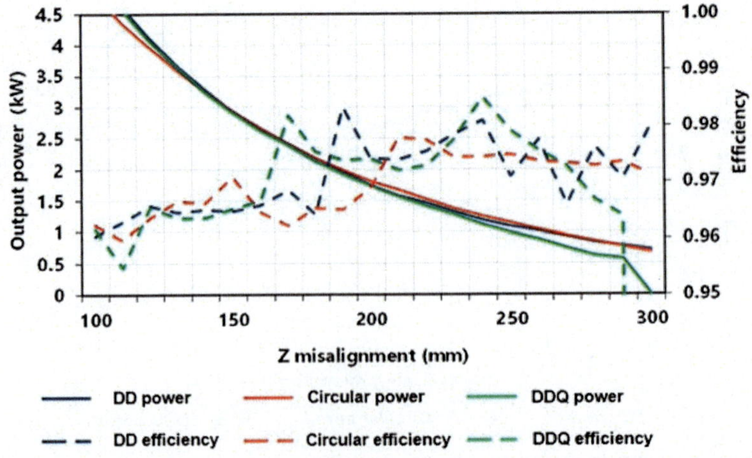

Fig. 5.25 DC-DC output power and efficiency versus Z-misalignment of the three coils

300 mm. In summary, DDQ coil shows better misalignment tolerance than DD coil and circular coil.

5.2.5 Prototype of DDQ Coil

From simulation results, DDQ coil is more promising than the other two types of coils facing misalignment. This part is to illustrate the performance of the DDQ prototype and compare with simulation results. Figures 5.26 and 5.27 show the structure of coils and ferrite of the DD-DDQ magnetic coupler prototype, which is built based on the simulation parameters. The shell of the prototype is made of nylon material in consideration of stress, weight reduction, heat dissipation, etc. The comparison between simulation results and experimental results is shown in Table 5.6.

Figures 5.28 and 5.29 demonstrate the experimental results of how DC-DC output power and efficiency change when misalignment happens in the *X*-direction and *Y*-direction, respectively. The impact on DC-DC output power and efficiency when both X-misalignment and *Y*-misalignment happen at the same time is shown in Fig. 5.30. The output power curve decreases as *X*-misalignment increases while the output power increases first and then decreases as the *Y*-misalignment becomes larger, which

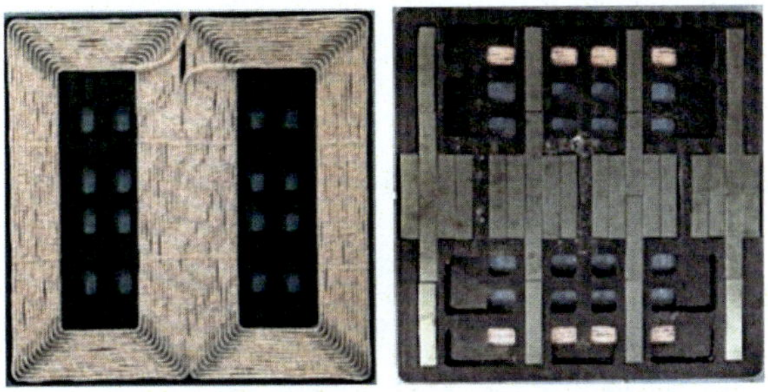

a. DD coil b. Ferrite

Fig. 5.26 Primary-side coils and ferrites

a. DD coil b. Q coil c. Ferrite

Fig. 5.27 Secondary-side coils and ferrites

Table 5.6 Coil parameters

Primary side			Secondary side					
			DD coil			Q coil		
	Simulation	Prototype		Simulation	Prototype		Simulation	Prototype
L_1(uH)	221.2	219.2	L_{2D}(uH)	136.2	134.3	L_{2O}(uH)	72.8	71.8
$R_1(\Omega)$	0.064	0.582	$R_{2D}(\Omega)$	0.032	0.332	$R_{2O}(\Omega)$	0.021	0.201
L_{f1}(uH)	65.65	64.03	L_{f2D}(uH)	43.00	44.26	L_{f2O}(uH)	22.00	21.51
$R_{Lf1}(\Omega)$	0.087	0.081	$R_{Lf2D}(\Omega)$	0.057	0.059	$R_{Lf2O}(\Omega)$	0.029	0.031
C_1(nF)	22.54	23.11	C_{2D}(nF)	37.62	36.8	C_{2O}(nF)	69.01	69.3
$R_{C1}(\Omega)$	0.017	0.033	$R_{C2D}(\Omega)$	0.010	0.087	$R_{C2O}(\Omega)$	0.054	0.056
C_{f1}(nF)	53.41	52.9	C_{f2D}(nF)	81.53	82.4	C_{f2O}(nF)	159.4	158.4
$R_{Cf1}(\Omega)$	0.070	0.110	$R_{Cf2D}(\Omega)$	0.046	0.105	$R_{Cf2D}(\Omega)$	0.024	0.025
			K_{12D}	0.301	0.289	K_{12O}	0.00	0.00

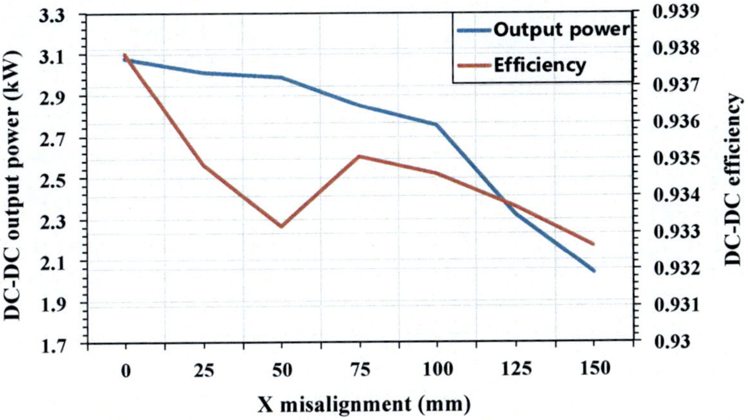

Fig. 5.28 Output power and efficiency of DC-DC versus *X*-misalignment

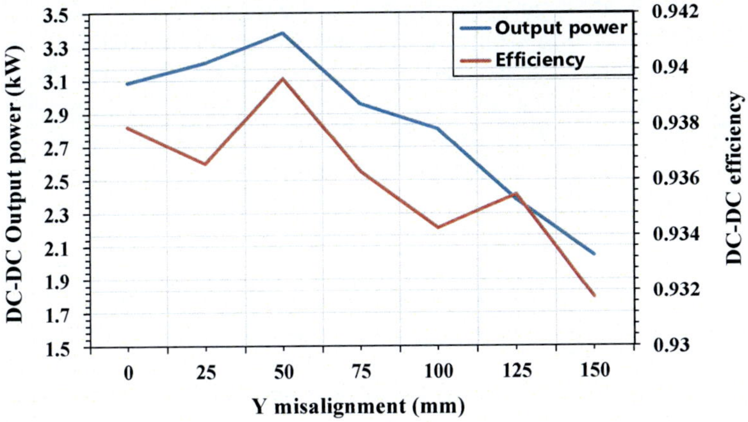

Fig. 5.29 Output power and efficiency of DC-DC versus *Y*-misalignment

are similar to simulation results shown in Figs. 5.23 and 5.24. When the misalignment is within 150 mm, the DC-DC output power can be more than 2 kW, and the efficiency is higher than 93% no matter the misalignment is in *X*- or *Y*-direction. When the misalignment is less than 100 mm in both *X* and *Y*-direction, the DC-DC output power and the efficiency can maintain 2.5 kW and 93%, respectively.

5.3 Quadrature Coils

In recent years, the demand for EVs has been growing due to the increase in greenhouse-gas emissions and the shortage of fossil fuels (Choi et al. 2014). However,

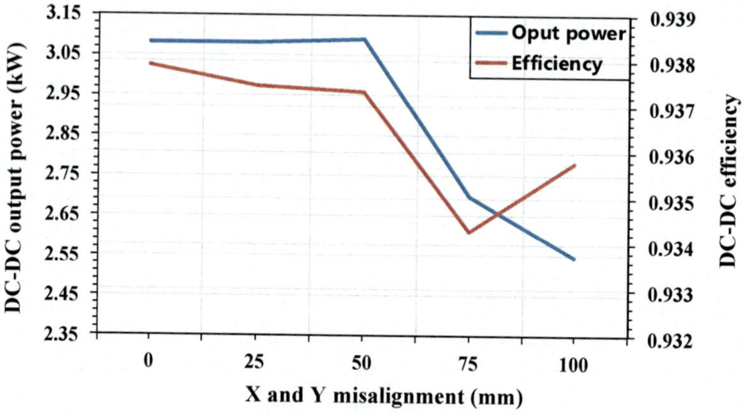

Fig. 5.30 Output power and efficiency of DC-DC versus *X*- and *Y*-misalignment

charging is one of the major barriers to the overall acceptance of EVs (Zhu et al. 2014). WPT technology is convenient, safe, and space-saving, which could be regarded as an ideal substitute for conductive charging (Huang et al. 2015; Ning et al. 2013). Therefore, WPT technology is applied to develop practical wireless charging systems for EVs. The relevant research in this field is of great variety, such as power electronic converters, resonant circuit topologies, coil design, safety, and foreign object detection (Wu et al. 2011; Jang et al. 2016). Among them, coil design and resonant circuit topologies are two key points (Li et al. 2015a; Deng et al. 2015; Kan et al. 2017).

The resonant circuit topology is essential in WPT as it helps to adjust the resonant frequency, minimize the VA rate of the power supply, improve power transfer capability and achieve high efficiency (Zhang et al. 2014). Four basic resonant networks labelled as "SS", "SP", "PS", and "PP" have been proposed and widely adopted in previous works. "P" or "S" denotes how the resonant capacitors are connected to the coupling coils, i.e., "P" represents parallel connection, and "S" represents series connection (Deng et al. 2015; Kan et al. 2017). The primary compensation capacitance of "SS" is independent of the coupling and load conditions. Nevertheless, it is difficult to adjust the output voltage (Cho et al. 2013). For "SP", the primary compensation capacitance could be affected by the coupling condition. For "PS" and "PP", the compensation capacitance is determined by both the coupling and load conditions. Therefore, the resonant tanks consisting of predetermined compensation capacitance may be broken by unexpected misalignment of couplers or load variation. In other words, the predetermined resonant frequency varies as coupling and load conditions change. Many advantageous resonant networks have been promoted to solve these problems. The LCC-LCC resonant network, shown in Fig. 5.31, is accepted as a suitable solution. The highlights of LCC-LCC resonant networks are listed as follows:

Determined by the predefined compensation capacitance and inductance, the resonant frequency of the LCC-LCC resonant network is irrelevant to the coupler's

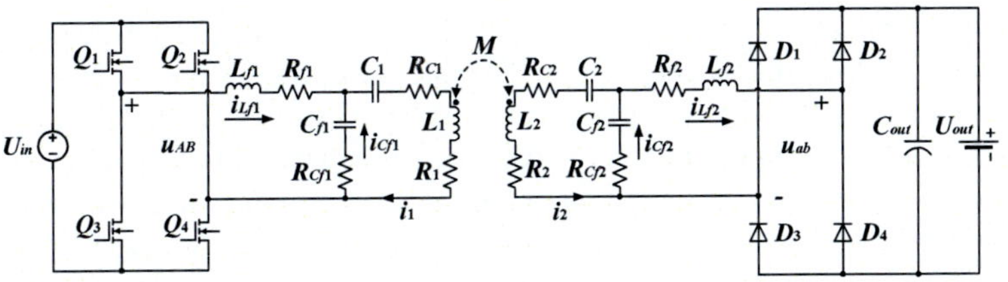

Fig. 5.31 LCC-LCC resonant network with its front-end and rear-end circuit

coupling coefficient and load condition. In addition, the current flowing into the rectifier could be considered as a constant current source, the amplitude of which is proportional to output voltage of the inverter. This simplifies the adjustment process of charging current. Moreover, part of the inductance of coupling coils is compensated by the series-connected compensation capacitance, which not only minimizes the inductance value and dimensions of compensation inductors, but also makes the tuning of resonant tanks more convenient compared with the LCL-T resonant network (Borage et al. 2005). Due to its robustness and effectiveness, LCC-LCC resonant network is utilized in this section.

Coil design is another essential area in WPT research and development, which greatly determines the power capacity and transmission efficiency of the whole WPT system. As essential design indexes, coupling coefficient, self-inductance, quality factor, and misalignment tolerance should be closely investigated (Zaheer et al. 2015). The magnetic coupler of the WPT system is loosely coupled, i.e., the primary coil and the secondary coil are linked across a large air gap, which could lead to poor coupling coefficient. In addition, it is difficult for two coupling coils to be fully aligned, and the misalignment may lead to a serious decline in transmission power and efficiency. In order to solve these problems, ferrite cores are arranged on the back of the coils to enhance the magnetic flux guidance and reduce the leakage loss (Budhia et al. 2013). On the other hand, the shape and turn number of coils are optimized to achieve a higher coupling coefficient and a larger stable charging area. In (Ning et al. 2013), a 2 kW 700 mm-diameter circular pad with a horizontal radial tolerance of 130 mm is proposed. It is too large in size, and the misalignment tolerance is very small. Reference (Takanashi et al. 2012) gives a flux-pipe coil that can transfer 3 kW with 90.1% efficiency at 100 mm *X*-axial offset. However, the efficiency drops dramatically when aluminum is utilized to shield the flux behind the pads. Reference (Budhia et al. 2013) proposes a bipolar single-sided flux coupler named DD pads. In (Nguyen et al. 2014), 8 kW DD pads are tested with 95.66% efficiency at the fully aligned position and 95.39% efficiency at 300 mm *X*-axial misalignment. However, the misalignment tolerance in the *Y*-axis is fairly small, which is totally different from that in the *X*-axis. DDQ pads are presented in Zaheer et al. (2015); Budhia et al. 2013; Covic et al. 2011), the misalignment tolerance of which is greatly improved, and the stable charging area also expands. However, the coupling coefficient of the coupler first rises and then declines rapidly with the increase of the *Y*-axial offset, which may lead to an overload of the charger.

The objective of this section is to propose a pair of quadrature coils with a converting circuit that enables the WPT system to transmit energy at a large horizontal offset. In this section, a mathematical description of the LCC-LCC resonant network is first established, and the calculation method of transmission power and efficiency is proposed. Then we propose the quadrature coils' structure. The variation in coupling coefficient with the horizontal misalignment is analyzed subsequently, and a converting circuit is proposed to avoid the rapid reduction of the coupling coefficient induced by the magnetic field cancellation. After that, a three-dimensional finite element analysis tool ANSYS MAXWELL is utilized to optimize the coil structure, and the misalignment tolerance of the optimized coils is compared with that of existing circular coils, DD-DD coupler, and DD-DDQ coupler. Finally, the correctness and effectiveness of the theoretical design are verified by experiments.

5.3.1 LCC-LCC Resonant Network

The structure of the LCC-LCC resonant network with its front-end inverter circuit and back-end rectifier circuit is shown in Fig. 5.31. The inverter circuit is composed of four MOSFETs (Q_1 to Q_4). The LCC-LCC resonant network is composed of compensation inductors L_{f1}, L_{f2}, compensation capacitors C_{f1}, C_{f2}, C_1, C_2, and coupling coils L_1, L_2. In addition, R_{f1}, R_{f2}, R_{Cf1}, R_{Cf2}, R_{C1}, R_{C2}, R_1, and R_2 are the parasitic resistance of L_{f1}, L_{f2}, C_{f1}, C_{f2}, C_1, C_2, L_1, and L_2 respectively. The rectifier circuit is composed of four fast recovery diodes D_1, D_2, D_3, D_4, and a filter capacitor C_{out}. u_{AB} is the input voltage applied to the LCC-LCC resonant network. u_{ab} is the output voltage before the rectifier diodes. i_{Lf1}, i_{Lf2}, i_{Cf1}, i_{Cf2}, i_1, and i_2 are the current on L_{f1}, L_{f2}, C_{f1}, C_{f2}, L_1, and L_2 respectively. To make the LCC-LCC resonant network work in the resonant state, the following relationships should be satisfied:

$$\begin{cases} (2\pi f_0)^2 = \frac{1}{L_{f1}\cdot C_{f1}} = \frac{1}{(L_1-L_{f1})\cdot C_1} \\ (2\pi f_0)^2 = \frac{1}{L_{f2}\cdot C_{f2}} = \frac{1}{(L_2-L_{f2})\cdot C_2} \end{cases} \tag{5.7}$$

where f_0 is the resonant frequency.

According to the analysis method proposed by Li et al. (2015b), the input and output voltages are replaced by the first-order harmonics of the square-waveform voltages at the switching frequency, and the resistance of all inductors and capacitors is neglected for simplicity. U_{AB}, U_{ab}, I_1, I_2, I_{Lf1}, I_{Lf2}, I_{Cf1}, and I_{Cf2} are adopted to represent the phasor form of the corresponding variables. The branch currents in the resonant state could be calculated by the superposition principle and Kirchhoff's law as follows:

$$\mathbf{I}_1 = \frac{\mathbf{U}_{AB}}{\mathrm{j}\omega_0 L_{f1}} \tag{5.8}$$

$$\mathbf{I}_{Lf1} = -\frac{K\sqrt{L_1L_2} \cdot \mathbf{U}_{ab}}{\mathrm{j}\omega_0 L_{f1} \cdot L_{f2}} \tag{5.9}$$

$$\mathbf{I}_{Cf1} = \frac{\mathbf{U}_{AB} \cdot L_{f2} + K\sqrt{L_1L_2} \cdot \mathbf{U}_{ab}}{\mathrm{j}\omega_0 L_{f1} \cdot L_{f2}} \tag{5.10}$$

$$\mathbf{I}_2 = \frac{\mathbf{U}_{ab}}{\mathrm{j}\omega_0 L_{f2}} \tag{5.11}$$

$$\mathbf{I}_{Lf2} = \frac{K\sqrt{L_1L_2} \cdot \mathbf{U}_{AB}}{\mathrm{j}\omega_0 L_{f1} \cdot L_{f2}} \tag{5.12}$$

$$\mathbf{I}_{Cf2} = \frac{\mathbf{U}_{ab} \cdot L_{f1} + K\sqrt{L_1L_2} \cdot U_{AB}}{\mathrm{j}\omega_0 L_{f1} \cdot L_{f2}} \tag{5.13}$$

where K is the coupling coefficient between the primary and the secondary coils.

U_{ab} should be in phase with I_{Lf2} because U_{ab} is a passive voltage generated according to the conduction mode of diodes D_1 ~ D_4 (Li et al. 2015b). Hence, U_{ab} lags U_{AB} by 90 degrees, and the transferred power could be obtained as follows:

$$P = \mathbf{U}_{ab} \cdot \mathbf{I}^*_{Lf2} = \frac{K\sqrt{L_1L_2} \cdot U_{ab} \cdot U_{AB}}{\omega_0 L_{f1} \cdot L_{f2}} = \frac{4K\sqrt{L_1L_2}}{\pi^3 f_0 \cdot L_{f1}L_{f2}} U_{in}U_{out} \tag{5.14}$$

U_{in} and U_{out} represent the input voltage and output voltage of the WPT system, respectively. For a given WPT system, U_{in}, U_{out}, L_1, L_2, L_{f1}, and L_{f2} are often determined. Therefore, the energy potential for the WPT system is directly determined by K. The stability of K under external disturbance directly affects the stability of output power. For WPT of EVs, parking deviation is the most important factor affecting K. Therefore, designing a pair of coupling coils with a high K which is insensitive to horizontal misalignment is crucial.

Since the parasitic resistance of all inductors and capacitors is extremely small, the overall power loss of the DC-DC system could be calculated approximately by using the branch currents obtained from Eqs. (5.8)–(5.13). The power loss of LCC-LCC resonant network Pres could be calculated as follows:

$$\begin{aligned} P_{res} = &\mathbf{I}_{Lf1} \cdot \mathbf{I}^*_{Lf1} \cdot R_{f1} + \mathbf{I}_{Cf1} \cdot \mathbf{I}^*_{Cf1} \cdot RCf1 + \mathbf{I}_1 \cdot \mathbf{I}^*_1 \cdot (R_{C1} + R_1) \\ &+ \mathbf{I}_{Lf2} \cdot \mathbf{I}^*_{Lf2} \cdot R_{f2} + \mathbf{I}_{Cf2} \cdot \mathbf{I}^*_{Cf2} \cdot R_{Cf2} + \mathbf{I}_2 \cdot \mathbf{I}^*_2 \cdot (R_{C2} + R_2) \end{aligned} \tag{5.15}$$

Hence, the efficiency of LCC-LCC resonant network η_{res} could be expressed as:

$$\eta_{res} = (1 - \frac{P_{res}}{P}) \times 100\% \tag{5.16}$$

Considering the working efficiency of the inverter and the rectifier circuits, the overall efficiency of the DC-DC system η could be expressed as:

$$\eta = \eta_{inv} \cdot \eta_{res} \cdot \eta_{rec} \tag{5.17}$$

where η_{inv} and η_{rec} indicate the efficiency of the inverter and the rectifier circuits, respectively.

In this section, we take the coupling coefficient as an evaluation index of the coupler's performance. Thus, it is reasonable to establish a numerical link between the DC-DC system's efficiency η and coupler's coupling coefficient K, so as to investigate the impact of coupler's design parameters on η. It is worth noting that P_{res} could be simplified into the following form according to Eq. (5.15):

$$\begin{aligned} P_{res} = {} & L_1 L_2 \frac{(R_{Cf1} + R_{f1})U_{ab}^2 + (R_{Cf2} + R_{f2})U_{AB}^2}{\omega_0^2 L_{f1}^2 L_{f2}^2} K^2 \\ & + \frac{L_{f1}^2(R_2 + R_{C2} + R_{Cf2})U_{ab}^2 + L_{f2}^2(R_1 + R_{C1} + R_{Cf1})U_{AB}^2}{\omega_0^2 L_{f1}^2 L_{f2}^2} \end{aligned} \tag{5.18}$$

Hence, the DC-DC system's efficiency could be arranged in the following form:

$$\begin{aligned} \eta = {} & \eta_{inv} \cdot \eta_{rec} - \eta_{inv} \cdot \eta_{rec} \frac{L_1 L_2[(R_{Cf1} + R_{f1})U_{ab}^2 + (R_{Cf2} + R_{f2})U_{AB}^2]}{\omega_0 \sqrt{L_1 L_2} \cdot L_{f1} L_{f2} U_{ab} U_{AB}} \cdot K \\ & - \eta_{inv} \cdot \eta_{rec} \frac{L_{f1}^2(R_2 + R_{C2} + R_{Cf2})U_{ab}^2 + L_{f2}^2(R_1 + R_{C1} + R_{Cf1})U_{AB}^2}{\omega_0 \sqrt{L_1 L_2} \cdot L_{f1} L_{f2} U_{ab} U_{AB}} \frac{1}{K} \end{aligned} \tag{5.19}$$

Equations (5.18) and (5.19) reveal the functional forms of K mapping to P_{res} and η, based on which the numerical relationship between K and η could be established by experimental data fitting. In addition, the theoretical relationship between K and η could also be conveniently determined by directly substituting the existing component parameters into Eq. (5.19).

5.3.2 *Quadrature Coils and Converting Circuit*

5.3.2.1 Structure of Quadrature Coils

According to Zhang et al. (2014); Borage et al. 2005; Takanashi et al. 2012), the DD-DD coupler has a large misalignment tolerance in the X-axis (X-axis is perpendicular to the main flux direction), whereas misalignment tolerance in the Y-axis (Y-axis is parallel to the main flux direction) is fairly small. An auxiliary unipolar coil named as quadrature coil (Q-coil) is usually assembled behind the DD-coil in the secondary side to compensate the poor Y-axial misalignment tolerance of DD-DD coupler. A DD-coil with a Q-coil is referred to as a DDQ pad (Budhia et al. 2013). When horizontal offset happens in the Y-axis, the coupling between the primary and the secondary DD-coils decreases significantly, but the coupling between the primary DD-coil and the secondary Q-coil increases, which compensates for the poor Y-axial misalignment tolerance of DD-DD coupler. Moreover, the Q-coil and the DD-coil are tuned independently, but their transmission power is superposed and then delivered to the output terminal.

In this section, a new type of coupler is proposed to improve the imbalance of misalignment tolerance in the X-axis and Y-axis. The proposed coupling coil is comprised of two identical DD-coils, which are overlapped at 90 degrees and connected in series to form one coupling coil as shown in Fig. 5.32. We name this new coupling coil as “quadrature coil”. The primary and the secondary coils share the same winding structure but different dimensions. When the primary and the secondary coils are fully aligned, all DD components are coupled, thus contributing to the overall power capacity enhancement. When there is a horizontal misalignment between two coupling coils, the tolerance ability in a specific direction could be promoted by adjusting the serial directions of quadrature coils based on the relative position of two coupling coils. Moreover, the secondary quadrature coil could be tuned and rectified by only one set of resonant network and rectifier, which reduces the design complexity and cost, simplifies the process of system power adjustment, and enhances the overall efficiency owing to fewer power components.

References (Budhia et al. 2013) provides guidance for the design of ferrite cores. Considering both the utilization efficiency of electromagnetic energy and cost, ferrite cores are arranged as square grids shown in Fig. 5.33. The longitudinal and lateral

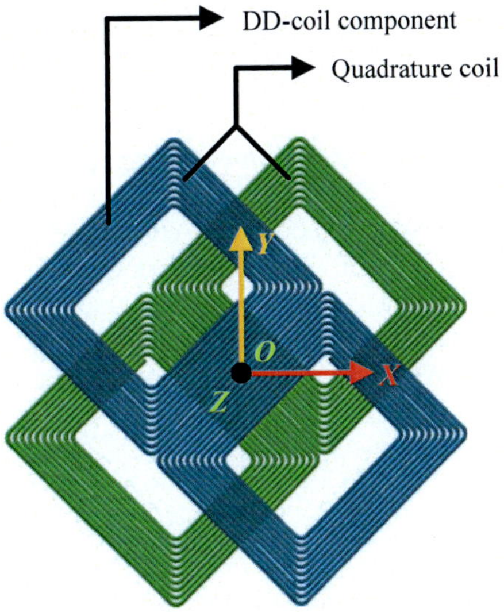

Fig. 5.32 Winding mode of quadrature coil

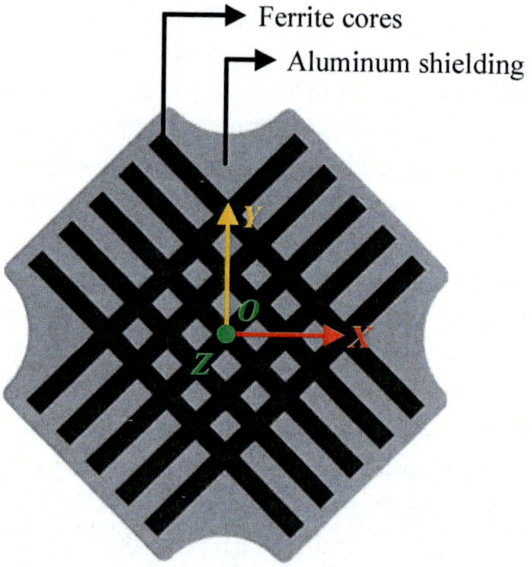

Fig. 5.33 Ferrite cores and aluminium shielding

directions of the square grids are parallel to two main flux directions of quadrature coils. Here we would like to clarify that there is only one group of ferrite cores on the primary or the secondary side matching the quadrature coil. Taking the primary side as an example, the vertical gap between the upper DD-coil component and ferrite cores is larger than that between the lower DD-coil component and ferrite cores by a height close to the diameter of the coil wire. However, the extra leakage caused by the height is negligible, and the power transfer performance would not be influenced by the minor difference of air gap. In the process of coupler design, the coil and ferrite cores should be kept as close as possible to minimize the leakage of coils. Moreover, the size of the aluminum shielding layer should be appropriately larger so as to shield the flux leakage and attenuate the radiation of magnetic flux to the surrounding environment. Quadrature coils and ferrite cores are combined to form a magnetic coupling device shown in Fig. 5.34.

5.3.2.2 Misalignment Tolerance of Quadrature Coils

A quadrature coil can be considered as a series configuration of two DD-coils. The two DD-coils constituting the primary coil are named P1 and P2 respectively, and the two DD coils constituting the secondary coil are named S1 and S2 respectively. The current flow diagram of the primary and secondary coils is shown in Fig. 5.35.

When the current flows through the primary coil, the flux linkage generated in the secondary coil could be expressed as:

$$\Psi_{P-S} = \Psi_{P1-S1} + \Psi_{P1-S2} + \Psi_{P2-S1} + \Psi_{P2-S2} \tag{5.20}$$

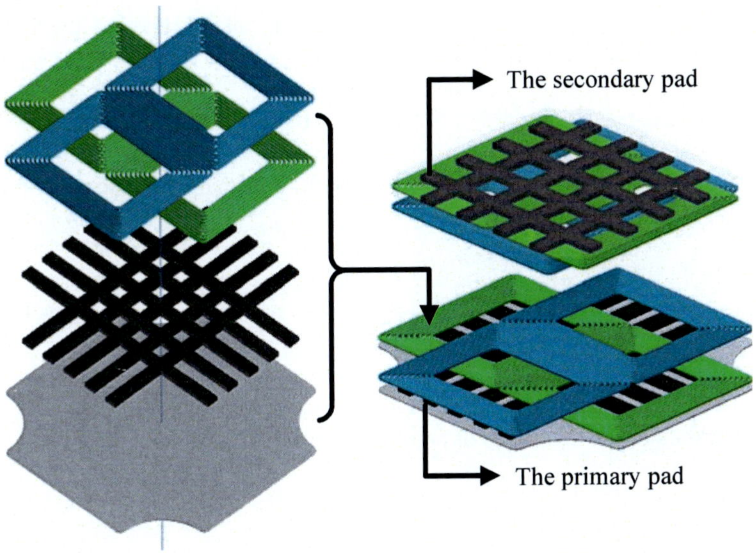

Fig. 5.34 Magnetic coupling device

Fig. 5.35 Current flow of the primary and the secondary coils

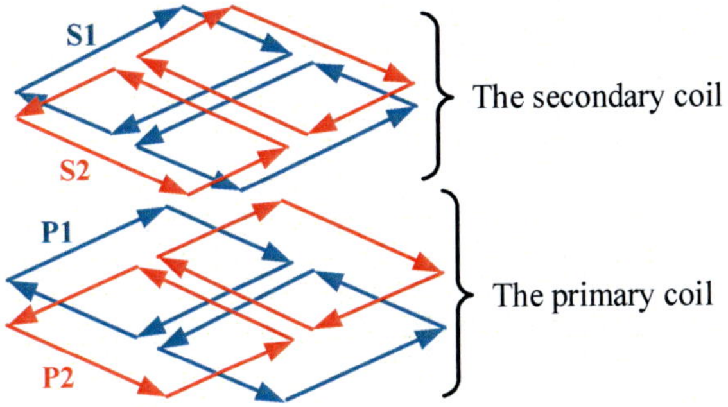

where $\Psi_{P\text{-}S}$ indicates the flux linkage generated in the secondary coil by the primary coil. $\Psi_{A\text{-}B}$ indicates the flux linkage generated in coil B by coil A (A, B represents P1, P2, S1, or S2). From the symmetrical structure of quadrature coils, we know that:

$$\begin{cases} \Psi_{P1-S1} \approx \Psi_{P2-S2} \\ \Psi_{P1-S2} \approx \Psi_{P2-S1} \end{cases} \tag{5.21}$$

Hence,

$$\Psi_{P-S} \approx 2(\Psi_{P1-S1} + \Psi_{P1-S2}) \tag{5.22}$$

When the primary and the secondary coils are completely aligned, the flux linkage of coil P1 and coil S2 is zero, namely:

$$\Psi_{P1-S2} = 0 \tag{5.23}$$

Therefore, the mutual inductance between the primary and the secondary coils could be calculated as follows:

$$M_{P-S} = \frac{\Psi_{P-S}}{I_P} \approx \frac{2\Psi_{P1-S1}}{I_P} = 2M_{P1-S1} \tag{5.24}$$

where I_P indicates the current in the primary coil. $M_{A\text{-}B}$ indicates the mutual inductance between coils A and B (A, B represents P1, P2, S1, or S2).

According to the definition of coupling coefficient, we know:

$$\begin{cases} K = \frac{M_{P-S}}{\sqrt{L_1 \cdot L_2}} \\ K_{P1-S1} = \frac{M_{P1-S1}}{\sqrt{L_{P1} \cdot L_{S1}}} \end{cases} \tag{5.25}$$

where L_{P1}, L_{P2}, L_{S1}, and L_{S2} indicate the self-inductances of coils P1, P2, S1, and S2, respectively. $K_{A\text{-}B}$ indicates the coupling coefficient between coils A and B (A, B represent P1, P2, S1 or S2).

Based on the calculation formula of series inductance and the symmetrical structure of quadrature coils, we get:

$$\begin{cases} L_1 = L_{P1} + L_{P2} + 2M_{P1-P2} \approx L_{P1} + L_{P1} + 0 = 2L_{P1} \\ L_2 = L_{S1} + L_{S2} + 2M_{S1-S2} \approx L_{S1} + L_{S1} + 0 = 2L_{S1} \end{cases} \tag{5.26}$$

The coupling coefficient of quadrature coils could be obtained by (5.24)–(5.26):

$$K \approx K_{P1-S1} \tag{5.27}$$

From Eq. 5.27, we observe that when the primary and the secondary coils are fully aligned, the coupling coefficient of quadrature coils is the same as that of coils P1 and S1. When there is a misalignment between the primary and the secondary coils, $\Psi_{P1\text{-}S1}$ and $\Psi_{P1\text{-}S2}$ vary with the increase of misalignment. Taking the offsets in the *X*-axis and *Y*-axis as examples, the coupling of quadrature coils is analyzed.

Figure 5.36 shows the schematic diagrams of quadrature coils obtained by looking down along the *Z*-axis. The arrows indicate the current flowing directions. S11 and S12 represent the two square sub-coils forming coil S1 in series. S21 and S22 represent the two square sub-coils forming coil S2 in series. Hence,

$$\begin{cases} \Psi_{P1-S1} = \Psi_{P1-11} + \Psi_{P1-12} \\ \Psi_{P1-S2} = \Psi_{P1-21} + \Psi_{P1-22} \end{cases} \tag{5.28}$$

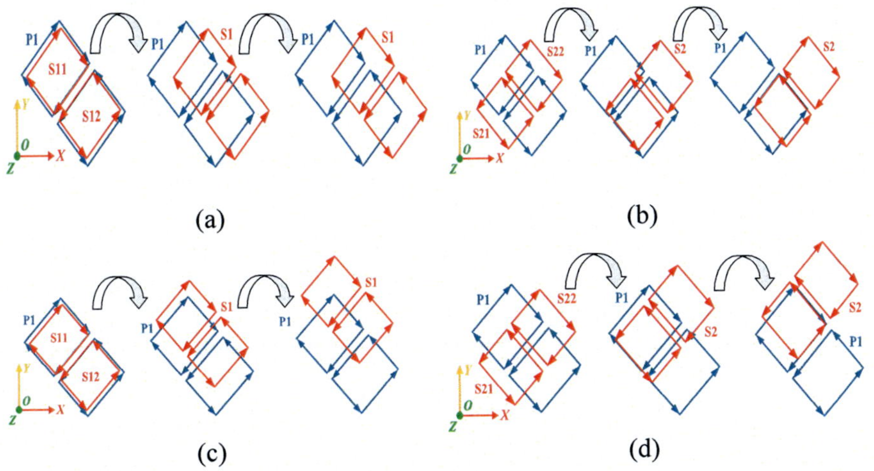

Fig. 5.36 Top view of quadrature coils: **a** Coil S1 is misaligned in the X-axis; **b** coil S2 is misaligned in the X-axis; **c** coil S1 is misaligned in the Y-axis.; **d** coil S2 is misaligned in the Y-axis

Substituting (5.28) into (5.22) yields:

$$\Psi_{P-S} \approx 2(\Psi_{P1-11} + \Psi_{P1-12} + \Psi_{P1-21} + \Psi_{P1-22}) \tag{5.29}$$

According to Fig. 5.36, when the primary and the secondary coils are fully aligned, we have the following numerical relationship:

$$\begin{cases} \Psi_{P1-11} = \Psi_{P1-12} > 0 \\ \Psi_{P1-21} = \Psi_{P1-22} = 0 \end{cases} \tag{5.30}$$

When the secondary coil is misaligned in the X-axis, $\Psi_{P1\text{-}11}$ and $\Psi_{P1\text{-}12}$ decrease at the same time, whereas the decline rate of $\Psi_{P1\text{-}11}$ is faster than that of $\Psi_{P1\text{-}12}$. $\Psi_{P1\text{-}21}$ increases from zero and then decreases, whereas $\Psi_{P1\text{-}22}$ declines first and rises subsequently. Figure 5.37 shows the variation of flux linkage with the X-axial misalignment. We can know that $|\Psi_{P1\text{-}21}|>|\Psi_{P1\text{-}22}|$.

Therefore, the mutual inductance between the primary and the secondary coils satisfies the following relationship:

$$M_{P-S} = \frac{\Psi_{P-S}}{I_P} > \frac{2(\Psi_{P1-11} + \Psi_{P1-12})}{I_P} = \frac{2\Psi_{P1-S1}}{I_P} = 2M_{P1-S1} \tag{5.31}$$

According to (5.25), (5.26), and (5.31), the relationship between K and $K_{P1\text{-}S1}$ could be obtained as follows:

$$K > K_{P1-S1} \tag{5.32}$$

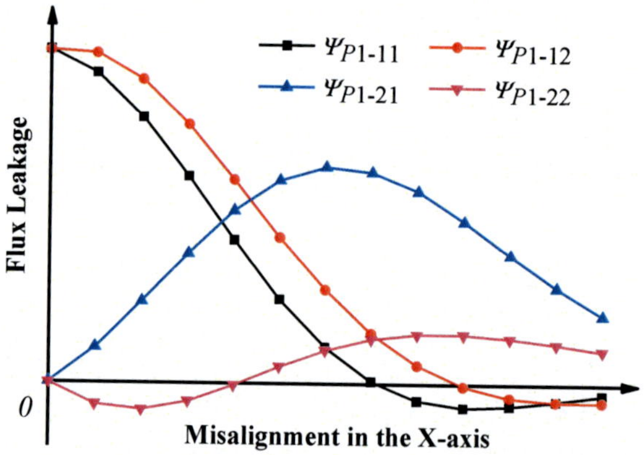

Fig. 5.37 Flux linkage versus the X-axial misalignment

When the secondary coil is misaligned in the Y-axis, $\Psi_{P1\text{-}11}$ and $\Psi_{P1\text{-}12}$ decrease at the same time, whereas the decline rate of $\Psi_{P1\text{-}11}$ is faster than that of $\Psi_{P1\text{-}12}$. $\Psi_{P1\text{-}21}$ decreases from zero and then increases, whereas $\Psi_{P1\text{-}22}$ rises first and declines subsequently. Figure 5.38 shows the variation of flux linkage with the Y-axial misalignment. We can know that $|\Psi_{P1\text{-}21}|>|\Psi_{P1\text{-}22}|$.

Therefore, the mutual inductance between the primary and the secondary coils satisfies the following relationship:

$$M_{P-S} = \frac{\Psi_{P-S}}{I_P} < \frac{2(\Psi_{P1-11} + \Psi_{P1-12})}{I_P} = \frac{2\Psi_{P1-S1}}{I_P} = 2M_{P1-S1} \tag{5.33}$$

According to (5.25), (5.26), and (5.33), the relationship between K and $K_{P1\text{-}S1}$ could be obtained as follows:

$$K < K_{P1-S1} \tag{5.34}$$

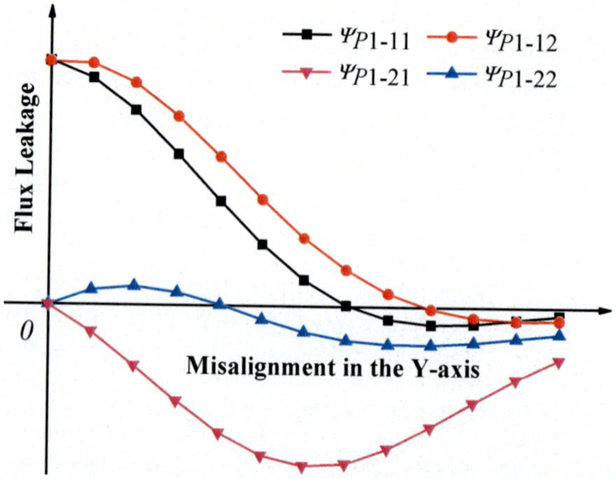

Fig. 5.38 Flux linkage versus the Y-axial misalignment

From (5.32) and (5.34), we observe that the coupling of quadrature coils in the X-axis is quite different from that in the Y-axis. When the secondary coil is misaligned in the X-axis, the coupling coefficient of quadrature coils is larger than those of coils P1 and S1. When the secondary coil is misaligned in the Y-axis, the coupling coefficient of quadrature coils is smaller than that of coils P1 and S1.

5.3.2.3 Configuration of Converting Circuit

In order to change the totally different coupling of quadrature coils in the X-axis and Y-axis and to obtain a higher coupling coefficient under the existing misalignment, a converting circuit is proposed, which can adjust the serial directions of quadrature coils based on the relative position between the primary and secondary coils. The configuration of the converting circuit is shown in Fig. 5.39.

In the wireless charging scene of EVs, the primary and the secondary coils are often not well aligned due to the limitation of driving skills and parking environment. The X-axial and Y-axial misalignments could be acquired based on the image recognition, or radar detection technology after the vehicle is parked. When the center of the secondary coil is detected to be located in Region A1 or A2, i.e., the X-axial misalignment is larger than the Y-axial misalignment, as shown in Fig. 5.40, the converting circuit will switch to Mode 1, i.e., SP1 and SS1 close the left endpoint and the right endpoint respectively, and both SP2 and SS2 close the lower endpoints. At this time, the coupling coefficient of the quadrature coils is insensitive to the X-axial misalignment, and a higher energy potential is achieved.

When the center of the secondary coil is detected to be located in Region B1 or B2, i.e., the Y-axial misalignment is greater than the X-axial misalignment, as shown in Fig. 5.41, the converting circuit will switch to Mode 2, i.e., SP1 and SS1 close the right endpoint and the left endpoint respectively, and both SP2 and SS2 close the upper endpoints. At this time, the coupling coefficient of the quadrature coils is insensitive to the Y-axial misalignment, and a higher energy potential is achieved.

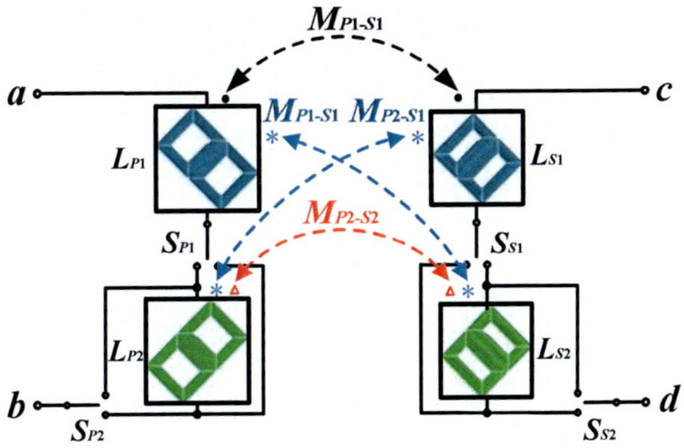

Fig. 5.39 Configuration of converting circuit

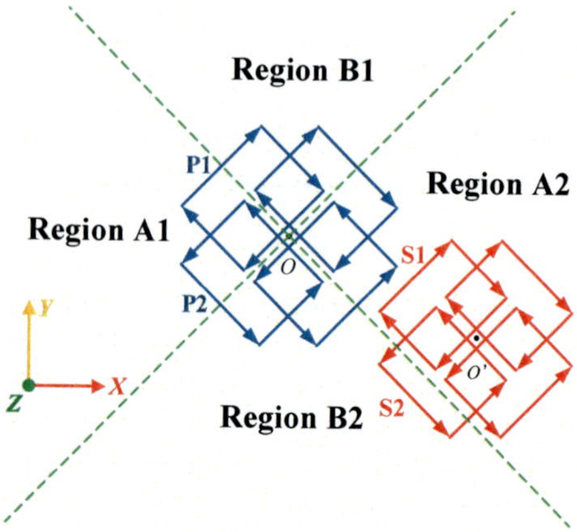

Fig. 5.40 Relative position of two quadrature coils in Mode 1

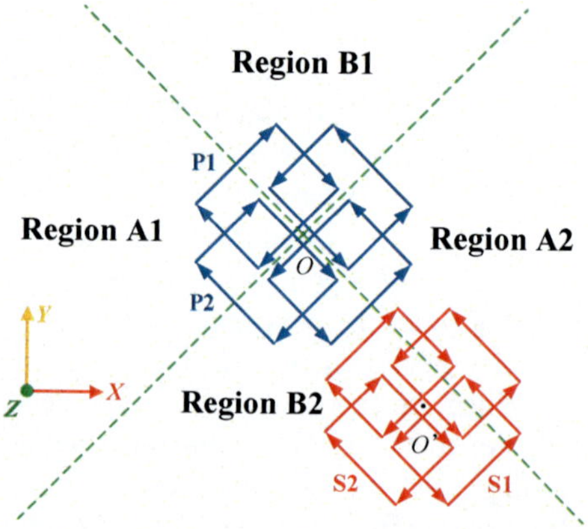

Fig. 5.41 Relative position of two quadrature coils in Mode 2

The mode conversion only changes the current flowing directions in coils P2 and S2 without changing the self-inductance of quadrature coils, so the resonant capacitances and inductances in the LCC-LCC resonant network do not need to be adjusted. In addition, coils are not energized when the mode switches, so there is no need to consider the effect of peak current at the moment of circuit conversion. Power MOSFETs are chosen as the switches in the converting circuit due to their high working frequency. To illustrate the switching process, the flow chart of the mode conversion is shown in Fig. 5.42.

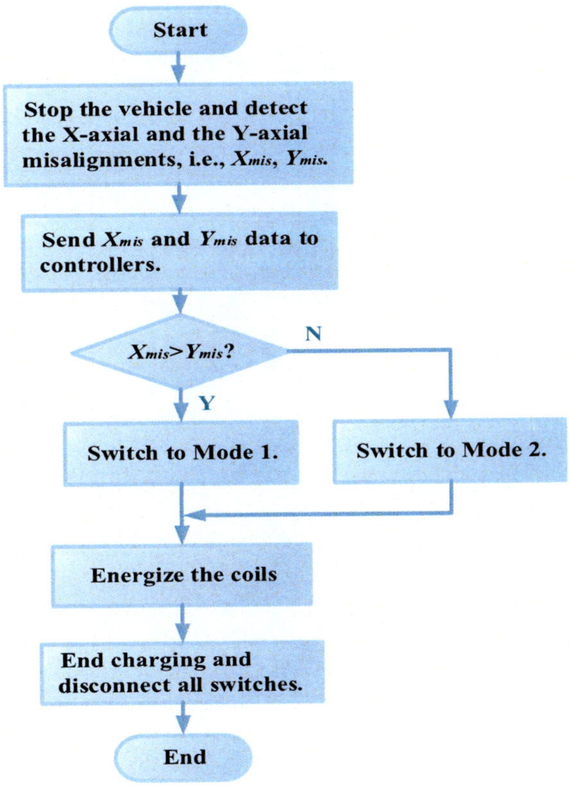

Fig. 5.42 Flow chart of mode conversion

5.3.3 *Optimization of Quadrature Coils*

5.3.3.1 Parametric Modelling

To analyze the performance of quadrature coils, a magnetic coupling model is built in ANSYS MAXWELL as shown in Fig. 5.43. Design specifications are listed in Table 5.7.

5.3.3.2 Structure Optimization

With the mode switching of the converting circuit, the coupling of quadrature coils in the X-axis is basically consistent with that in the Y-axis. Thus, only the X-axial misalignment is investigated here. The objective of structure optimization is to achieve high coupling coefficients at both the fully aligned position and the position offset by 150 mm in the X-axis. Set $N_{pri} = N_{sec} = 10$ initially, and select W_{pri},

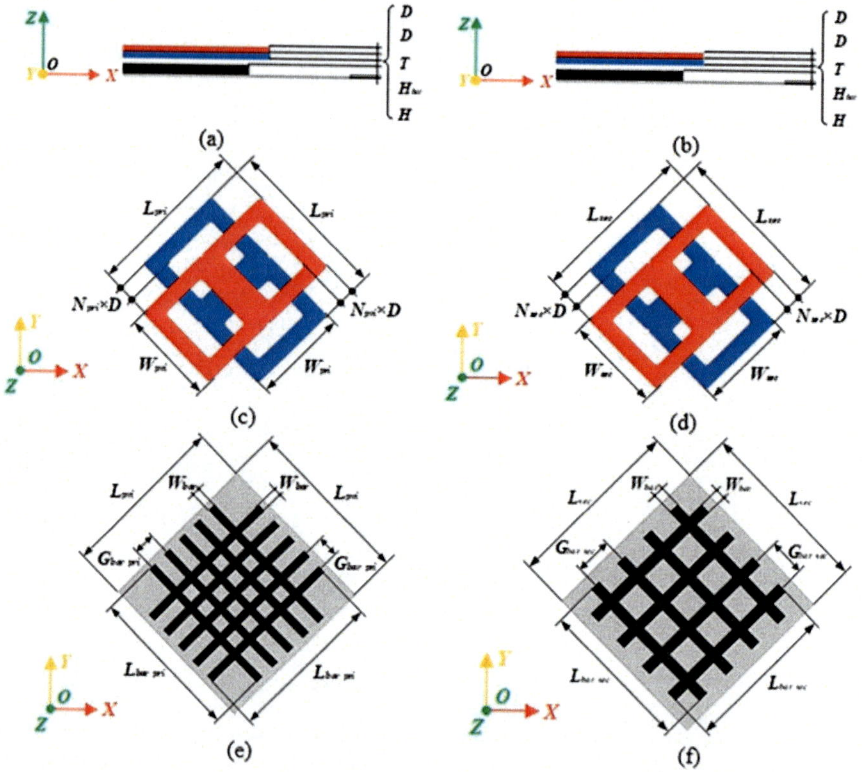

Fig. 5.43 Parametric model of quadrature coils: **a** the primary pad in vertical direction; **b** the secondary pad in vertical direction; **c** the primary coil; **d** the secondary coil; **e** the primary ferrite cores and aluminium shielding; **f** the secondary ferrite cores and aluminium shielding

W_{sec} as independent variables. The variations of coupling coefficients are investigated at both the fully aligned position and the position offset by 150 mm in the X-axis with ANSYS MAXWELL, which is shown in Fig. 5.44.

Define K_0 and K_{150} as the coupling coefficients at the fully aligned position and the position offset by 150 mm in the X-axis, respectively. In Fig. 5.44, when W_{pri} and W_{sec} achieve 300 mm and 320 mm respectively, K_{150} takes the maximum value, and a high K_0 value is also obtained. In order to get an optimal result, smaller steps are used in the FEA simulation. We observe that when W_{pri} and W_{sec} are 290 mm and 330 mm respectively, a better result is achieved, i.e., $K_0 = 0.2727$, $K_{150} = 0.1854$.

As the turn number increases, both the self-inductance and the mutual inductance of the coupler increase, and a higher energy potential is obtained. However, the increased parasitic resistance leads to a decline in efficiency, so the turn number of the coils should be limited within a certain range. For a 3.5 kW WPT system, the self-inductance of the coil is generally required to be about 200 μH. The relationship between the self-inductance and the turn number of the primary coil is shown in

Table 5.7 Specifications of quadrature coils

Symbol	Parameter	Value
P_{in}	Input power	3.5 kW
U_{in}	Input voltage	355 V
U_{out}	Output voltage	306 V
h	Air gap between coupling coils	110 mm
f_0	Resonant frequency	85 kHz
$L_{pri} \times L_{pri}$	Installation area of the primary coil	500 mm × 500 mm
$L_{sec} \times L_{sec}$	Installation area of the secondary coil	400 mm × 400 mm
D	Diameter of Litz wire	5.5 mm
L_{bar_pri}	Length of the primary ferrite cores	462 mm
L_{bar_sec}	Length of the secondary ferrite cores	346.5 mm
W_{bar}	width of ferrite cores	24.5 mm
H_{bar}	Height of ferrite cores	9.5 mm
G_{bar_pri}	Interval of the primary ferrite cores	$(W_{pri}\text{-}N_{pri} \times D)/4$ [a]
G_{bar_sec}	Interval of the secondary ferrite cores	$(W_{sec}\text{-}N_{sec} \times D)/3$ [a]
T	Vertical gap between coil and ferrite cores	5 mm
H	Height of aluminum shielding	2 mm

[a] W_{pri} and W_{sec} denote the width of the primary DD-coil components and the secondary DD-coil components, respectively. N_{pri} and N_{sec} indicate the turn number of the primary and the secondary coils, respectively

Fig. 5.45. When the turn number of the primary coil achieves 9, the self-inductance of the primary coil is 218.63 μH, which meets the design requirements.

Fix the turn number of the primary coil, and take the turn number of the secondary coil as the independent variable. The trends of both the coupling coefficient and self-inductance of the secondary coil with the variation of N_{sec} are investigated at the fully aligned position and the position offset by 150 mm in the *X*-axis. Simulation results are shown in Fig. 5.46.

From Fig. 5.46, we observe that when N_{sec} reaches 11, K_0 and K_{150} are close to the maximum value, i.e., 0.2715 and 0.1844, and the secondary coil obtains a self-inductance value of 248.5959 μH, which is close to 200 μH.

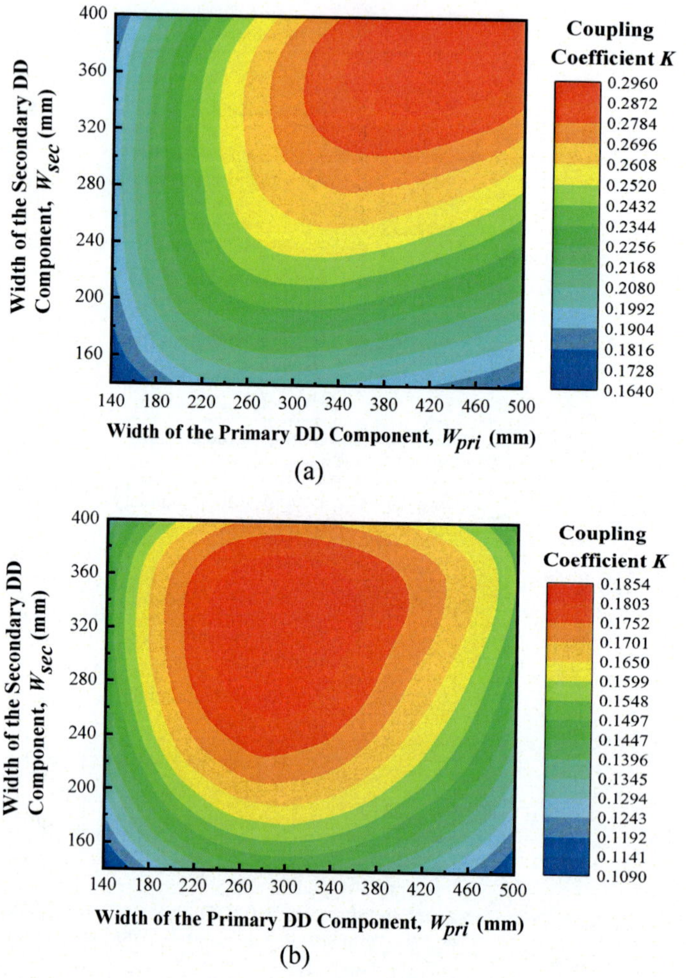

Fig. 5.44 *K* profiles with the variation of W_{pri} and W_{sec}: **a** *K* profiles at the fully aligned position; **b** *K* profiles at the position offset by 150 mm in the *X*-axis

5.3.3.3 Misalignment Tolerance of Optimized Coils

ANSYS MAXWELL is employed to analyze the misalignment tolerance of the optimized coils. The coupling coefficient profiles of the quadrature coils with a converting circuit and without a converting circuit are shown in Fig. 5.47. As can be seen, the converting circuit greatly enlarges the charging area. With the converting circuit, the stability of the coupling coefficient is effectively improved.

In order to compare and analyze the misalignment tolerances of different types of couplers, the same structure optimization method is applied to the circular coupler, DD-DD coupler, and DD-DDQ coupler. The parameters of these magnetic couplers

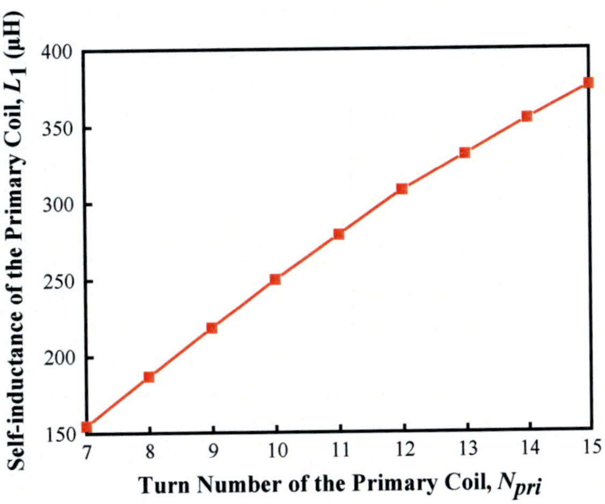

Fig. 5.45 L_1 curve with the variation of N_{pri}

are listed in Table 5.8. Simulated in ANSYS MAXWELL, the coupling coefficient curves of different couplers with the X-axis and Y-axis misalignments are shown in Fig. 5.48.

As shown in Fig. 5.48, since the circular coils have a symmetrical structure, the misalignment tolerance of circular coils in the X-direction is consistent with that in the Y-direction. The coupling coefficient of circular coils decreases rapidly with the increase of horizontal offset. When the horizontal offset is ~130 mm, the coupling coefficient decreases to 0.15. When the horizontal offset reaches ~230 mm, the coupling coefficient decreases to 0. The X-axial and Y-axial misalignment tolerances of the DD-DD coupler are seriously unbalanced. The X-axial misalignment tolerance seems quite large. When the X-axial offset reaches ~220 mm, the coupling coefficient decreases to 0.15. When the X-axial offset reaches 360 mm, the coupling coefficient remains at 0.06. However, the Y-axial misalignment tolerance of the DD-DD coupler is very small. When the Y-axial offset reaches ~90 mm, the coupling coefficient drops to 0.15. When the Y-axial offset reaches 150 mm, the coupling coefficient decreases to 0. The X-axial and Y-axial misalignment tolerances of quadrature coils are consistent under the action of the converting circuit. When the X-axial or Y-axial offset reaches ~190 mm, the coupling coefficient decreases to 0.15. When the X-axial or Y-axial offset comes to 360 mm, the coupling coefficient remains at 0.03. Therefore, quadrature coils with a converting circuit could achieve large misalignment tolerances in both the X-axial and the Y-axial directions.

Figure 5.48 also shows the coupling coefficient curves of the DD-Q coupler and DD-DD coupler, which constitute the DD-DDQ coupler, varying with the misalignment distance in the X-axis and Y-axis. When there is an X-axial misalignment, the coupling coefficient of the DD-Q coupler is 0 approximately. At this time, the DD-DDQ coupler is equivalent to the DD-DD coupler, which has excellent misalignment characteristics. When there is a Y-axial misalignment, the coupling coefficient of the DD-DD coupler decreases rapidly, whereas that of DD-Q coupler increases first and

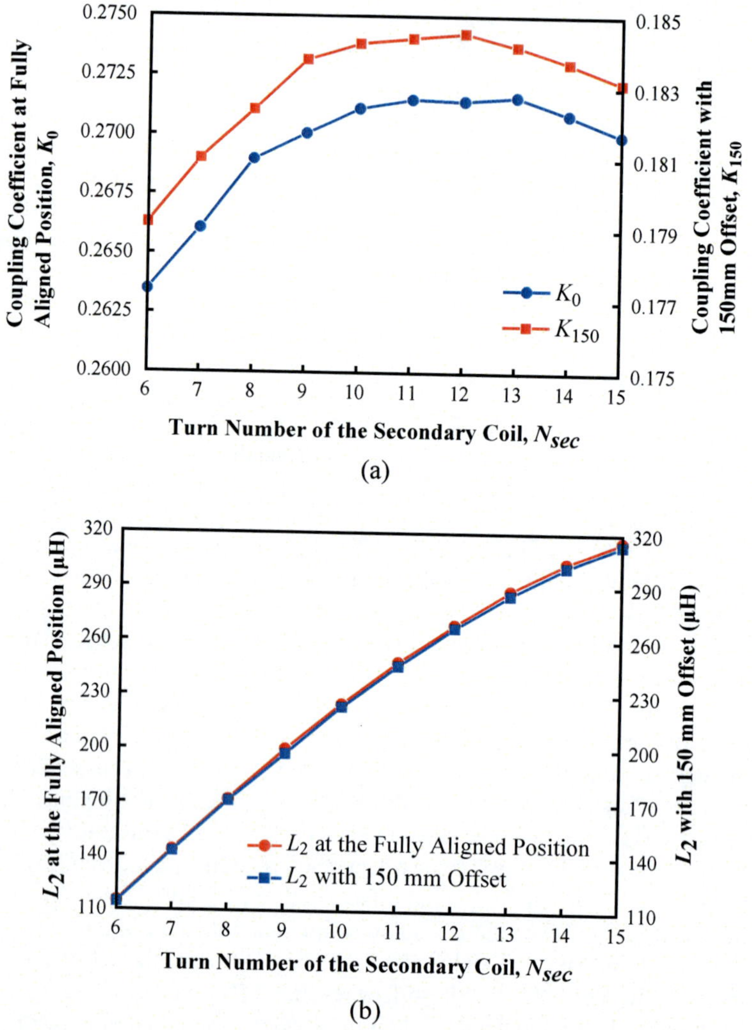

Fig. 5.46 K and L_2 curves with the variation of N_{sec}: **a** K curves with the variation of N_{sec}; **b** L_2 curves with the variation of N_{sec}

decreases subsequently, which compensates for the Y-axial misalignment tolerance. Since the secondary DD-coil and Q-coil are tuned independently, the power calculation method proposed in previous section would not be appropriate for evaluating the misalignment characteristics of the DD-DDQ coupler. According to Budhia et al. (2013), the transferred power of couplers is analyzed and compared when there is a Y-axial misalignment between two coupling coils. The transferred power curves varying with the Y-axial misalignment are shown in Fig. 5.49. We can observe that with the increase of the Y-axial offset, the transferred power of the DD-DDQ coupler

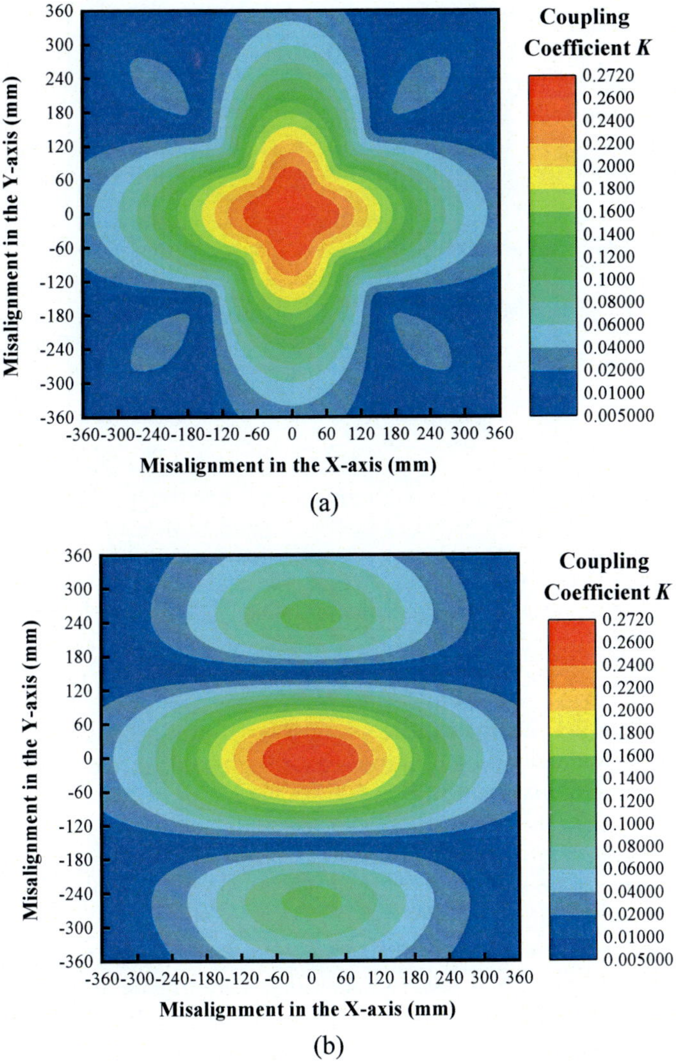

Fig. 5.47 *K* profiles of quadrature coils with a converting circuit and without a converting circuit: **a** with a converting circuit; **b** without a converting circuit

first increases to the peak value of 3.8 kW and then decreases rapidly. When the *Y*-axial offset is ~180 mm, the power drops to 0. Subsequently, the power increases again but is much lower than the rated value. The power curve of quadrature coils decreases slowly with the increase of the *Y*-axial offset. If 2 kW is set as the minimum power limit of the WPT system, the *Y*-axial offset range of quadrature coils is 0–180 mm approximately, whereas that of the DD-DDQ coupler is about 0–130 mm. Therefore, the *Y*-axial misalignment tolerance of quadrature coils is obviously larger

Table 5.8 Parameters of three types of coupling coils

Categories	Circular coils	DD-DD coupling coils	DD-DDQ coupling coils
Structure diagram			
Dimension of the primary side	Diameter: 500 mm	Length × Width: 500 mm × 500 mm	Length × Width: 500 mm × 500 mm
Dimension of the secondary side	Diameter: 400 mm	Length × Width: 400 mm × 400 mm	Length × Width: 400 mm × 400 mm
Turn number of the primary coil	20	12	12
Turn number of the secondary coil	28	16	DD → 16; Q → 13
Self-inductance of the primary coil	220.70 μH	220.10 μH	218.40 μH
Self-inductance of the secondary coil	202.22 μH	180.16 μH	DD → 170.57 μH; Q → 77.49 μH

than that of the DD-DDQ coupler. As the *Y*-axial offset is less than 60 mm, the transferred power of the DD-DDQ coupler would exceed the rated value, which may increase the risk of overload.

In order to make the comparison more explicit, we give a bar chart in Fig. 5.50 to show the 2 kW charging ranges of various couplers in the *X*-axis and *Y*-axis. From Fig. 5.50, we can observe that the *X*-axial and *Y*-axial charging ranges of quadrature coils are consistent and generally large. Though the *X*-axial charging range of the DD-DD coupler is slightly larger than that of quadrature coils, the *Y*-axial charging range of DD-DD coupler is much smaller than that of quadrature coils. DD-DDQ coupler shares almost the same *X*-axial charging range with DD-DD coupler, and the *Y*-axial charging range extends owing to the extra Q-coil, but the *Y*-axial charging range of DD-DDQ coupler is still a lot smaller than that of quadrature coils. The *X*-axial and *Y*-axial charging ranges of the circular coils are consistent, but much smaller than those of quadrature coils. In short, the generally large and symmetrical misalignment tolerance in the *X*-axis and *Y*-axis is the most obvious advantage of quadrature coils with a converting circuit.

As a prominent advantage of quadrature coils, the large *Y*-axial misalignment tolerance is fairly significant for the stable and reliable wireless power transmission of EVs. For the wireless charging equipment of EVs, the *Y*-axial alignment of the coupler is often not as simple to adjust as mentioned in Budhia et al. (2013), Nguyen et al. (2014). One way to limit the *Y*-axial misalignment of the coupler is to set a parking limit block on the parking space. However, when the vehicle parks with a

Fig. 5.48 *K* curves of different couplers with horizontal misalignments: **a** with the *X*-axial misalignment; **b** with the *Y*-axial misalignment

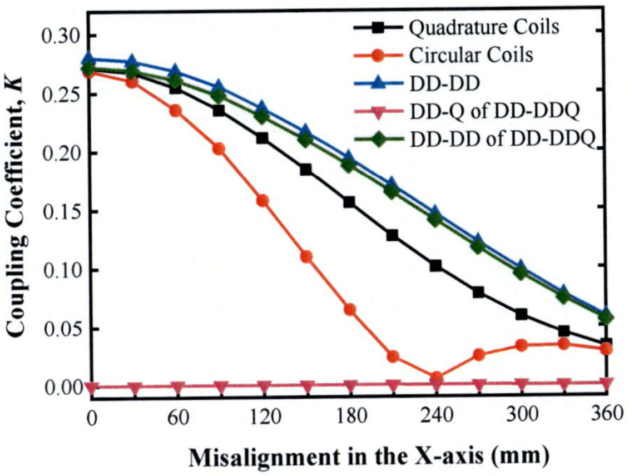

(a)

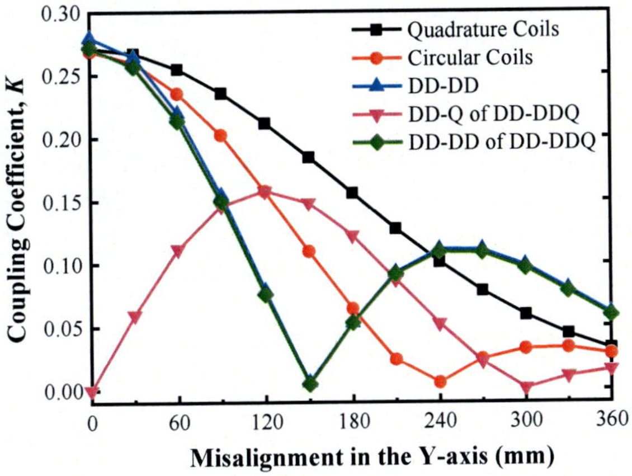

(b)

yaw misalignment, the *Y*-axial misalignment of the coupler will increase. The profile of vehicle parking with a yaw misalignment is shown in Fig. 5.51a. θ is the vehicle's yaw angle, and *A* is the contact point between the wheel and the parking limit block. *O* and *O'* are the center points of the primary and the secondary coils respectively, *B* and *B'* are the footpoints of the perpendiculars from *O* and *O'* to the parking limit block respectively, *r* is the distance between point *A* and *O*, *h* and *h'* are the vertical distances from *O* and *O'* to the parking limit block respectively, *l* and *l'* are the lengths of line segments *AB* and *AB'* respectively, α is the angle between *AO* and *AB*, Δh is the *Y*-axial misalignment of the coupler. According to the geometric relationship in Fig. 5.51a, the following formula could be obtained:

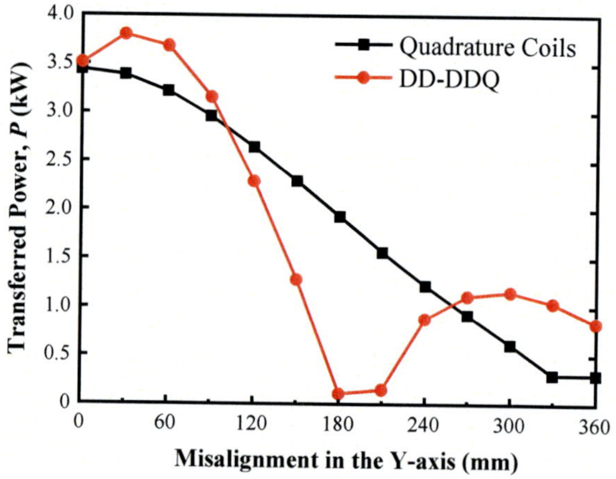

Fig. 5.49 Transferred power versus the *Y*-axial misalignment

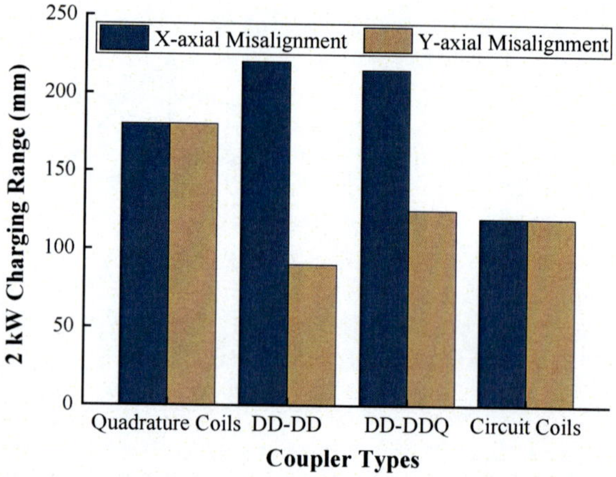

Fig. 5.50 2 kW charging range of various couplers in the *X*-axis and *Y*-axis

$$\begin{cases} \Delta h = r(\sin(\alpha + \theta) - \sin\alpha) \\ h = r\ \sin\alpha \\ l = r\ \cos\alpha \end{cases} \tag{5.35}$$

After simplification, we have:

$$\Delta h = h(\cos\theta - 1) + l\ \sin\theta \tag{5.36}$$

When θ takes a small value, the above equation could be simplified as follows:

$$\Delta h \approx l \sin\theta \tag{5.37}$$

For a 1.8 m-wide vehicle, the value of l is 0.9 m. When θ equals to 5°, Δh is 78.44 mm approximately. We can observe that a small yaw misalignment leads to an

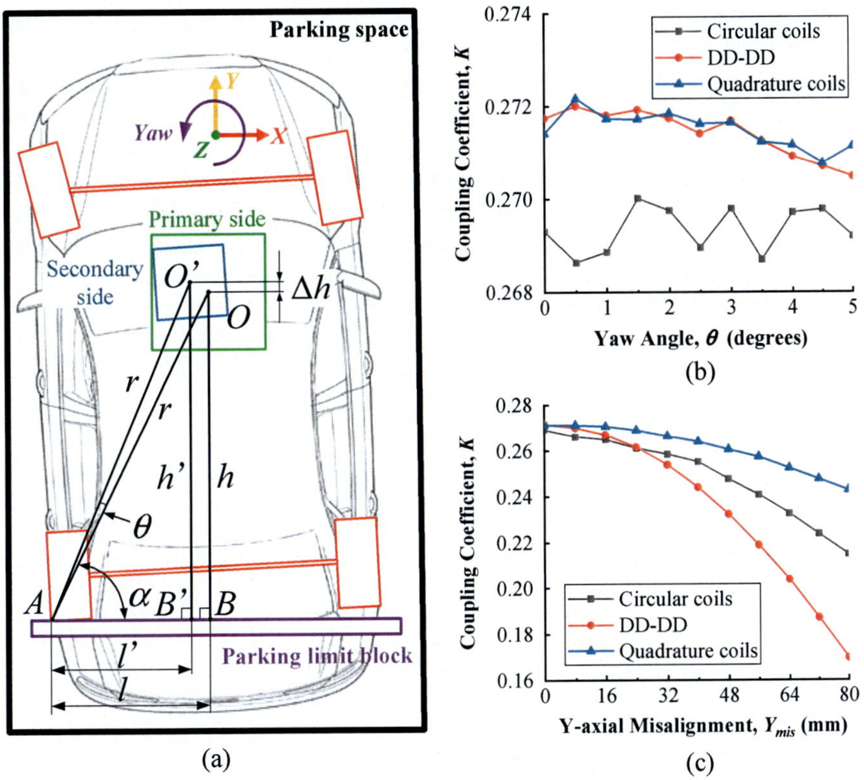

Fig. 5.51 Yaw misalignment of the vehicle and its effect on the coupling of coupler: **a** Vehicle parking with a yaw misalignment; **b** coupling coefficient versus the yaw angle; **c** coupling coefficient versus the *Y*-axial misalignment

un-negligible *Y*-axial misalignment of the coupler. The coupling coefficient curves varying with the yaw angle and the *Y*-axial misalignment are shown in Fig. 5.51b, c, respectively. With the increase of θ, K does not decrease obviously, but with the increase of Y_{mis}, K drops rapidly. Considering the assembly error of the coupler and parking limit block, as well as the different chassis sizes of various vehicles, we find it difficult to limit the *Y*-axial misalignment of the coupler to a small range by merely a parking limit block. Budhia et al. (2013) and Nguyen et al. (2014) mentioned that visual detection signals could be utilized to guide the driver to adjust the coupler's *Y*-axial position, but this process requires highly sensitive and accurate sensors with good dynamic characteristics and it is unreliable to ensure that all drivers can limit the *Y*-axial misalignment of the coupler to a few centimeters.

Quadrature coils could adjust the modes of the converting circuit flexibly according to the coupler's static alignment state. When the *X*-axial misalignment distance is larger than the *Y*-axial misalignment distance, the converting circuit switches to Mode 1. At this time, the coupling coefficient of quadrature coils is insensitive to the X-axial misalignment, and a higher energy potential is achieved. When the X-axial misalignment distance is less than the Y-axial misalignment distance, the converting circuit switches to Mode 2. At this time, the coupling coefficient of quadrature coils is insensitive to the Y-axial misalignment, and a higher energy potential is achieved. Trigger information for mode conversion could be provided by the small current detection or sensor feedbacks of the vehicle's static parking position (Jeong et al. 2018). Under the action of the converting circuit, the X-axial and Y-axial misalignment tolerances of quadrature coils are exactly consistent and generally large.

5.3.3.4 Misalignment Tolerance of Optimized Coils

In order to verify the effectiveness of the structural optimization process, we compare the optimized quadrature coils with a set of quadrature coils with original design parameters. The original quadrature coils' parameters are set as follows: both N_{pri} and N_{sec} are set to 10; W_{pri} and W_{sec} are set to 500 mm and 400 mm, respectively. The flux distribution profiles of two sets of quadrature coils with different design parameters are shown in Fig. 5.52. We can observe that the magnetic flux of the optimized quadrature coils is more intensively distributed between the coils on both sides. In other words, the optimized quadrature coils have a smaller flux leakage, larger power transmission capacity, and higher power transmission efficiency.

Safety is also an essential element in the design of wireless charging couplers for electric vehicles. We have reviewed the SAE J2954 standard and found the restrictive conditions concerning leakage of wireless charging couplers. In Clause 10.1, three physical regions are defined to facilitate EMF safety management of WPT systems. Region 1 is defined as the area underneath the vehicle. Region 2 is defined as the region around the periphery of the vehicle from the edge of the sheet metal. Region 3 is the vehicle interior. The EMF mainly distributes in Region 1 due to the normal assembly position of magnetic couplers. Therefore, prolonged exposure in Region 1 exceeding the EMF limits shall be strictly prohibited. Several methods are taken into account, such as active or passive access control, detection, and shut down before exposure occurs, which may be beyond the scope of this paper. In addition, the EMF in Regions 2 and 3 shall comply with the guidelines for General Public EMF exposure referenced in ICNIRP 2010. With the standard operating frequency of 81.38 to 90 kHz, the General Public Reference Levels of ICNIRP 2010 stipulates that the magnetic flux density in Regions 2 and 3 shall be below 27.0 μT RMS. Besides, the magnetic flux in Regions 2 and 3 shall be constrained to a more stringent level based on the AAMI/ISO 12117–2012 Appendix M because the pacemakers and implanted neuro-stimulators can only operate as designed below 15.0 μT rms in 81.38 to 90 kHz magnetic fields.

ANSYS MAXWELL is employed to study the leakage distribution of quadrature coils and verify application security. The flux distribution profiles are shown in

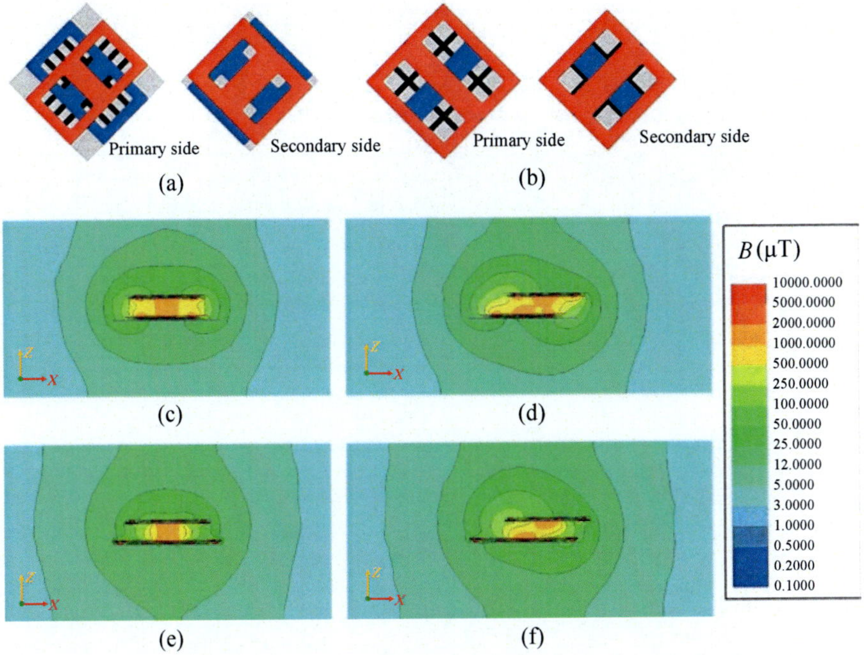

Fig. 5.52 **a, b** Different coupler designs where: **a** optimized; **b** original; **c–f** flux distribution profiles where: **c** optimized coils fully aligned; **d** optimized coils with *X*-axial offset: 150 mm; **e** original coils fully aligned; **f** original coils with *X*-axial offset: 150 mm

Fig. 5.53. The optimized quadrature coils are assembled under a vehicle, the dimension of which is 1.8 m × 4 m. With the shielding of steel chassis, the leakage intensity is greatly weakened in Regions 2 and 3, where the maximum leakage reaches 1.13 μT as the power consumption is set as 3.5 kW. We can observe that the maximum leakage is entirely within the standard requirements mentioned above for the fully aligned and 150 mm offset in the *X*-axis scenarios.

In order to compare the emissions of various couplers, we give a measured line originating from the center of the couplers' air gap to the position beyond the vehicle's edge as shown in Fig. 5.54a–h and detect the flux density along the measured line by simulation. From Fig. 5.54i, j, we observe that the leakage does not change substantially with a 150 mm horizontal offset of couplers. The DD-DD curves and the quadrature coils curves are quite close to each other. The flux density of circular coils is higher than that of the DD-DD coupler and quadrature coils within the edge of the steel chassis. As the measured distance exceeds the edge of steel chassis, the flux density of circular coils is much lower than that of the DD-DD coupler and quadrature coils, though all couplers' emissions are extremely low (below 0.5 μT RMS) at this time. Generally, all the couplers' leakages outside the edge of steel chassis satisfy the requirements of SAE J2954 standard for the fully aligned and 150 mm offset in the *X*-axis scenarios.

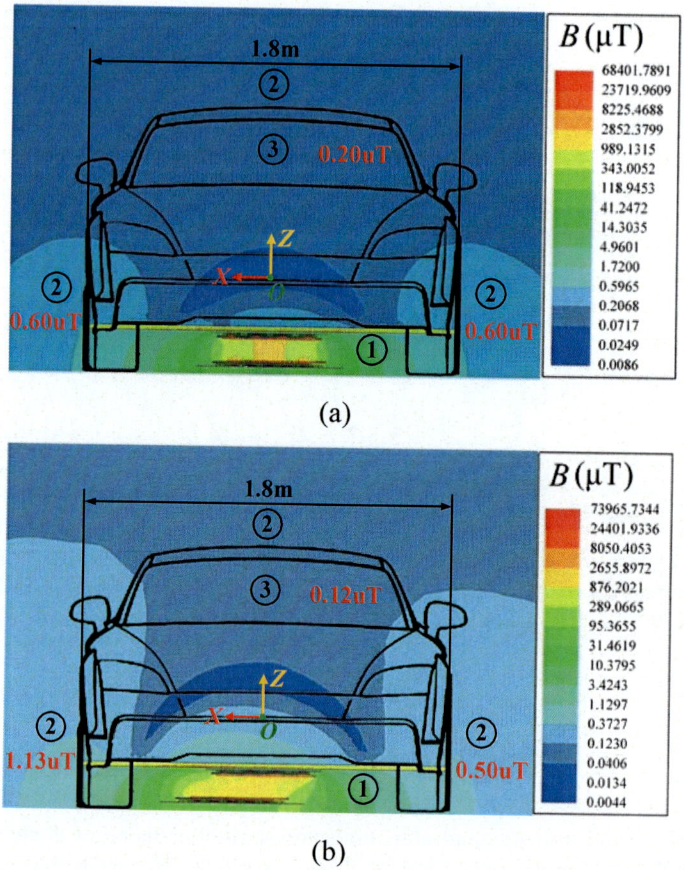

Fig. 5.53 Flux distribution profiles of quadrature coils mounted on a vehicle: **a** coupler fully aligned; **b** coupler horizontally offset by 150 mm

5.3.4 Experiments Verification

In order to verify the misalignment characteristics of quadrature coils, a 3.5 kW WPT experiment platform is built as shown in Fig. 5.55a. Chroma 63200A is adopted as an electric load, and the constant voltage mode is used to simulate the charging process of power batteries. Coupling coils, ferrite cores and, aluminium shielding are also shown in Fig. 5.55 b, c.

According to Eqs. (5.31) and (5.48), and the parameters listed in Tables 5.7 and 5.8, the theoretical values of resonant capacitances and inductances could be calculated. At the same time, the actual values of the resonant capacitances and inductances could also be measured. All data are listed in Table 5.9, from which we observe that the measured values are very close to the corresponding design values. To compare and analyze the simulation and experimental results, PSPICE is utilized to obtain the

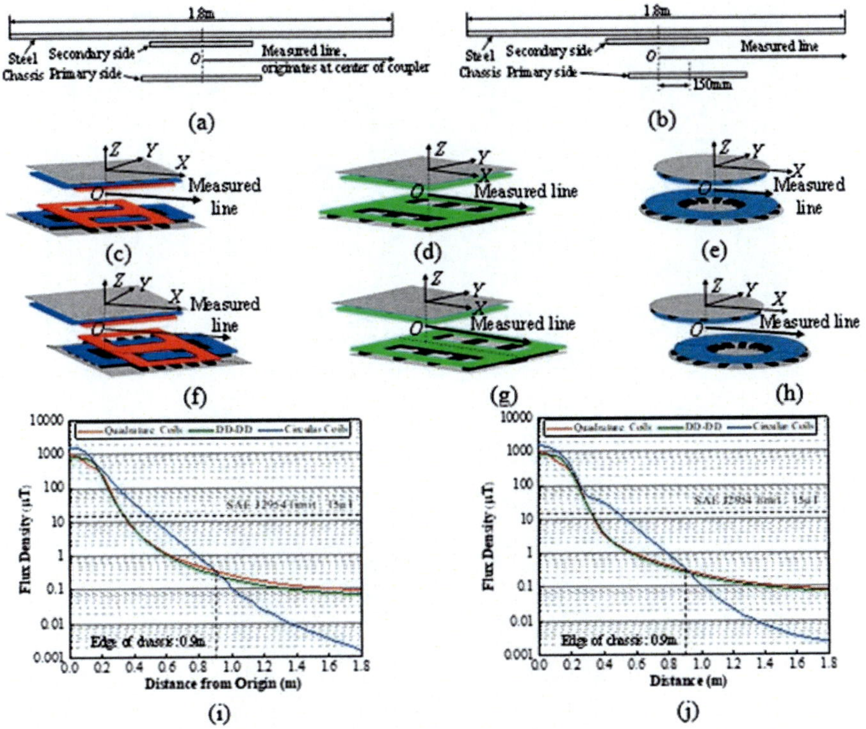

Fig. 5.54 **a**, **b** measured lines for: **a** coupler fully aligned; **b** coupler horizontally offset by 150 mm; **c–e** various couplers fully aligned: **c** quadrature coils; **d** DD-DD; **e** circular coils; **f–h** various couplers horizontally offset by 150 mm: **f** quadrature coils; **g** DD-DD; **h** circular coils; **I**, **j** Flux density along measured lines: **i** couplers fully aligned; **j** couplers horizontally offset by 150 mm

power curves as the *X*-axial and *Y*-axial misalignments increase. Subsequently, the input and output power, as well as the power transmission efficiency of the quadrature coils are experimentally measured. The simulation and experimental results are shown in Fig. 5.56.

As Fig. 5.56 shows, the experimental results are fairly consistent with the theoretical analysis and simulation results. The power capacity of quadrature coils is extremely high under the *X*-axial misalignment, but quite low under the *Y*-axial misalignment. When the secondary coil is misaligned along the angle bisector direction of *XOY*, where the *X*-axial misalignment equals to the *Y*-axial misalignment, both the power and efficiency curves are in the region surrounded by the *X*-axial misalignment curves and the *Y*-axial misalignment curves. The imbalanced misalignment tolerances in the *X*-axis and *Y*-axis can be overcome by the converting circuit. When the quadrature coils are fully aligned, the input and output power of the WPT system is 3.534 kW and 3.349 kW, respectively, and the efficiency reaches 94.78%. When the secondary coil is misaligned by 150 mm in the *X*-axis or *Y*-axis (with converting circuit transferring to another mode), the input and output power of the

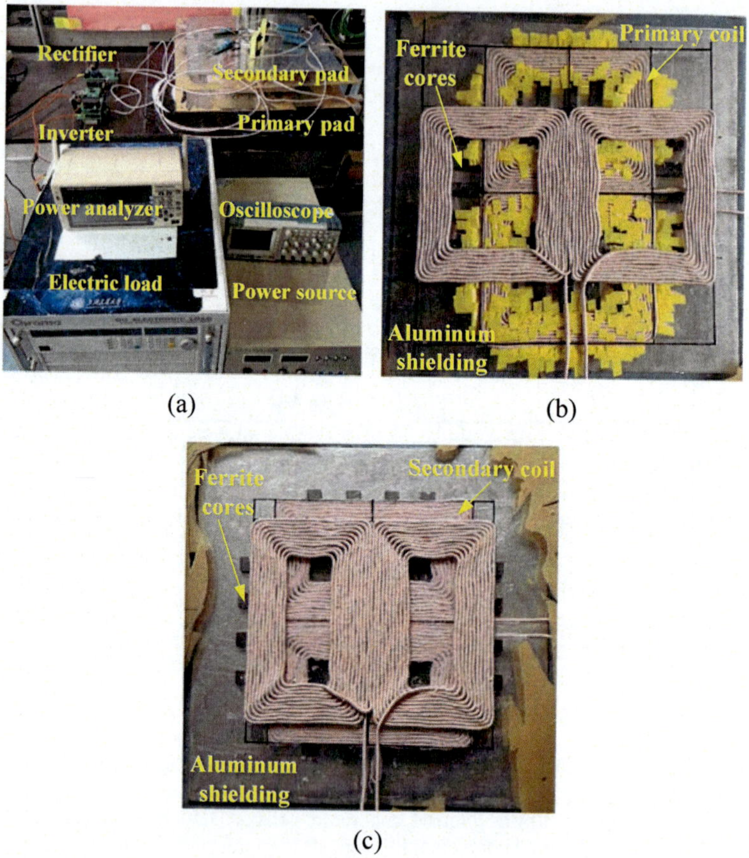

Fig. 5.55 Experimental setup: **a** 3.5 kW WPT experiment platform; **b** primary coil with ferrite cores and aluminum shielding; **c** secondary coil with ferrite cores and aluminum shielding

Table 5.9 Design value and measured value

Parameter	Design value	Measured value
$L1$	218.63 μH	218.63 μH
$L2$	248.60 μH	218.63 μH
$Lf1$	58.00 μH	218.63 μH
$Lf2$	54.00 μH	218.63 μH
$C1$	22.825 nF	22.93 nF
$C2$	20.184 nF	20.68 nF
$Cf1$	60.447 nF	60.42 nF
$Cf2$	66.149 nF	66.77 nF
K	0.2715	0.2737

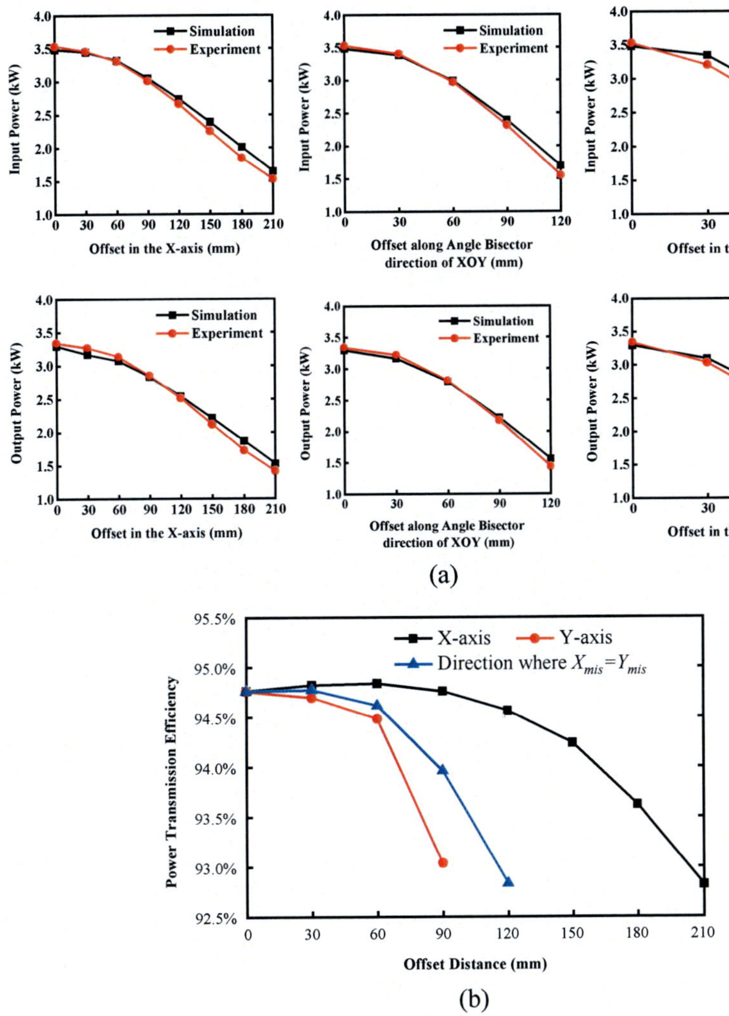

Fig. 5.56 The power and efficiency curves of the WPT system with a horizontal offset: **a** Input and output power versus offset distance; **b** power transmission efficiency versus offset distance

WPT system is 2.256 kW and 2.126 kW, respectively, and the efficiency is still as high as 94.22%. The waveforms of the input/output current and voltage in the experiment are also observed during the offset process, as shown in Fig. 5.57. From Fig. 5.57, we can observe that the equivalent impedance of the inverter load is slightly inductive, which contributes to the soft switching of power MOSFETs in the inverter. Hence an efficient and stable operation of the inverter is obtained. Moreover, the input current and voltage of the rectifier are in phase and successive, which indicates that the power is smoothly transferred to the secondary side. When there is an X-axial misalignment

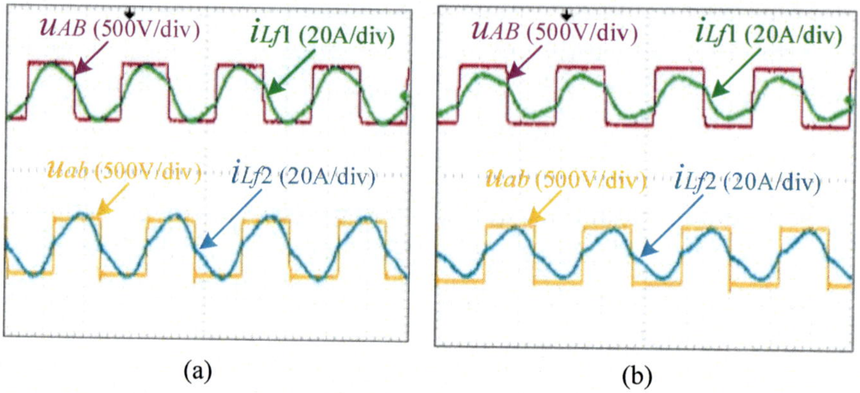

Fig. 5.57 Experimental waveforms: **a** Coupler fully aligned. **b** Coupler horizontally offset by 150 mm

between two coupling coils, the inverter and rectifier still operate in an efficient state, yet the fundamental amplitudes of the input and output currents decrease.

This section proposes a novel type of coupler named as quadrature coils for the WPT system. This coupler improves the stability of energy transmission when there is a horizontal misalignment between two coupling coils. In this work, the variation of the coupling coefficient with the increasing offsets in the *X*-axis and *Y*-axis is analyzed theoretically first. Then a converting circuit is proposed, which could change the topology of quadrature coils based on the relative position of two coupling coils. With this converting circuit, large misalignment tolerances are achieved both in the *X*-axis and *Y*-axis. Finally, ANSYS MAXWELL is utilized to optimize the structure of quadrature coils, and a 3.5 kW WPT experiment platform is established according to the optimization parameters. Both the simulation and experimental results demonstrate the effectiveness of this new coupler.

5.4 Coil Integration for Volume Reduction

In recent years, the study of EV's WPT has made great progress. Research has shown that this technique has broad market prospects (Li and Mi 2015). At present, the study of WPT focuses on power electronic converters, resonant circuit topologies, coil design, the control method, and safety (Li and Mi 2015). Among these varieties, coil design and resonant circuit topologies are two essential points in WPT.

The resonant circuit topology is the key element in WPT, because it helps to adjust the resonant frequency, minimize the VA rating of power supply, improve power transfer capability, and achieve high efficiency (Yungtaek and Jovanovic 2003). There are four basic compensation topologies labeled as SS, SP, PS, and PP. "S" or "P" means the way capacitors how to connect with coils. In reference (Li et al. 2015b), a double-sided LCC topology is presented. With the proposed topology, the resonant frequency is irrelevant to the coupling coefficient and load condition. And the topology has high efficiency. Reference (Deng et al. 2015) shows that double-sided LCC topology is more appropriate in EV's application.

Coil design is another important element in WPT. It greatly determines the power capacity and transmission efficiency. When designing coils, essential elements should be considered, such as the coupling coefficient, self-inductance of coils, quality factor of coils, and misalignment. In order to increase the coupling coefficient and quality factors, ferrites with high permeability are usually added, which bring problems of calculating the inductance of coils, coupling coefficient, and core loss. A 3D finite element method using ANSYS Maxwell can be used for coil design and analysis (Nguyen et al. 2014). Reference (Ning et al. 2013) presents a 2 kW 700-mm-diameter circular pad having a horizontal radial tolerance of 130 mm. The sizes of circular pads are too large, transferring high power and have poor tolerance of misalignment. Reference (Takanashi et al. 2012) gives a flux-pipe coil which can transmit 3 kW with 90% efficiency across a 200 mm air gap. However, the flux-pipe has two flux conductive paths, and one of the paths is wasted, which decreases the efficiency. In order to solve this problem, a DD pad is presented in reference (Budhia et al. 2013). In reference (Nguyen et al. 2014), an 8 kW DD pad with 200 mm air gap and ± 300 mm horizontal misalignment tolerance is tested. The charger transfers energy with an efficiency of 95.66% (DC to DC) at the desired position and still maintains at 95.39% efficiency with a ± 300 mm horizontal misalignment. But the area of the pad is large. The above pads can be tested using SS, LCL or LCC topology. If using LCL or LCC topology, there is an extra large compensation coil. Hence, the system is not appropriate in EV's application. Reference (Deng et al. 2015) presents integrated coils using LCC topology, and it can transfer 5.7 kW with a peak efficiency of 95.36% from DC to DC at rated operation condition. More research about this structure can be found in Reference (Li et al. 2015a; Deng et al. 2015). However, the mutual inductance between the main coil and the compensation coil on the same side makes the parameter design difficult (Deng et al. 2015). And the extra cross-side mutual inductance increases the reactive power and influences the efficiency (Deng et al. 2015).

In order to simplify the design procedure and eliminate or reduce the influence of cross-side mutual inductance of integrated coils using double-sided LCC compensation topology, this section gives a new method of integrating compensation coils and main coils. This new integration method theoretically eliminates the mutual inductance between the main coil and the compensation coil on the same side to simplify the design procedure and also eliminates the cross-side mutual inductance to reduce the reactive power. A method of optimizing the integrated coils is also presented. Finally, the system shows good performance and can deliver 3 kW with 92.16% efficiency at an air gap of 150 mm.

5.4.1 Analysis of Circuit Model

Research shows that double-sided LCC compensation topology has a better performance and is more appropriate in EV's application. With this topology, the resonant frequency is irrelevant to the coupling coefficient and load condition. And it can provide a unity power factor both in the input port and the output port of the topology. Figure 1 shows the double-sided LCC compensation topology with the front-end circuit and the rear-end circuit. In Fig. 5.58, U_{in} and U_{out} mean the input DC voltage and battery voltage, respectively. On the primary side, AC power is provided by a high-frequency full-bridge inverter composed of 4 MOSFETs: S_1 ~S_4. L_{f1}, C_{f1}, L_1 and C_1 constitute the primary resonant tank, and its resonant frequency is tuned the same as the frequency of the inverter. The high-frequency current in the primary coil (L_1) can stimulate the same frequency alternating magnetic field, which induces an alternating voltage in the secondary coil L_2. On the secondary side, L_{f2}, C_{f2}, L_2 and C_2 constitute the secondary resonant tank, and its resonant frequency is tuned the same as the primary side. Then, high-frequency AC power is rectified to DC power by a rectifier composed of 4 fast recovery diodes. Finally, the DC power is injected into batteries. If integrating main coils and compensation coils, there are 6 coupling coefficients. Among them, the cross-side coupling coefficient K_{2f1}, K_{1f2}, and K_{f1f2} increase the reactive power and influence the efficiency according to Deng et al. (2015). Although the same side coupling coefficient K_{1f1} and K_{2f2} have no influence on the efficiency, they make the parameter design difficult. Hence it is a goal to eliminate or minimize the five extra coupling coefficients.

The new integrated coils are shown in Fig. 5.59. The primary side and secondary side have the same size (600 mm × 600 mm) and structure. Main coil $L_1(L_2)$ and compensation coil $L_{f1}(L_{f2})$ both choose DD pad. The compensation coil is placed under the center of the main coil on the same side. But the current direction of the main coil and the compensation coil on the same side is different. Figure 5.3 shows the current direction of each coil. A DD pad is composed of two series rectangular coils named *P1*(*S1*) and *P2*(*S2*) respectively in main coils and $C_{p1}(C_{s1})$ and $C_{p2}(C_{s2})$ respectively in compensation coils just as shown in Fig. 5.60.

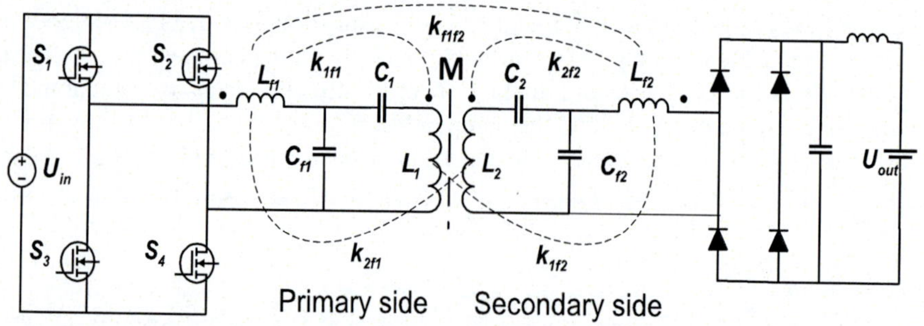

Fig. 5.58 Double-sided LCC compensation topology with front-end circuit and rear-end circuit

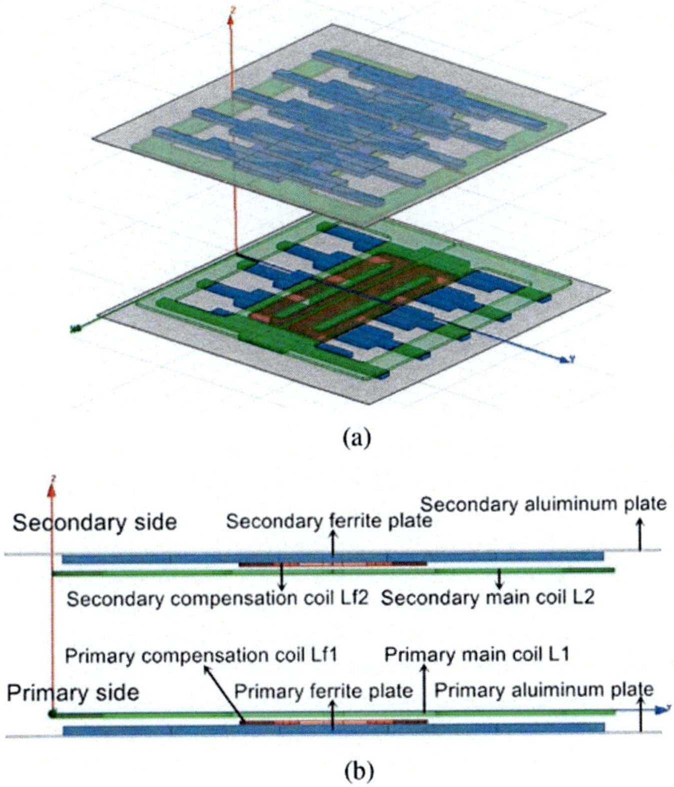

Fig. 5.59 The new integrated coils: **a** Overview; **b** side view

Fig. 5.60 The current direction of each coil

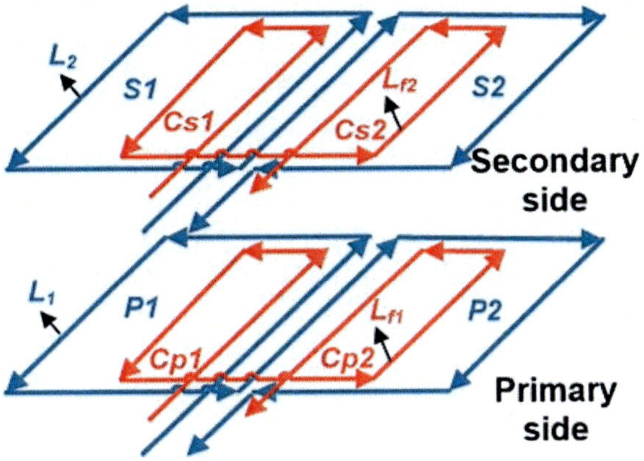

On the primary side, the flux linkage in the compensation coil generated by the main coil is:

$$\psi_{P-Cp} = \psi_{P1-Cp1} + \psi_{P1-Cp2} + \psi_{P2-Cp1} + \psi_{P2-Cp2} \tag{5.38}$$

where ψ_{P1-Cp1} stands for the flux linkage in rectangular coil C_{p1} of L_{f1} generated by rectangular coil $P1$ of L_1. ψ_{P1-Cp2}, ψ_{P2-Cp1} and ψ_{P2-Cp2} have similar definitions. Because the compensation coil is placed under the center of the main coil, there is the following relationship:

$$\psi_{P1-Cp1} = -\psi_{P1-Cp2},\ \psi_{P2-Cp1} = -\psi_{P2-Cp2} \tag{5.39}$$

Thus,

$$\psi_{P-Cp} = 0 \tag{5.40}$$

So, there is no mutual inductance between the main coil and the compensation coil on the same side and $K_{1f1} = 0$ theoretically. Similarly, there is $K_{2f2} = 0$, $K_{1f2} = 0$, $K_{2f1} = 0$ theoretically. The mutual inductance between L_{f1} and L_{f2} is not eliminated though, the value is so small that can be neglected because the size of compensation coils is small, and the distance between compensation coils is large.

According to reference (Mohammad and Choi 2018), the resonant equations of double-sided LCC compensation topology are expressed as:

$$\begin{cases} (2\pi f)^2 = \frac{1}{L_{f1} \cdot C_{f1}} = \frac{1}{(L_1 - L_{f1}) \cdot C_1} \\ (2\pi f)^2 = \frac{1}{L_{f2} \cdot C_{f2}} = \frac{1}{(L_2 - L_{f2}) \cdot C_2} \end{cases}, \tag{5.41}$$

where f is the tuned frequency. According to SAE TIR J2954, f is chosen as 95 kHz. when the circuit parameters are tuned to resonant values and circuit loss is neglected, the output power is expressed as:

$$P = \frac{4K\sqrt{L_1 L_2}}{\pi^3 f_0 \cdot L_{f1} L_{f2}} U_{in} U_{out}. \tag{5.42}$$

5.4.2 *Optimization of the Coil*

5.4.2.1 Optimization of Main Coils

Based on EV's application, we decide to choose the dimension of the primary main coil and secondary main coil both as 600 mm × 600 mm. Figure 5.61 shows the

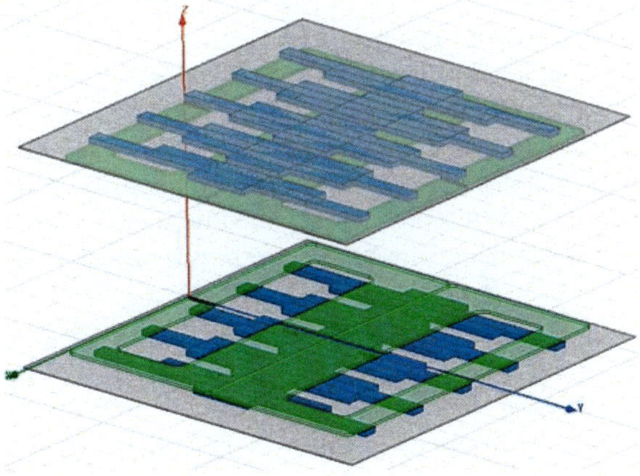

Fig. 5.61 Overview of main coils without compensation coils

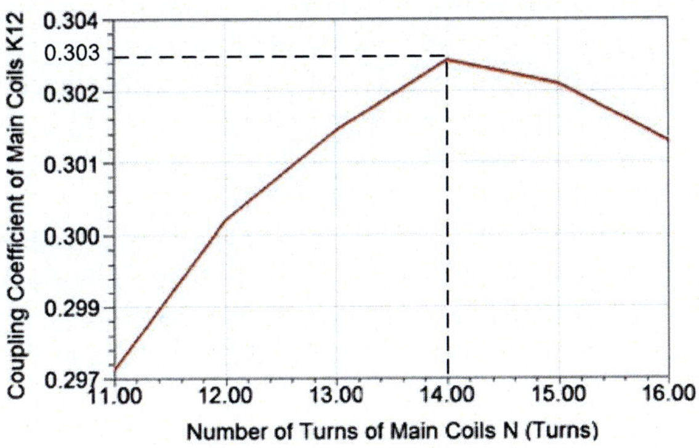

Fig. 5.62 Simulation results of coupling coefficient of main coils K12 with number of turns of main coils N

overview of main coils without compensation coils. Based on Eq. (2.5), the coupling coefficient K12 is related to output power. The value is desired to be high. In DD pad, coupling coefficient K12 is only determined by the width of main coils if other parameters of the pad are fixed. So, it is desired to simulate the relationship between the width of main coils and the coupling coefficient K_{12} without compensation coils. Figure 5.62 shows the simulation results of the coupling coefficient with the width of main coils determined by the number of turns of main coils N.

According to the simulation, when the turn number of main coils is 14, the coupling coefficient of main coils is maximal. Meanwhile, the inductance of main coils is determined.

5.4.2.2 Optimization of Compensation Coils

Once the inductance and coupling coefficient of main coils, input voltage, output voltage, and output power are determined, the product of L_{f1} and L_{f2} can be obtained according to Eq. (2.5). Next step, L_{f1} and L_{f2} should be optimized to ensure the efficiency is maximal theoretically.

In the circuit, there is inevitable loss as coils and capacitors have resistance and ferrites also expend energy. It is necessary to consider the resistance while optimizing the compensation coils. Figure 5.6 shows the circuit model ignoring the inverter with resistance and equivalent load resistor. Based on part 3.1, main coils have been constructed. The measured results are shown in Table 5.9. The resistance of compensation coils and capacitors can be calculated by:

$$R_1=\frac{2\pi f L_1}{Q_1},\ R_2=\frac{2\pi f L_2}{Q_2},\ R_{Lf1}=\frac{2\pi f L_{f1}}{Q_{f1}},\ R_{Lf2}=\frac{2\pi f L_{f2}}{Q_{f2}} \tag{5.43}$$

$$R_{C1}=\frac{D}{2\pi f C_1},\ R_{C2}=\frac{D}{2\pi f C_2},\ R_{Cf1}=\frac{D}{2\pi f C_{f1}},\ R_{Cf2}=\frac{D}{2\pi f C_{f2}} \tag{5.44}$$

Q is the quality factor of the compensation coil. Its value is estimated as 1000 according to the previous version of the coils we have tested. D is the dissipation factor of capacitors. We use EPCOS film capacitors, and their dissipation factor is claimed as 0.002. Then values of C_1, C_2, C_{f1}, C_{f2} can be calculated by resonant Eqs. (5.41), and values of $R_1, R_2, R_{Lf1}, R_{Lf2}, R_{C1}, R_{C2}, R_{Cf1}$, and R_{Cf2} can be estimated by Eqs. (2.6) and (2.7).

In Fig. 5.63, the circuit is divided into six blocks. η_1 represents the efficiency of block1, which is the ratio of the real power obtained by the load over the real power entering into block 1. η_2 represents the efficiency of block 2, which is the ratio of the real power obtained by block 1 over the real power entering into block 2. $\eta 3$ represents the efficiency of block 3, which is the ratio of the real power obtained by block 2 over the real power entering into block 3. $\eta 4$ represents the efficiency of block 4, which is the ratio of the real power obtained by block 3 over the real power entering into block 4. $\eta 5$ represents the efficiency of block 5, which is the ratio of the real power obtained by block 4 over the real power entering into block 5. $\eta 6$

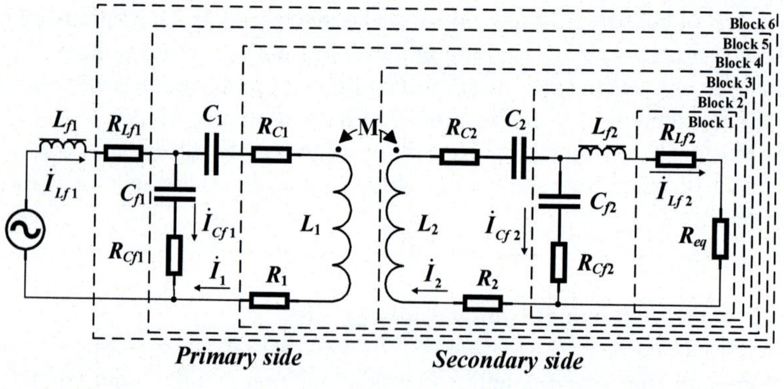

Fig. 5.63 Circuit model ignoring the inverter with resistance and equivalent load resistor

represents the efficiency of block 6, which is the ratio of the real power obtained by block 5 over the real power entering into block 6. $\eta 1$, $\eta 2$, $\eta 3$, $\eta 4$, $\eta 5$, $\eta 6$ can be expressed as:

$$\eta_1 = \frac{R_{eq}}{R_{eq} + R_{f2}} \tag{5.45}$$

$$\eta_2 = \frac{\left| R_{Cf2} + \frac{1}{j2\pi f C_{f2}} \right|^2 (R_{eq} + R_{Lf2})}{\left| R_{Cf2} + \frac{1}{j2\pi f C_{f2}} \right|^2 (R_{eq} + R_{Lf2}) + \left| R_{eq} + R_{Lf2} + j2\pi f L_2 \right|^2 R_{Cf2}} \tag{5.46}$$

$$\eta_3 = \frac{Re(Z_{block2})}{Re(Z_{block2}) + R_2 + R_{C2}} \tag{5.47}$$

$$\eta_4 = \frac{Re(Z_{refl})}{Re(Z_{refl}) + R_{C1} + R_1} \tag{5.48}$$

$$\eta_5 = \frac{\left| R_{Cf1} + \frac{1}{j2\pi f C_{f1}} \right|^2 \left[R_{C1} + R_1 + Re(Z_{refl}) \right]}{\left| R_{Cf1} + \frac{1}{j2\pi f C_{f1}} \right|^2 \left[R_{C1} + R_1 + Re(Z_{refl}) \right] + \left| R_1 + R_{C1} + j2\pi f L_1 + \frac{1}{j2\pi C_1} + Z_{refl} \right|^2 R_{Cf1}} \tag{5.49}$$

$$\eta_6 = \frac{Re(Z_{block5})}{Re(Z_{block5}) + R_{Lf1}} \tag{5.50}$$

The efficiency of the overall circuit is:

$$\eta = \eta_1 \cdot \eta_2 \cdot \eta_3 \cdot \eta_4 \cdot \eta_5 \cdot \eta_6 \tag{5.51}$$

The efficiency η is only related to L_{f2}. We can plot the efficiency over L_{f2} curve by Eq. (5.48) using MATLAB. The curve shows in Fig. 5.64. When L_{f2} is 59 μH, the efficiency is maximal. Thus, L_{f1} can be calculated according to Eq. (2.5). Then

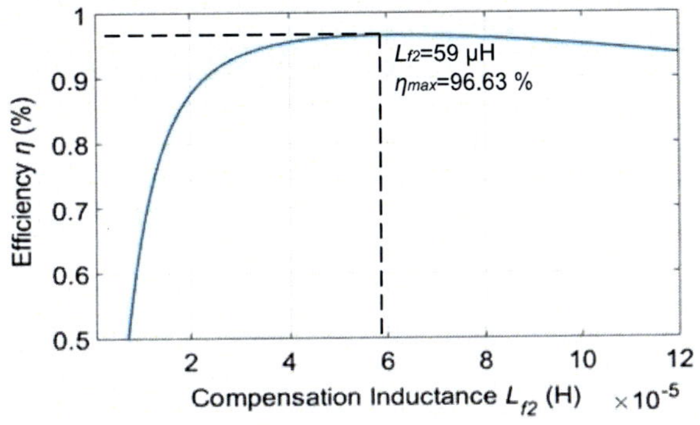

Fig. 5.64 Efficiency η over $Lf2$ curve

Table 5.10 Designed values and measured values of circuit parameters

Parameters	Design value	Measured value
$L1$	292μH	296.8μH
$L2$	292μH	284.2μH
$Lf1$	56.19μH	57.11μH
$Lf2$	59μH	59.35μH
$C1$	11.82nF	11.645nF
$C2$	13.7nF	13.75nF
$Cf1$	47.29nF	46.59nF
$Cf2$	49.93nF	50.14nF
$K12$	0.303	0.301
$K2f1$	0	0.00121
$K1f2$	0	0.00123
$K1f1$	0	0.0113
$K2f2$	0	0.00638
$Kf1f2$	0	0.0549

compensation coils are added into the previous main coil model in MAXWELL just as shown in Fig. 5.59. Compensation coils are smaller than main coils and are placed under the center of main coils on the same side. According to simulation, compensation coils are established and the measured values are shown in Table 5.10.

After the determination of main coils and compensation coils, other values of resonant elements can be calculated by previous equations.

5.4.3 *Experiment Verification*

After simulation and calculation, a wireless power system is established. Figure 5.65 shows the experimental setup of main coils, compensation coils and ferrite bars with an aluminum plate. Figure 5.66 shows the experimental workspace. Table 5.10 shows the designed values and measured values of circuit parameters. The measured values are within allowed error limits.

Figure 5.67a gives the system power characteristics with input voltage 300 V and output voltage 300 V. Figure 5.67b gives the waveform, including the voltage of one MOSFET V_{MOSFET}, drive voltage V_{drive}, and the current of L_{f1} I_{f1} when fully aligned. The input power is 3.2599 kW, and output power is 3.0044 kW. With the designed power, efficiency from DC-DC is 92.16%. The peak efficiency can reach 94.1%.

This section provides a new method of integrating main coils and compensation coils using double-sided LCC compensation topology, which can eliminate or reduce the extra coupling coefficient. This method simplifies the parameter design procedure and reduces the reactive power. A method of optimizing the integrated coils is also

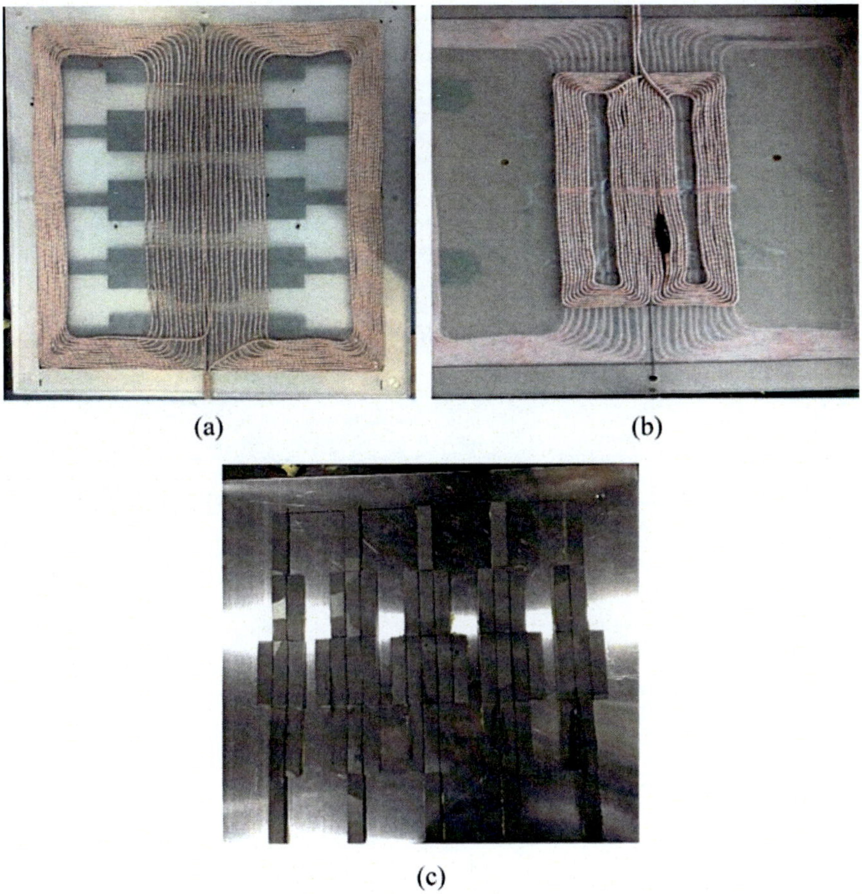

Fig. 5.65 **a** Main coils; **b** compensation coils; **c** Ferrite bars with an aluminum plate

used. Finally, a wireless power transfer system with 3 kW output power and 92.16% efficiency (from DC-DC) at an air gap of 150 mm is presented.

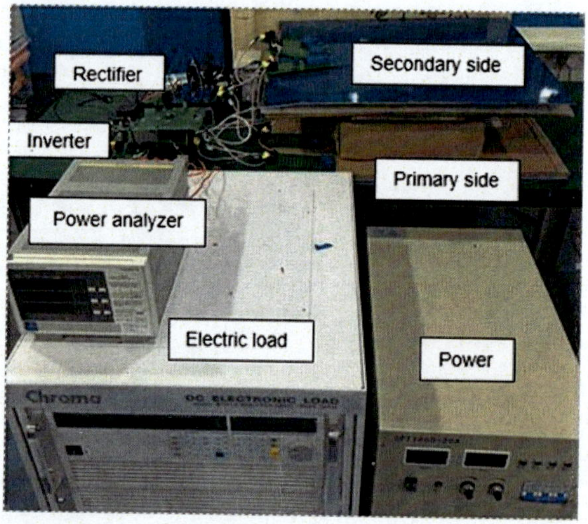

Fig. 5.66 Experimental workspace

Fig. 5.67 **a** System power characteristics; **b** waveform when fully aligned

(a)

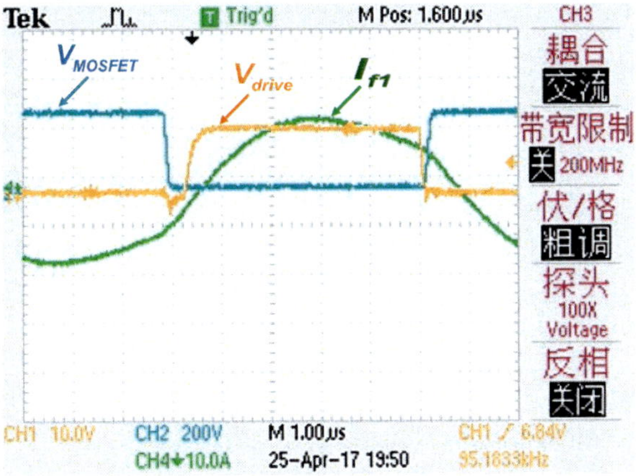

(b)

References

Borage M, Tiwari S, Kotaiah S (2005) Analysis and design of an LCL-T resonant converter as a constant-current power supply. IEEE Trans Ind Electron 52:1547–1554

Budhia M, Covic G and Boys JT (2010) A new IPT magnetic coupler for electric vehicle charging systems. In: 36th annual conference on IEEE industrial electronics society (IECON), Glendale, AZ, pp 2487–2492

Budhia M, Covic GA, Boys JT (2011) Design and optimization of circular magnetic structures for lumped inductive power transfer systems. IEEE T Power Electron 26(11):3096–3108

Budhia M, Boys JT, Covic GA et al (2013) Development of a single-sided flux magnetic coupler for electric vehicle IPT charging systems. IEEE Trans Ind Electron 60:318–328

Covic GA, Kissin MLG, Kacprzak D et al (2011) A bipolar primary pad topology for EV stationary charging and highway power by inductive coupling. In: Energy conversion congress and exposition, pp 1832–1838
Choi SY, Huh J, Lee WY et al (2014) Asymmetric coil sets for wireless stationary EV chargers with large lateral tolerance by dominant field analysis. IEEE Trans Power Electron 29:6406–6420
Cho S, Lee I, Moon S et al (2013) Series-series compensated wireless power transfer at two different resonant frequencies. In: 2013 IEEE ECCE Asia Downunder, Melbourne 1052–1058.
Deng J, Li W, Nguyen TD et al (2015) Compact and efficient bipolar pads for wireless power chargers: design and analysis. IEEE Trans Power Electron 30:6130–6140
Huang CY, James JE, Covic GA (2015) Design considerations for variable coupling lumped coil systems. IEEE Trans Power Electron 30:680–689
Jang GC, Jeong SY, Kwak HG et al (2016) Metal object detection circuit with non-overlapped coils for wireless EV chargers. In: Power electronics conference, pp 1–6
Jeong SY, Kwak HG, Jang GC et al (2018) Dual-purpose nonoverlapping coil sets as metal object and vehicle position detections for wireless stationary EV chargers. IEEE Trans Power Electron 33:7387–7397
Kan T, Nguyen TD, White JC et al (2017) A new integration method for an electric vehicle wireless charging system using lCC compensation topology. IEEE Trans Power Electron 32:1638–1650
Li W, Zhao H, Li S et al (2015a) Integrated LCC compensation topology for wireless charger in electric and plug-in electric vehicles. IEEE Trans Ind Electron 62(4215):4225
Li S, Li W, Deng J et al (2015b) A double-sided lcc compensation network and its tuning method for wireless power transfer. IEEE Trans Veh Technol 64:2261–2273
Li S, Mi CC (2015) Wireless power transfer for electric vehicle applications. IEEE J Em Sel Top P 3(1):4–17
Mohammad M, Choi S (2018) Optimization of ferrite core to reduce the core loss in double-D pad of wireless charging system for electric vehicles. In: 2018 IEEE applied power electronics conference and exposition (APEC), San Antonio, TX, pp 1350–1356
Ning P, Miller JM, Onar OC et al (2013) A compact wireless charging system for electric vehicles. In: Energy conversion congress and exposition, pp 3629–3634
Nguyen TD, Li S, Li W et al (2014) Feasibility study on bipolar pads for efficient wireless power chargers. In: Applied power electronics conference and exposition, pp 1676–1682
Takanashi H, Sato Y, Kaneko Y et al (2012) A large air gap 3 kW wireless power transfer system for electric vehicles. In: Energy conversion congress and exposition, pp 269–274
Wu HH, Gilchrist A, Sealy K et al (2011) A review on inductive charging for electric vehicles. In: Electric machines and drives conference, pp 143–147
Yungtaek J, Jovanovic MM (2003) A contactless electrical energy transmission system for portable-telephone battery chargers. IEEE Trans Ind Electron 50:520–527
Zaheer A, Hao H, Covic GA et al (2015) Investigation of multiple decoupled coil primary pad topologies in lumped IPT systems for interoperable electric vehicle charging. IEEE Trans Power Electron 30:1937–1955
Zhang W, Wong SC, Tse CK et al (2014) Design for efficiency optimization and voltage controllability of series–series compensated inductive power transfer systems. IEEE Tran Power Electron 29:191–200
Zhu Q, Wang L, Liao C (2014) Compensate capacitor optimization for kilowatt-level magnetically resonant wireless charging system. IEEE Trans Ind Electron 61:6758–6768

Chapter 6
Soft-Switching Converters

6.1 Introduction

In converters, a high operating frequency is utilized to satisfy the requirements for high power conversion density, small size, and light weight. However, when the switching frequency rises, the switching loss and EMI problem of the converter become more evident. Soft switching technology is a solution to the difficulties mentioned above.

Typical EVs WPT system consists of AC/DC (PFC and BUCK circuit), DC/AC (inverter circuit), resonant compensation network, transmitting coil and receiving coil, rectifier, and battery (Li et al. 2015). The PFC circuit is split into two parts: rectification and boost. Switching losses will arise in the boost, buck, and inverter circuits of the wireless charging system due to the presence of the power switch. Because the transmission power of the wireless charging technology is relatively high, the power switches will create a lot of heat if the soft switch is not utilized, even if a heat sink is supplied. This will impair transmission efficiency and even lower the transmission power level. As a result, the use of soft switching technology in wireless charging systems is critical.

Soft switching can be done in several ways. Many researchers have used the concept for designing additional resonant branches (Sharifi and Jabbari 2014; Pattnaik et al. 2010; Kumar et al. 2014; Khalilian et al. 2014; Amini and Farzanehfard 2014; Adib and Farzanehfard 2010; Zhang et al. 2010, 2017; Emrani and H. Farzanehfard 2012; Mousavi et al. 2011). Capacitors, inductors, and diodes make up the passive auxiliary circuit, which has a simple structure. zero-voltage-switching (ZVS) or zero-current-switching (ZCS) is achieved using a passive resonant circuit due to the immutable characteristic of the current or voltage on the inductor or capacitor. Auxiliary switch and passive components (capacitor, inductor, and diode) make up the active auxiliary resonant circuit. The circuiting energy is decreased by controlling the energy stored in the resonant tank with an auxiliary switch to achieve zero-voltage-transition (ZVT) or zero-current-transition (ZCT).

X. Zhang et al., *Wireless Power Transfer Technologies for Electric Vehicles*,
Key Technologies on New Energy Vehicles,
https://doi.org/10.1007/978-981-16-8348-0_6

Another technique for achieving soft switching is interleaved soft switching topology (Tsai et al. 2011; Abbasi et al. 2014; Jung et al. Apr. 2011; Hsieh et al. Apr. 2009). To construct an interleaved framework, these converters are connected in parallel. An inductor is employed as a symmetrical structure bridge to achieve ZVS. The capacitor can be charged or discharged depending on the positive and negative current in the inductor. The current and voltage of the switch can be buffered. The inherent feature of the interleaved architecture is that each sub-structure serves as an auxiliary circuit for another sub-structure while also providing power to the load.

Soft switching can also be accomplished using a coupled inductor, as seen in Chen et al. (2016), Aksoy et al. (2105), Yu et al. (2010), Bodur and Bakan (2004), Zhang et al. (2017), Lin et al. (1998). The weight and volume of the converter are reduced by using a coupled inductor as a resonant element with the same magnetic core. To reduce voltage or current shock, the electromagnetic field transfers the impact energy of the current and voltage to other branches.

In the wireless charging system, the boost converter adopts the method of coupling inductance to realize the zero voltage zero current transition (ZVZCT) turn-on and ZVS turn-off of the main switch; the main switch of the BUCK converter realizes the ZCS turn-on and ZCT turn-off by adding an auxiliary circuit; The ZVS condition for the inverter is achieved by adjusting the secondary series capacitor. The following is an analysis of the three modules mentioned above.

6.2 Zero Voltage Zero Current Transition Boost Converter

Boost converters are widely utilized in hybrid energy storage systems, solar power production systems, electric drive systems, and other applications as one of the most regularly used DC–DC converters. In the wireless charging system, a ZVZCT boost converter with coupled inductors is employed. The main switch of the converter is turned on for ZVZCT and off for ZVS. When the auxiliary switch is turned off, the turn-off current and voltage are lower than the input current and output voltage, respectively. As a result, the auxiliary switch is comparable to being used in a low-power application, which is one of the benefits of this circuit.

6.2.1 The Structure and Principle of Topology

The resonant sub-circuit is made up of a coupled inductor, a resonant capacitor, three diodes, and an auxiliary switch, as illustrated in Fig. 6.1a. The bypass capacitor of the main switch S_1 is in parallel with the stray capacitance in C_{S1}.

To make the functioning principle of the converter more understandable, the mutual inductance decoupling equivalent circuit is presented in Fig. 6.1b. Because of the large output capacitor C_1 and inductor L_1, the output voltage V_o and main

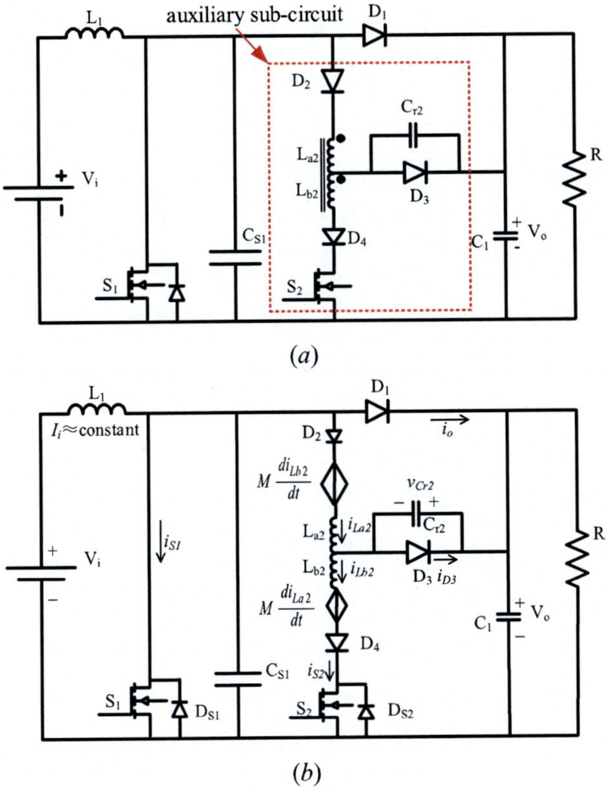

Fig. 6.1 Boost converter. **a** Proposed circuit. **b** Equivalent circuit

inductor current I_i are expected to remain constant. Figure 6.2 shows the main waveforms of the boost converter. A PWM cycle is split into 10 operational modes in this diagram. Figure 6.3 depicts the operation modes of the topology:

(1) *Mode 1* $[t_0 \leq t < t_1]$:

The auxiliary switch S_2 is turned on in mode 1. The current in D_1 progressively drops, whereas the current in D_2 gradually rises from zero. S_2, D_2, and D_4 are all ZCS turn-on due to coupled inductors, while C_{r2} is charging. Therefore, the auxiliary switch realizes ZCS turn-on.

(2) *Mode 2* $[t_1 \leq t < t_2]$:

In mode 2, the current in D_1 decreases to zero, and C_{S1} begins to discharge for the resonance with the auxiliary circuit. The voltage of C_{S1} drops. The current in D_1 becomes 0.

(3) *Mode 3* $[t_2 \leq t < t_3]$:

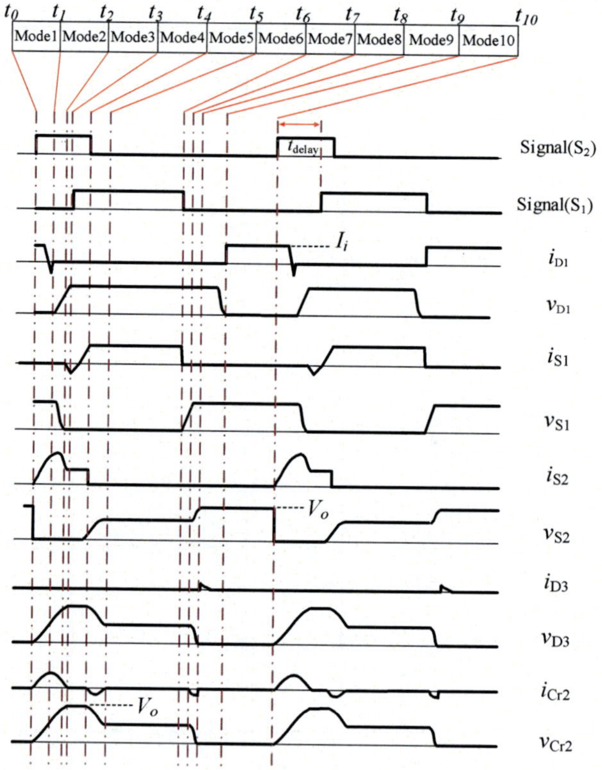

Fig. 6.2 Magnetization key waveforms and switching performance

When the bypass capacitor of S_1 is entirely discharged in this mode, the voltage of the main switch S_1 drops to zero. The antiparallel diode of the S_1 starts to conduct current, providing a time interval for the ZVT to turn-on. Current i_{S2} are the total currents of i_{Cr2} and i_{D2} in this mode.

(4) *Mode 4 [$t_3 \leq t < t_4$]:*

C_{r2} is entirely charged at the beginning of this mode. It is known by KVL, $V_{C1}=v_{Cr2} + v_{Lb2} - Mdi_{La}/dt$. As a result, after the charge is finished, the final two terms will be zero, resulting in a voltage of approximately V_{C1} for v_{Cr2}. The antiparallel diode of S_1 is held in the conduction condition throughout this time, and the voltage crossing S_1 is zero. At this time, S_1 will be applied by the active signal. As a result, S_1 is turned on by ZVT. The freewheeling current of i_{La2}-$I_i$persists. The ZVZCT turn-on margin for the main switch is determined by the time before i_{La2}-I_i goes to zero.

(5) *Mode 5 [$t_4 \leq t < t_5$]:*

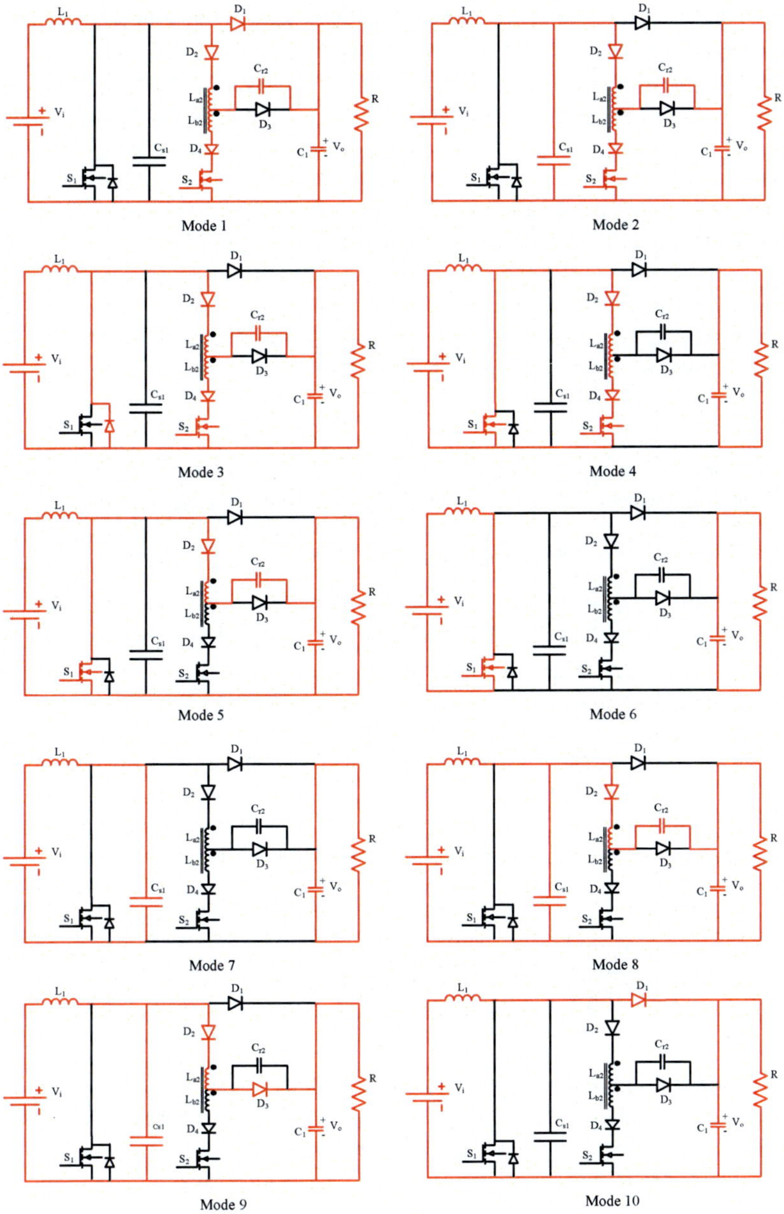

Fig. 6.3 Topology of operations stages. Mode 1 ($t_0 \leq t < t_1$). Mode 2 ($t_1 \leq t < t_2$). Mode 3 ($t_2 \leq t < t_3$). Mode 4. ($t_3 \leq t < t_4$). Mode 5 ($t_4 \leq t < t_5$). Mode 6 ($t_5 \leq t < t_6$). Mode 7 ($t_6 \leq t < t_7$). Mode 8 ($t_7 \leq t < t_8$). Mode 9 ($t_8 \leq t < t_9$). Mode 10 ($t_9 \leq t < t_{10}$)

When the auxiliary switch is switched off at the start of this mode, L_{b2} will transmit electromagnetic energy to L_{a2} through the winding. L_{a2} and C_{r2} start to resonate. At this point, C_{r2} will release power to L_{a2}. The L_{a2} and L_{b2} induction voltages reverse instantly, and this reverse voltage can reduce the current influence on the S_2.

(6) *Mode 6* $[t_5 \leq t < t_6]$:

This is a standard PWM mode. The energy from the power source will be stored in the inductor L_1, and the output capacitor C_1 will supply the load.

(7) *Mode 7* $[t_6 \leq t < t_7]$:

Through the main inductor L_1, the power supply V_i begins to charge C_{S1}. As a result, the voltage of S_1 will gradually rise, and S_1 will be able to turn off in ZVS.

(8) *Mode 8* $[t_7 \leq t < t_8]$:

In mode 8, the residual energy in the C_{r2} will be released owing to the input current of I_i as the voltage across the C_{S1} climbs to a specific amount.

(9) *Mode 9* $[t_8 \leq t < t_9]$:

Mode 9 is the process of releasing residual energy in L_{a2} by releasing current through D_2 and D_3. At the end of this mode, C_{S1} will be charged to the output voltage V_o.

(10) *Mode 1* $[t_9 \leq t < t_{10}]$:

This traditional mode is the main switching off-stage of the boost converter. Both S_1 and S_2 are in the off-state throughout this time. The main inductor will store energy from the power source in this mode since the DC power source supplies energy to both the load and the main inductor.

6.2.2 *The Conditional Constraint of ZVT Turned-On and ZVS Turned-Off for Main Switch*

S_1 always meets the ZVS turn-off requirement due to the existence of C_{S1}. However, before removing the signal from S_2, C_{S1} must be fully discharged. When the body diode of S_1 is in the active state, the drive signal must be applied to the main switch for ZVZCT turn-on. In this scenario, the C_{S1} discharge interval should not exceed t_2-t_1. For the main switch ZVZCT, the topology must meet the following requirements:

$$i_{Lb2\_3} - I_i \geq 0 \tag{6.1}$$

$$C_{S1} \leq \frac{I_i}{V_o}(t_2 - t_1) \tag{6.2}$$

6.2.3 Configuration of Coupled Inductor for Soft Switching

The major benefit of this topology is that the required freewheeling current for the main switch ZVT can be configured flexibly by adjusting the coupled inductor ratio. The resonant peak current of the auxiliary sub-circuit provides the majority of the freewheeling current. A portion of the resonant current is transmitted to the load side in the auxiliary subcircuit, while another portion is engaged in the resonance of the C_{S1}. The coupled inductor plays a critical function in the distribution of resonance energy. S_1 requires a significant freewheeling current to accomplish ZVT under heavy load, however, the need for freewheeling current is extremely low under light load. According to various loads, coupled inductors with various ratios are built to identify the resonant energy distribution that is sufficient for S_1 to switch on with ZVT while wasting no more energy, resulting in the desired effect. The ratio will not alter after the design in real experiments.

Setting the correct coupled inductor ratio can effectively optimize the proportional distribution for the two portions of the current. It is obvious that the load determines I_i. As a result, ZVZCT can adapt to varied loads by adjusting the freewheeling current through a coupled inductor ratio.

$$coupling\ ratio = \frac{L_{b2}}{L_{a2}} \tag{6.3}$$

Freewheeling current increases when I_o (or I_i) falls, as seen in Fig. 6.4. The circulation loss of the auxiliary circuit steadily increases as the freewheeling current increases. This is also why most soft switching converters have low efficiency when used with a light load. If this topology can implement ZVZCT under heavy load with

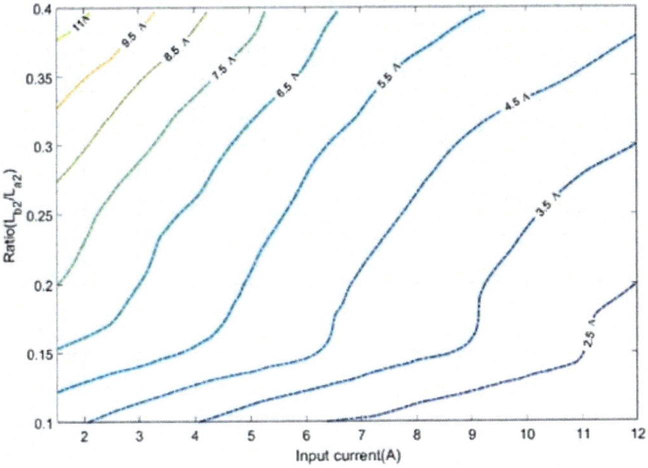

Fig. 6.4 S_1 freewheeling current influence from coupling ratio and input current

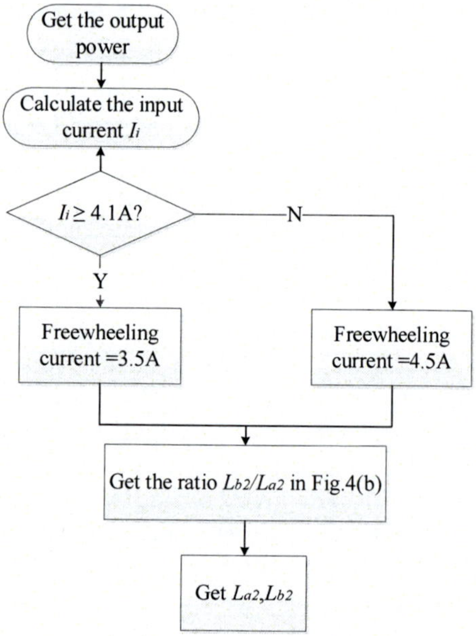

Fig. 6.5 Proposed soft switching optimization flow chart

the same parameters, it must also be able to implement ZVZCT under light load. Soft switching is significantly less efficient with light loads.

It is only necessary to change the ratio of the connected inductor when designing a soft switch with varied loads. The design has been simplified and enhanced significantly. The connection between the freewheeling current, the coupled inductor ratio, and the input current is shown in Fig. 6.4. The flow chart of the suggested soft switching optimization is shown in Fig. 6.5, and the approach for optimizing the coupled inductor ratio is as follows:

1. Get the output power in practical application, and calculate the input current I_i.
2. Choose the right freewheeling current. Excessive freewheeling currents would increase circulation loss, whereas too little freewheeling currents would increase the likelihood of ZVZCT failure. As a result of theoretical study and empirical findings, the optimum freewheeling currents for $I_i \geq 4.1$A and $I_i < 4.1$A are 3.5A and 4.5A, respectively.
3. In Fig. 6.4, draw a vertical line with an abscissa equal to the computed input current. This line intersects with the curve that depicts the freewheeling current that has been optimized. The optimum ratio of the coupled inductor is then the ordinate of the crossover point.

6.2.4 Voltage and Current Waveforms of Boost Converter

The main switching voltage and current waveform for turning on and off are shown in Fig. 6.6. The voltage and the current of S_1 progressively fall to zero, after which the current reverses and rises from zero, resulting in the ZVZCT of S_1. The current is swiftly decreased to zero when the signal is withdrawn. Then, in V_{S1}, there is no leaping. As a result, ZVS is possible to achieve during the off process.

The S_2 switching mechanism and conduction properties are depicted in Fig. 6.7. The current continuously increases as the active signal of S_2 is applied, and S_2 switches on ZCS. It can also be observed that the turn-off current and voltage of S_2 are lower than I_i and V_o, implying that S_2 is equal to operating in low-power applications. The capacitor C_{r2} starts discharging mode again after a while, and the

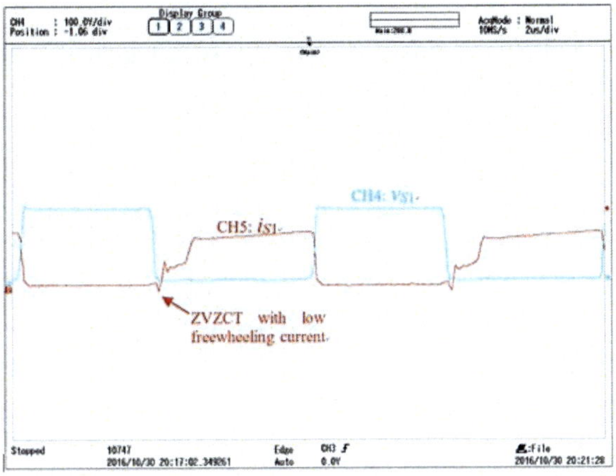

Fig. 6.6 Voltage and current of S_1

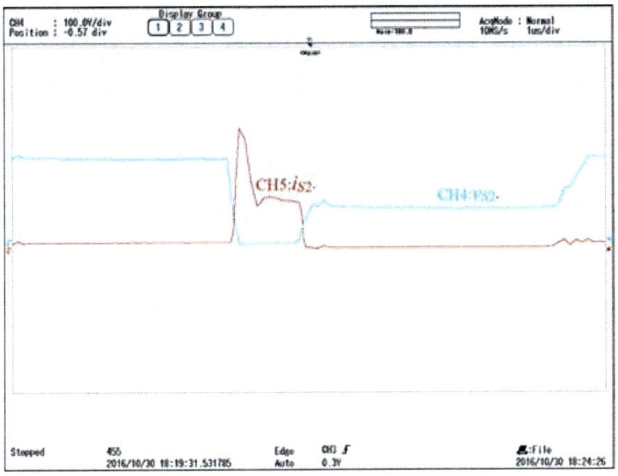

Fig. 6.7 Voltage and current of S_2

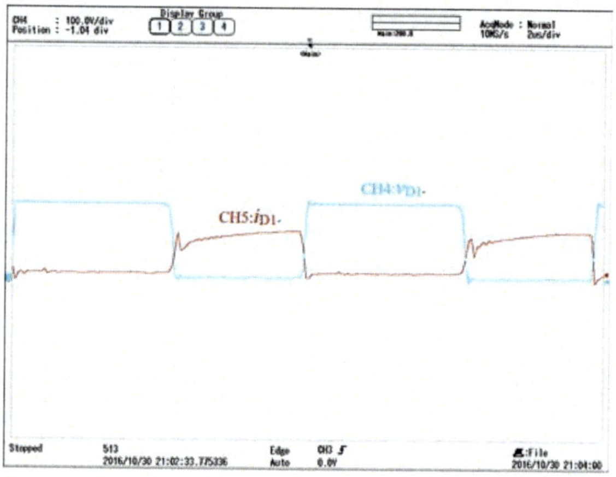

Fig. 6.8 Voltage and current of D_1

auxiliary switch voltage finally rises. The main diode operates in soft switching, as illustrated in Fig. 6.8.

6.3 Zero-Current-Switching PWM Buck Converter

Buck converters, which are a sort of basic DC-DC converter, are frequently employed in industrial applications. In this case, a ZCS PWM buck converter with magnetic coupling effect is employed. The auxiliary circuit, which includes the auxiliary switch, auxiliary capacitor, and coupled inductor, distinguishes this buck converter from typical buck converters. The main inductor is the main coil of the coupled inductor. One auxiliary coil is linked to the auxiliary circuit, while the other is connected in series with the main diode. The control method of the main switch does not need to be modified, while the main switch is closed in the ZCT state by adjusting the auxiliary switch. Furthermore, the main switch is turned on with ZCS condition, and the auxiliary switch is turned on and off with ZCS condition.

6.3.1 *Converter Topology and Operation Principles*

The proposed ZCS buck converter is made up of a standard buck converter and an additional resonant circuit, as shown in Fig. 6.9a. The windings L_1, L_2, and L_a are on a single toroidal ferrite core, and the coupling coefficient is less than 0.75. The main inductor is L_1, whereas the auxiliary coupled inductor is L_2. The input and output filtering capacitors are C_1 and C_2, respectively. And in the auxiliary circuit, C_a resonates with L_a. D, D_1, and D_2 are the main diode, the antiparallel body of S_1 and S_2, respectively.

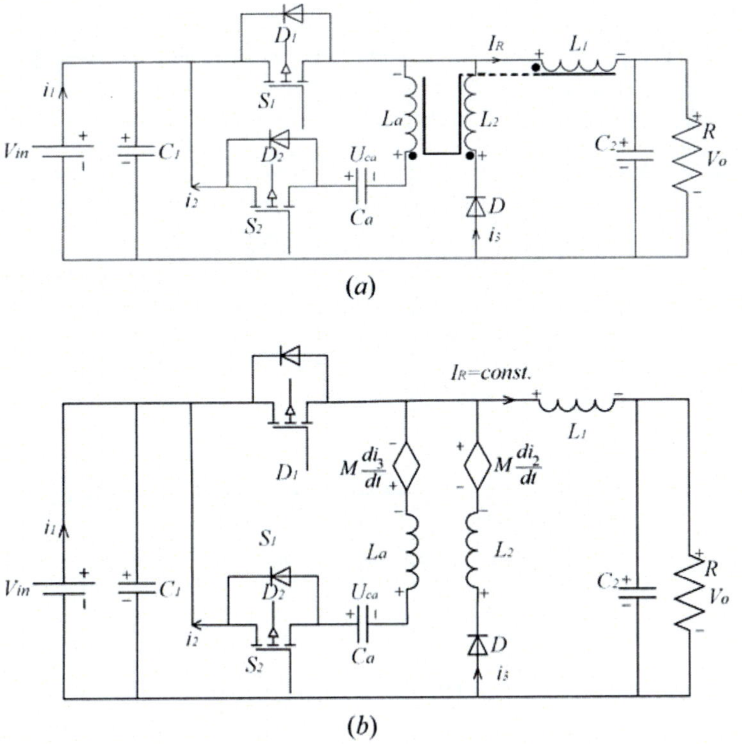

Fig. 6.9 Topology of the proposed ZCS buck converter. **a** Schematic of the proposed buck converter. **b** Decoupling equivalent circuit

Because the main inductor and output capacitance are large enough, it is assumed that the main inductor current I_R and the output voltage V_o remain constant to make analysis and derivation easier. Then, for mode analysis with $di_R/dt \approx 0$, the coupling between L_1 and L_a, L_1 and L_2, and L_1 and L_2 may be ignored. As a result, the decoupling equivalent circuit could be seen in Fig. 6.9b.

One PWM period is split into six modes in the study of operating principles, and the corresponding circuits for each mode are illustrated in Fig. 6.10, where the red solid arrows represent the real current direction. Figure 6.11 depicts the theoretical waveforms.

(1) *Mode 1*$[t_0 \le t < t_1]$:

The initial values at t_0 are $i_1 = 0$, $i_2 = 0$, $U_{Ca} = U_{Ca+}$, $i_3 = I_R$. The resonance between C_a and L_a, C_1, and L_2, begins when the active signal is supplied to the switch S_1. S_1 ZCS turn-on is achieved for these resonances. Simultaneously, the auxiliary capacitor C_a discharges00A0through the antiparallel body diode of S_2, causing the main diode current i_3 to fall while i_1 increases. Because the current in the main diode falls to zero when this mode finishes at t_1, ZCS turn-off of the main diode is achieved.

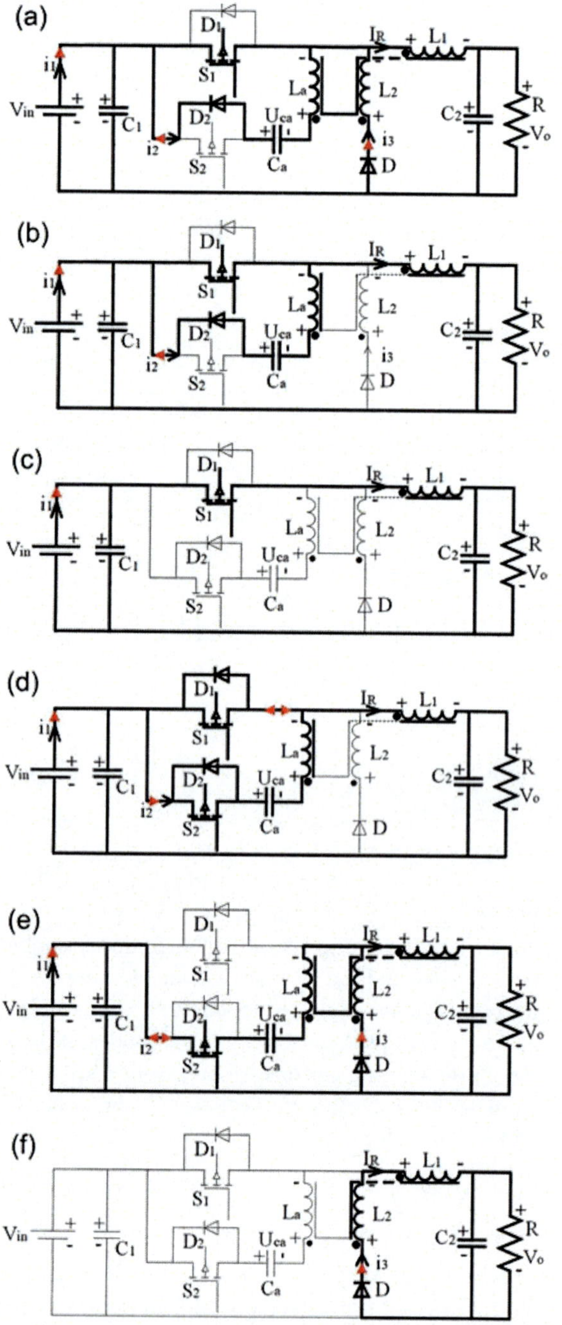

Fig. 6.10 Equivalent circuits for each mode. **a** Mode 1 ($t_0 \leq t < t_1$). **b** Mode 2 ($t_1 \leq t < t_2$). **c** Mode 3 ($t_2 \leq t < t_3$). **d** Mode 4. ($t_3 \leq t < t_4$). **e** Mode 5 ($t_4 \leq t < t_5$). **f** Mode 6 ($t_5 \leq t < t_6$)

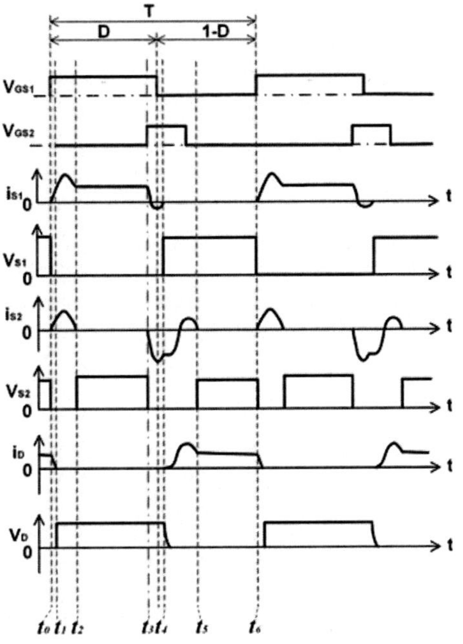

Fig. 6.11 Theoretical waveforms of operation modes

(2) *Mode 2[$t_1 \leq t < t_2$]:*

C_a continues to resonate with L_a after the main diode D shuts off in mode 2. Meanwhile, as i_2 reaches its maximum value, U_{Ca} falls to zero. Then, as U_{Ca} continues to discharge from a positive to a negative voltage, i_2 begins to fall. At t_2, i_2 equals zero, and U_{Ca} reaches the minimum voltage U_{min}. In this way, ZCS turn-off of the antiparallel body diode of S2 is achieved.

(3) *Mode 3[$t_2 \leq t < t_3$]:*

At time t_2, i_2 is zero, and S_2 clamps the minimal voltage U_{min} on the auxiliary capacitor C_a. i_1 equals I_R and remains constant in this mode. Mode 3 is the same as the operating mode of a conventional buck converter.

(4) *Mode 4[$t_3 \leq t < t_4$]:*

At time t_3, the starting circumstances are $U_{Ca} = U_{min}$, $i_1 = I_R$, and $i_2 = 0$. At t_3, the auxiliary switch S_2 is activated, and L_a begins to resonate with C_a. Because L_a is in series with S_2 and the initial value of i_2 is zero, S_2 begins with ZCS. i_2 rises in the negative direction during this phase, whereas i_{S1} decreases from positive to negative current. D_1 is turned on once i_{S1} is negative. Because all of the current on S_1 flows through D_1, the ZCT turn-off of S_1 is achieved. C_a continues to be charged by L_a after S_1 has been turned off. When the current on S_1 reaches zero, the mode ends.

(5) *Mode 5[$t_4 \leq t < t_5$]:*

This mode is split into two parts. C_a is charged in the first segment, and i_1 is positive and declines from a positive value to zero. ZVS turn-on of the main diode D is achieved and i_3 begins to rise from zero. In the second section, C_a begins to discharge when i_1 reaches zero, and i_1 continues to fall when U_{Ca} exceeds V_{in}. Because the current of S_2 travels through the antiparallel diode D_2, the ZCS turned-off state of S_2 could be achieved as soon as i_1 turns negative.

(5) Mode 6$[t_5 \leq t < t_6]$:

The main inductor L_1 provides the load through the main diode D in this mode. This mode is known as the freewheeling current mode, and it is the same as in a traditional buck converter. The converter will begin the next PWM period when the active signal is applied to S_1 at t_6.

6.3.2 ZCS Turn-On Condition and ZCT Turn-Off Condition for the Main Switch

The ZCS turn-on condition of S_1 is always achieved due to the presence of L_a and L_1. C_a begins to discharge when an active signal is supplied to S_1 at t_0, and the discharging current is inhibited by L_a. If there is no inductor in series with the main diode, the freewheeling current will pass through it before t_0, resulting in a massive short current on S_1 at t_0 due to the reverse recovery time of the main diode. The main switch can then be switched off with ZCT condition as long as the current on S_1 turns negative and the antiparallel body diode D_1 is conductive. To satisfy these requirements, the energy stored in the auxiliary capacitor must be sufficient, and the C_a/L_a ratio must be adequate. As a result, the following condition should be satisfied:

$$U_{\min}\sqrt{\frac{C_a}{L_a}} \leq -I_R \tag{6.4}$$

When the current on the main switch is negative, the active signal of the main switch should be removed. The following equation is used to compute the turn-off time of S_1:

$$\begin{cases} t_{S_1\_off} \geq t_3 + \sqrt{L_a C_a} \arcsin \frac{I_R}{U_{\min}} \sqrt{\frac{L_a}{C_a}} \\ t_{S_1\_off} \leq t_3 + \sqrt{L_a C_a}\pi - \sqrt{L_a C_a} \arcsin \frac{I_R}{U_{\min}} \sqrt{\frac{L_a}{C_a}} \end{cases} \tag{6.5}$$

6.3.3 ZCS Turn-On and Turn-Off Conditions for the Auxiliary Switch

Because the increasing rate of i_2 is restricted by the coupled inductors, the auxiliary switch is turned on with the ZCS condition. When S_1 is turned off, the resonance between L_a, C_a, and L_2 begins. When the voltage of C_a exceeds the input voltage and current flows through the body diode of S_2, and the auxiliary switch is turned off using ZCS. The turn-off time meets the following criteria:

$$\begin{cases} t_{S2\_off} \geq t_6 + \frac{2}{\omega_2}\arctan\left[\frac{C_a V_{in}\omega_2 - \sqrt{C_a^2 V_{in}^2 \omega_2^2 + I_R^2}}{I_R}\right] \\ t_{S2\_off} \leq t_6 + \frac{2}{\omega_2}\arctan\left[\frac{C_a V_{in}\omega_2 + \sqrt{C_a^2 V_{in}^2 \omega_2^2 + I_R^2}}{I_R}\right] \end{cases} \quad (6.6)$$

Because the resonance between L_a, C_a, and L_2 should be completed before the next switching cycle, the following conclusion is derived:

$$t_6 - t_4 \leq (1 - D)T \quad (6.7)$$

6.3.4 Voltage and Current Waveforms of Buck Converter

Figures 6.12 and 6.13 illustrate the waveforms of the main and auxiliary switches, respectively. When V_{GS1} reaches 12 V, current i_{S1} steadily rises while V_{S1} rapidly drops to zero. Despite the minimal overlap between i_{S1} and V_{S1}, it is possible to

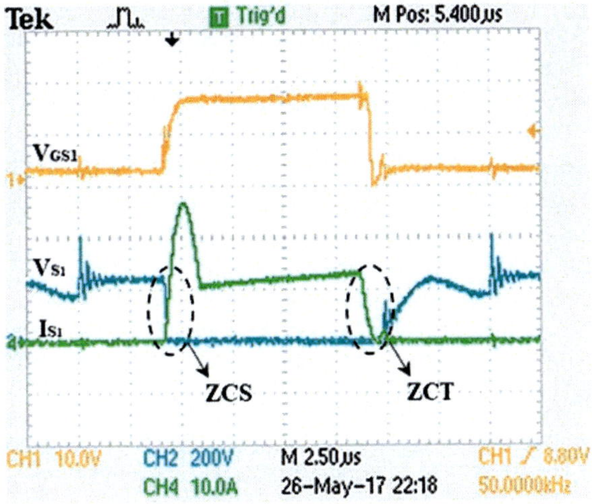

Fig. 6.12 The waveforms of the main switch S_1

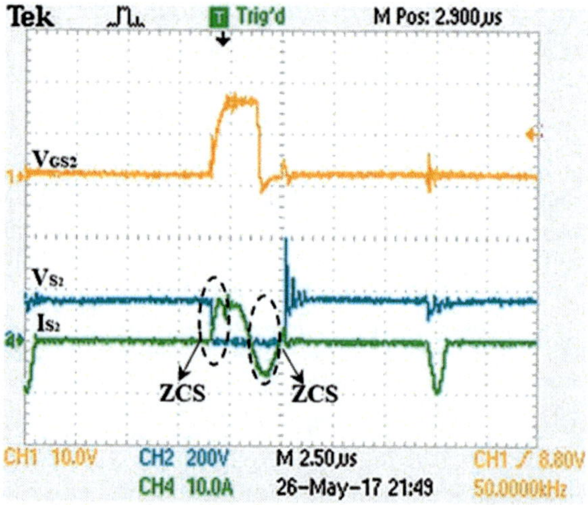

Fig. 6.13 The waveforms of the auxiliary switch S_2

conclude that significant reductions in turn-on losses have been achieved. As a result, S_1 is turned on with ZCS condition. S_1 is switched off with ZCT condition because current i_{S1} falls to negative current before V_{GS1} goes to low and turn-off losses are avoided. Furthermore, the current overshoots of the main switch during turn-on and turn-off are within permissible limits. The conclusion drawn from the foregoing analysis is that the ZCS and ZVS conditions of S_2 and D can be achieved.

6.4 Zero Voltage Switching Inverter

The inverter circuit converts DC power to a high-frequency square wave voltage, which is essential for the compensation network to achieve resonance. For the MOSFET to accomplish soft switching, the wireless charging system operates in the inductive region. Furthermore, if the MOSFETs in the upper and lower legs are turned on at the same time, a short circuit will occur, and the MOSFET will be destroyed if the input voltage is too high. As a result, achieving soft switching requires the proper dead time (Zhang et al. 2015). The ZVS condition for inverter MOSFETs is achieved here by adjusting the secondary series capacitor, and the optimal dead time for soft switching is described.

6.4.1 *Double-Sided LCC Resonant Compensation Network*

Figure 6.14 depicts the configuration of the double-sided LCC resonant WPT system. The inverter is made up of four MOSFETs (S_1–S_4), parasitic capacitors (C_{s1}–C_{s4}), and parasitic antiparallel diodes (D_1–D_4). The self-inductances of the transmitting

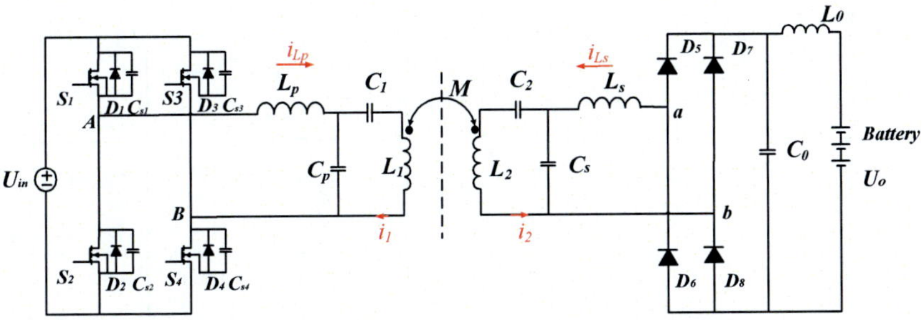

Fig. 6.14 Schematic of double-sided LCC resonant compensation network

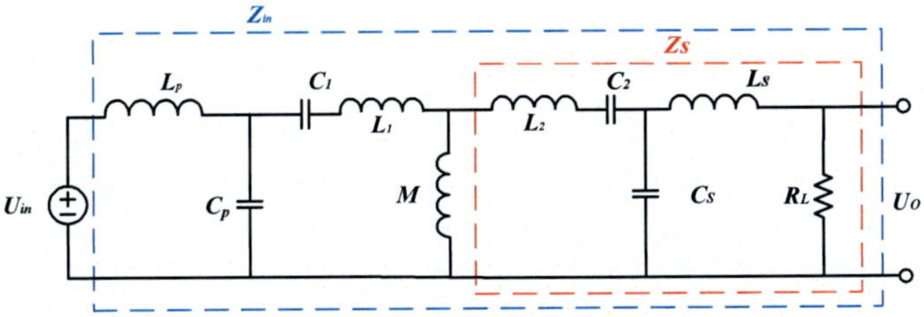

Fig. 6.15 Equivalent schematic of double-sided LCC resonant compensation network

and receiving coils are denoted by L_1 and L_2, respectively. The primary side compensation inductor and capacitors are denoted by L_p, C_p, and C_1. The secondary side compensation components are L_s, C_s, and C_2. M is the coefficient inductance of the two main coils. In addition, the rectifier is made up of rapid recovery diodes (D_5–D_8), a filtering capacitor C_o, and a filtering inductor L_o.

The parasitic resistance of the capacitors, inductors, and coils is disregarded for the sake of simplicity. Figure 6.15 depicts the primary and secondary resonant compensation equivalent network. Where R_L denotes the equivalent impedance of the battery and rectifier. According to Li et al. (2015), the corresponding impedance can be represented as a pure resistance, which is defined as $R_L = \pi^2 U_O^2 / 8P_{out}$, Pout is the output power of the WPT system.

The parameters of a double-sided LCC resonant compensation network are built using the equations in Li et al. (2015):

$$\begin{cases} L_p \cdot C_p = \dfrac{1}{\omega_0^2}, \ L_s \cdot C_s = \dfrac{1}{\omega_0^2}, \\ L_1 - L_p = \dfrac{1}{\omega_0^2 C_1}, \ L_2 - L_s = \dfrac{1}{\omega_0^2 C_2} \circ \end{cases} \tag{6.8}$$

The equivalent impedance of the equivalent double-sided LCC compensation network is given by:

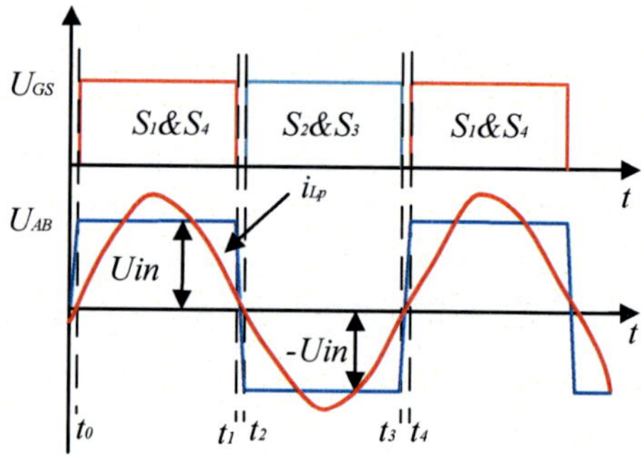

Fig. 6.16 Operation waveform of the inverter

$$Z_{in} = \frac{\omega_0^2 L_p^2 L_s^2}{M^2 R_L} \quad (6.9)$$

It is a completely resistant load. The inverter output voltage is in phase with the output current in this scenario, resulting in a unity power factor, however, this is not a good situation for MOSFETs to produce ZVS. By tuning the secondary series capacitor C_2, the circuit becomes inductive, allowing the MOSFET to achieve ZVS.

6.4.2 *Time Domain Analysis of Switching Mode*

The primary operational waveform is depicted in Fig. 6.16. Four phases can be found in a switching cycle. The initial part of the switching cycle is investigated.

Mode 1 $[t_0, t_1]$: the inverter output current i_{Lp} is freewheeling through D_1 and D_4, and C_{s1} and C_{s4} have been discharged to zero voltage before the S_1 and S_4 are turned on. At the moment t_1, S_1, and S_4 achieve zero voltage turn-on. U_{AB}=U_{in} at this time.

Mode 2 $[t_1, t_2]$: at t_1, both the S_1 and S_4 are switched off, and the inverter output current i_{Lp} (also known as cut-off current) is charging the capacitors C_{s1} and C_{s4} to voltage U_{in}, while the capacitors C_{s2} and C_{s3} are discharged to voltage zero.

Modes 3 and 4 are similar to mode 1 and mode 2, respectively. The turn-off current may be decreased and the MOSFET turn-off loss reduced by adjusting the capacitor C_2 and dead-time optimization.

6.4.3 *Optimization Method of Dead-Time of Inverter MOSFET*

The inverter output current charges and discharges the parasitic capacitors of the MOSFET during the dead-time, according to the preceding study. If the dead-time is adjusted incorrectly, the transmitting power and efficiency will suffer. As a result, for MOSFETs to accomplish ZVS, the dead-time setting is critical.

To guarantee ZVS and lower switching loss in MOSFETs, On the one hand, the dead-time must be long enough to charge or discharge the parasitic capacitors of both the MOSFETs and the PCB; on the other hand, the inverter current must be near to zero at the switching point, which implies the dead-time must be shorter than the diode freewheeling duration. As a result, the dead-time must satisfy the following expression:

$$t_c + t_{off} < t_{DT} < t_d \tag{6.10}$$

where t_c is the time required to charge and discharge parasitic capacitors, t_{off} is the time it takes for a MOSFET to turn off, and t_d is the time it takes for a diode to freewheel.

6.4.4 Voltage and Current Waveforms of Inverter

As shown in Fig. 6.17a, there is a negative current on the MOSFET due to diode freewheeling before the MOSFET turns on. When the drive voltage U_{GS} rises to 90%, the MOSFET has forward current (I_D) and the MOSFET turns on at zero voltage. According to Fig. 6.17b, because of the long dead-time, there is positive current on the MOSFET before it is turned on, and even after the MOSFET is turned off, the voltage of the switch decreases quickly to zero.

According to Fig. 6.18b, if the dead-time is too long, the inverter will oscillate at a high frequency. According to the testing results, a dead-time setting that is too short or too big would cause distortion in the output voltage and current of the inverter, resulting in a decreased efficiency and transmitting power of the WPT system. As shown in Fig. 6.18a, by choosing the appropriate dead-time, MOSFETs

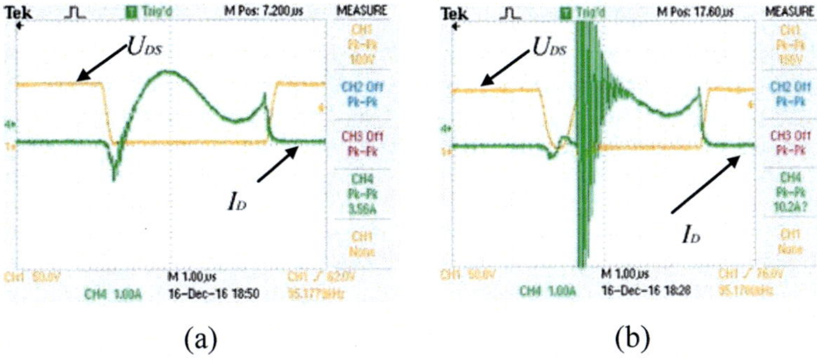

Fig. 6.17 Voltage and current of MOSFET. **a** = 500 ns; **b** = 1.2 us

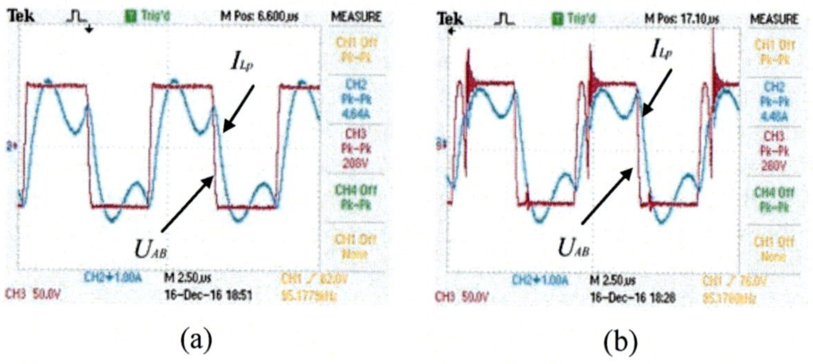

Fig. 6.18 Voltage and current of inverter. **a** =500 ns; **b** = 1.2 us

can accomplish ZVS and reduce voltage and current distortion, therefore increasing transmission efficiency.

References

Abbasi M, Mortazavi N, Rahmati A (2014) A novel ZVS interleaved boost converter. In: The 5th annual international power electronics, drive systems and technologies conference (PEDSTC 2014), pp 535–538

Adib E, Farzanehfard H (2010) Zero-voltage-transition PWM converters with synchronous rectifier. IEEE Trans Power Electron 25(1):105–110

Aksoy I, Bodur H, Faruk Bakan A (2010) A New ZVT-ZCT-PWM DC–DC converter. IEEE Trans Power Electron 25(8):2093–2105

Amini MR, Farzanehfard H (2014) Switched resonator DC/DC converter with a single switch and small inductors. Power Electron Iet 7(6):1331–1339

Bodur H, Bakan AF (2004) A new ZVT-ZCT-PWM DC-DC converter. IEEE Trans Power Electron 19(4):676–684

Chen YT, Shiu SM, Liang RH (2016) A new family of zero-voltage-transition nonisolated bidirectional converters with simple auxiliary circuit. IEEE Trans Industr Electron 63(3):1519–1527

Emrani, Farzanehfard H (2012) Zero-current switching resonant buck converters with small inductors. Iet Power Electron 5(6):710–718

Hsieh YC, Hsueh TC, Yen HC (2009) An interleaved boost converter with zero-voltage transition. IEEE Trans Power Electron 24(4):973–978

Jung DY, Ji YH, Park SH, Jung YC, Won CY (2011) Interleaved soft-switching boost converter for photovoltaic power-generation system. IEEE Trans Power Electron 26(4):1137–1145

Khalilian M, Adib E, Farzanehfard H (2014) Family of single-switch soft-switching pulse-width modulation DC-DC converters with reduced switch stress. Iet Power Electron 7(8):2182–2189

Kumar SS, Panda AK, Ramesh T (2014) A simple passive auxiliary circuit for reduction of losses and efficiency enhancement of synchronous buck converter. In: India Conference, pp 1–6.

Li S, Li W, Deng J et al (2015) A Double-sided LCC compensation network and its tuning method for wireless power transfer. IEEE Trans Veh Technol 64(6):2261–2273

Lin RL, Zhao Y, Lee FC (1998) Improved soft-switching ZVT converters with active snubber. In: 13th APEC, pp 1063–1069

Mousavi M, Pahlevaninezhad P, Das, Jain P (2011) ZCS PWM bidirectional converter with one auxiliary switch. In: Energy conversion congress and exposition, pp 1175–1180

Pattnaik S, Panda AK, Mahapatra K (2010) Efficiency improvement of synchronous buck converter by passive auxiliary circuit. IEEE Trans Ind Appl 46(6):2511–2517
Sharifi S, Jabbari M (2014) Family of single-switch quasi-resonant converters with reduced inductor size. Power Electron Iet 7(10):2544–2554
Tsai C-T, Shen C-L, Su J-C, Kuo Y-C (2011) Analysis and implementation of an interleaved ZVS buck converter with coupled-inductors. In: 6th IEEE conference on industrial electronics and applications, pp 1392–1397
Yu W, Qian H, Lai JS (2010) Design of high-efficiency bidirectional DC–DC converter and high-precision efficiency measurement. IEEE Trans Power Electron 25(3):650–658
Zhang Xi, Qian W, Li Z (2017) Design and analysis of a novel ZVZCT boost converter with coupling effect. IEEE Trans Power Electron 32(12):8992–9000
Zhang W, Zhang X, Xiao S (2010) A novel soft switch for buck converter. In: IEEE international symposium on power electronics for distributed generation systems, pp 180–184
Zhang Z, Wang F, Costinett DJ, Tolbert LM, Blalock BJ, Lu H (2015) Dead-time optimization of SiC devices for voltage source converter. In: 2015 IEEE applied power electronics conference and exposition (APEC), Charlotte, NC, pp 1145–1152
Zhang Y, Cheng XF, Yin C, Cheng S (2017) Analysis and research of a soft-switching bidirectional DC-DC converter without auxiliary switches. IEEE Trans Ind Electron 99:1–1

Chapter 7
Foreign Object Detection/LOD and Protections

7.1 Introduction

When an inductive power transfer system functions, electric energy is transmitted between the transmitter and the receiver via magnetic flux generated by a high-frequency AC current. Although shielding layers are utilized to reduce the radiation of the magnetic field outside the magnetic couplers, the magnetic field inside the magnetic couplers is still strong. An eddy current will be generated if a metal object (MO), such as a key, a bolt, a coin, a screw, a paper clip, or a tinfoil, is present between the transmitting and receiving coils during power transmission, resulting in overheating and potential fire risks (Jafari et al. 2019; Moghaddami and Sarwat 2018; Lu et al. 2021; Zhang et al. 2019; Thai et al. 2020b). In addition to the issues caused by metal objects, living objects (LOs), like humans, dogs, cats, and birds, exposed to the magnetic field generated by the magnetic couplers may experience many health symptoms, such as nausea, vomiting, dyspnea, dizziness, and blood pressure changes (Jeong et al. 2016a).

Therefore, it is essential to equip the wireless charging system with special designs to detect metal objects and living objects to ensure the stability, reliability, and safety of the wireless power transfer system. The technology and research on solving these issues is called foreign object detection (FOD), which is divided into metal object detection (MOD) and living object detection (LOD).

7.2 Lod

Living object detection makes wireless power transfer systems able to immediately shut down when humans or animals enter between the two transmitting magnetic couplers or when there is a living object close to the wireless charging system.

X. Zhang et al., *Wireless Power Transfer Technologies for Electric Vehicles*,
Key Technologies on New Energy Vehicles,
https://doi.org/10.1007/978-981-16-8348-0_7

There are mainly two strategies of LOD, which are field-based detection method and sensor-based detection method.

7.2.1 Field-Based Detection Method

The field-based detection method can infer whether there is a foreign object or a living object by directly monitoring the characteristic parameters of the wireless charging system without using an external sensor. The capacitive detection method is the most promising field-based detection method as the inductance and equivalent resistance are negligibly affected by living objects while the capacitance around the living objects can be affected. Thus, the detection system's equivalent capacitance variation is utilized for living object detection (Jeong et al. 2016a; Novak et al. 2017; Thai et al. 2018; Katz et al. 2015).

Patent (Novak et al. 2017) promoted a capacitive living object detection method using multiple metal plates assembled around the transmitting coil. The principle of this living object detection system is depicted in Fig. 7.1. When a living object is present within the detecting distance, there will be an extra capacitance in the equivalent circuits as shown in Fig. 7.1b compared to Fig. 7.1a. This extra capacitance will lead to a variation of the sensing voltage, which can be detected by the presence detector.

In 2015, Jeong et al. proposed a novel comb pattern capacitive living object detection system for wireless power transfer systems (Jeong et al. 2016a). The schematic diagram is shown in Fig. 7.2a. This comb pattern capacitive detecting system is placed above the transmitter coil, whose capacitance will change when a living object, such as a cat or a dog, is within the detecting height. To obtain uniform capacitance capabilities throughout the entire sensing pattern, the comb design was employed as the distance *d* between two electrodes can be constant throughout the detecting area. In addition, the pattern should be as narrow as possible, compared with other capacitive sensors, to ensure the efficiency of the wireless power transfer system would not be affected. Normally, the detecting distance of capacitive sensors is only several millimeters. However, since the detection pattern should be assembled inside the transmitter pad for living object detection for wireless power transfer system, the detection height of the capacitive sensors should be modified. As the distance between the two electrodes of the capacitor increases, the effective detection height *h* will increase (Jeong et al. 2016a). However, as mentioned above the comb pattern should be as narrow as possible to keep the efficiency of the wireless power transfer system, and therefore the capacitance capability of the comb pattern is small, which is sensitive to noise. Hence, the sensing circuit for this comb pattern should involve an active RC integrator to detect even the smallest voltage change, an RC low pass filter for continuous capacitance detecting and a reset switch circuit. This comb pattern capacitive LOD sensor showed relatively high reliability, and its simple structure also means the cost can be relatively low. However, due to the single comb structure, the sensitivity of this type of sensor is quite low. therefore, in 2018, they improved

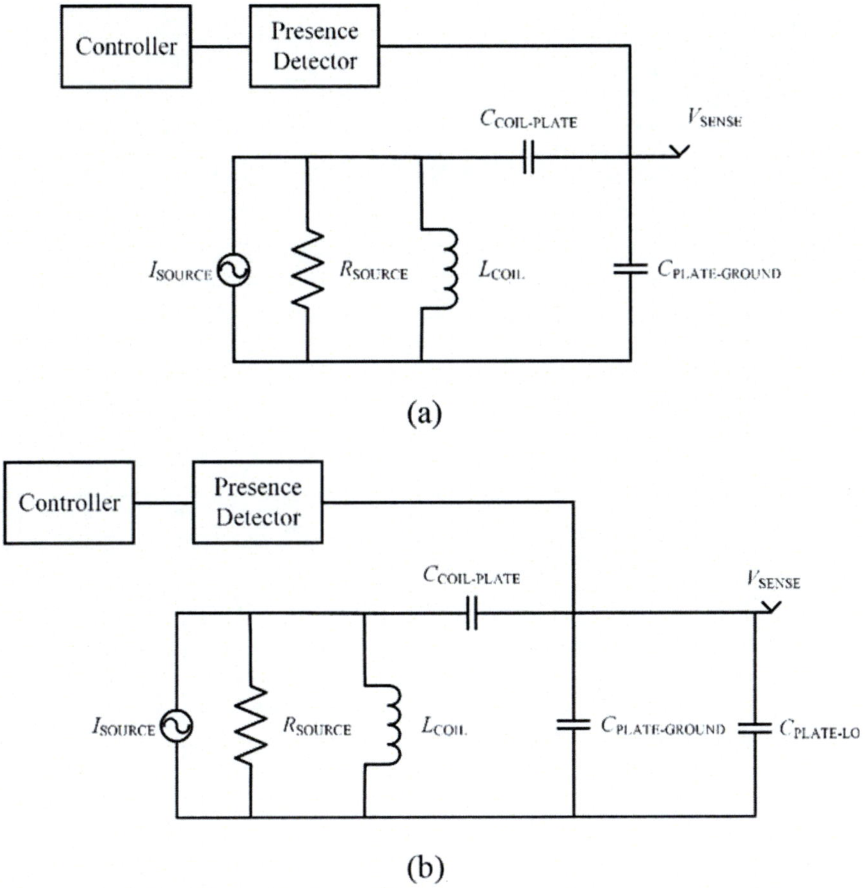

Fig. 7.1 Equivalent circuit of capacitive LOD system: **a** without a living object and **b** with a living object (Novak et al. 2017)

this system and proposed a multiple comb pattern capacitive living object detection system, as shown in Fig. 7.2b (Thai et al. 2018). They established the physical model of different comb pattern capacitive living object detection systems, analyzed the deformation characteristics of different patterns by simulation, and derived the relationship between the capacitance change and the circuit output voltage. Then, they concluded that multiple comb pattern capacitive living object detection systems show much higher sensitivity than single comb pattern capacitive living object detection systems.

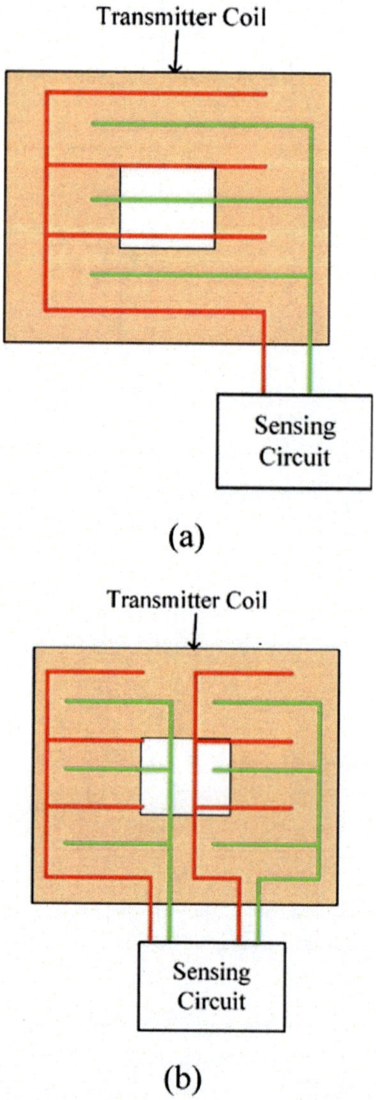

Fig 7.2 **a** Comb pattern capacitive LOD sensor and **b** Multiple comb pattern capacitive LOD sensor (Jeong et al. 2016a; Thai et al. 2018)

7.2.2 Sensor-Based Detection Method

The sensor-based living object detection method is the most available solution for wireless power transfer systems. Radar sensors, pressure sensors, infrared sensors, ultrasonic sensors (Novak et al. 2017), thermal sensors, and visual sensors, like image cameras, can be utilized in living object detection (Ote et al. 2020; Hoffman et al. 2015; Widmer et al. 2016; Sieber et al. 2018; Stillig et al. 2020; Poguntke et al. 2017; Wang et al. 2019; Dimke et al. 2019; Ombach 2014). Figure 7.3 shows several types

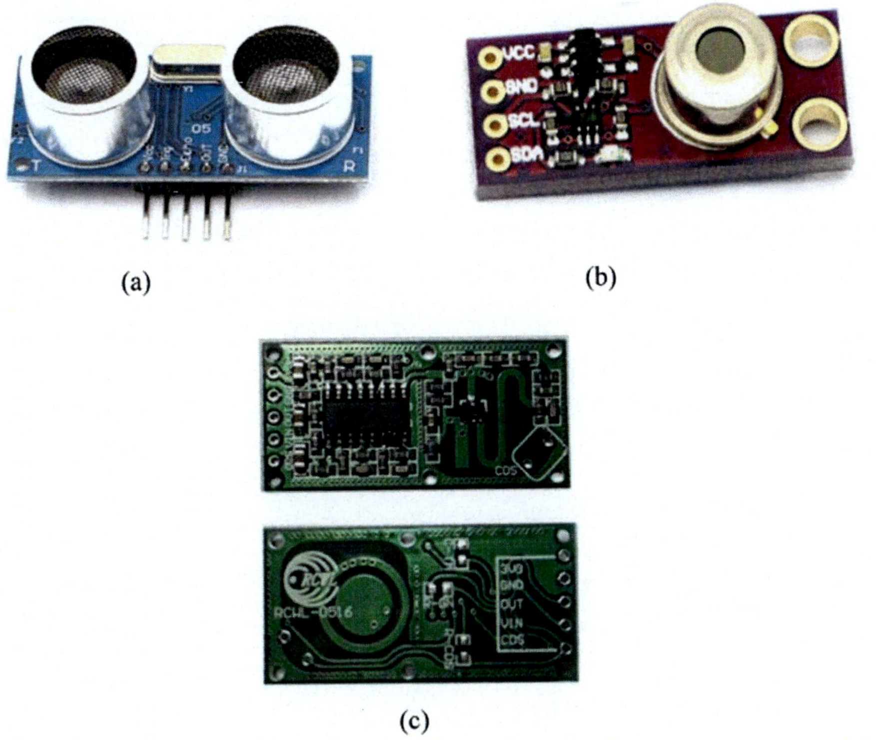

Fig. 7.3 Sensors that can be used in living object detection system: **a** ultrasonic module, **b** infrared module, and microwave module

of sensors that may be utilized in the LOD system. Moreover, multiple sensors can be combined to form a living object detection system (Hoffman et al. 2015). The sensors are usually installed on the ground. The sensors and LOD controller can be installed in the parking area. Another type of on-ground installation is that the sensors and LOD controller are integrated with the transmitter pad of the wireless power transfer system, as shown in Fig. 7.4.

The infrared-based approach is to use infrared sensors to detect the infrared rays emitted by living objects, therefore detecting their presence. The principle of infrared sensors shown in Fig. 7.5. The Fresnel lens is utilized to focus and concentrate the infrared rays emitted by living objects to promote the sensitivity of the infrared sensor, and the detection distance. Then, the pyroelectric sensor transfers the infrared signal into electrical signals. The bandpass amplifier is an ultra-low frequency signal since the amplitude of the electric pulse signal output by the pyroelectric sensor is relatively low, and its frequency is around 0.3–10 Hz, which depends on the moving speed of the living objects. Therefore, high gain, low noise, and low-frequency bandpass amplifiers are required for high gain amplification before the signal can be sent to the next circuit. To effectively suppress noise interference, improve the reliability of the module, and reduce the probability of misoperation, a voltage comparator is set

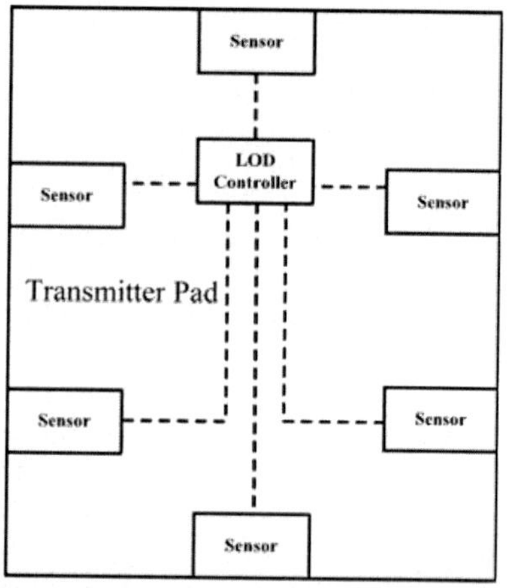

Fig. 7.4 Schematic of sensor-based living object detection integrated with the transmitter pad

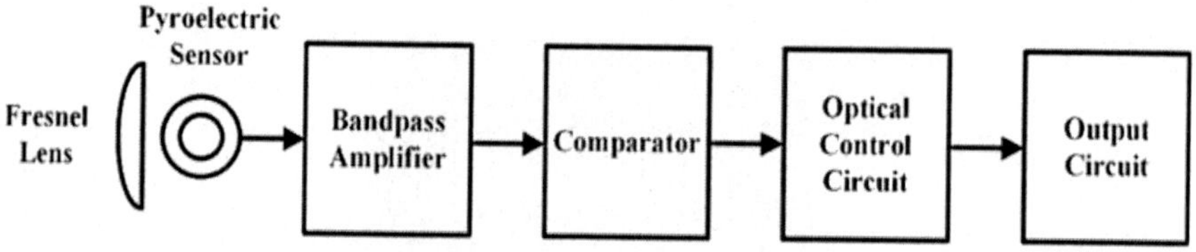

Fig. 7.5 Principle of infrared sensor

in the infrared-based module. The function of the optical control circuit is to control the signal input to the comparator by using the light-sensitive characteristics of the photoresist. In the daytime, the photoresist is irradiated by light, and the resistance value becomes very small. If the small resistance value is connected to the input of the comparator, the input signal amplitude of the comparator would never reach the jump value required by the threshold signal. Therefore, there would not be any output from the comparator. On the contrary, at night, the photoresist is not irradiated by light, and the resistance becomes very large, which almost does not affect the input signal of the comparator. Therefore, optical control can be achieved both day and night. The output circuit outputs the information of living objects to the controller to judge the presence of living objects. The frared-sensor-based LOD system is simple and straightforward, with minimal cost but low reliability. However, it is prone to missed detection and cannot handle the situations when a small organism enters the detecting region.

The vision-based approach is to use a camera to capture images in the immediate surroundings of the wireless power transfer system and then use computer processing to identify the presence of living objects. This approach is expensive and time-consuming since it necessitates the creation of a reference library for diverse living objects. Moreover, camera maintenance and cleaning should be taken into consideration.

Ultrasonic sensing is widely used in non-contact measurement methods, because the ultrasonic sensing system and circuit structure are easier to implement than other sensing methods, with cheaper cost and quicker sensing. Ultrasonic energy has a concentration of energy, low attenuation, good directivity, great reflectivity, and it may travel at various speeds in solids, liquids, and gases. Living species such as bats and whales have evolved an echolocation system that uses ultrasound to replace eyesight in low-light situations, based on the above-mentioned outstanding properties of ultrasound. According to different judging criteria, there are now three techniques to realize ultrasonic distance measurement, which can be broadly split into phase detection method, acoustic wave amplitude detection method, and transit time detection method. The phase-detection approach is the most accurate of these three techniques, but it has a restricted detection range. The reflected wave has a big impact on the amplitude detecting method. The transit time technique identifies the difference in time between the echo and the emission. The transit time technique is the most commonly utilized, and the hardware control and software design are the simplest. The ultrasonic-wave-based living object detection system utilizing the transit time technique is shown in Fig. 7.6. To start the detecting process, the controller sends the excitation pulse via the pulse transmitting circuit to the ultrasonic wave transmitter, transferring the electrical signal into the ultrasonic wave. Then, the ultrasonic wave transmitter sends the detection ultrasonic wave to detect the presence of living objects, and the reflecting wave is received by the ultrasonic wave receiver, transferring the ultrasonic wave into an electrical signal. The ultrasonic wave traveling time can be calculated via the time interval of the detection wave and the echo wave. The distance of living objects then can be acquired by multiplying the ultrasonic wave traveling time and the traveling speed of the ultrasonic wave. Moreover, the temperature compensating module can be utilized to promote the accuracy of the ultrasonic-sensor-based LOD system as the traveling speed of ultrasonic waves varies at different temperatures.

There are several types of radar sensors, of which microwave one is promising in LOD detection due to its wavelength. The principle of the microwave sensor is based on the Doppler effect. The frequency difference formula caused by the relative

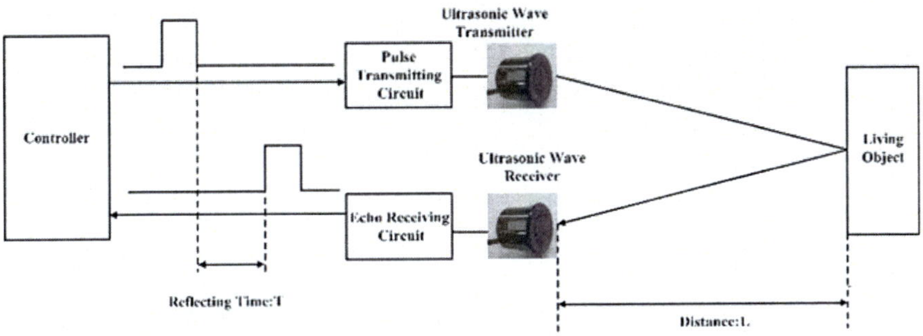

Fig. 7.6 Principle of ultrasonic wave sensor for living object detection

motion of the living object and the wave source is shown in the following formula:

$$f^{'}=\left(\frac{v \pm v_0}{v \mp v_s}\right) \cdot f \tag{7.1}$$

where $f^{'}$ is the frequency of the returned microwave and f is the frequency of the emitted microwave. v is the speed of the microwave. v_s is the moving speed of the wave source relative to the medium. If the wave source is moving towards the living object, the symbol is a " + " sign. Otherwise, it is the "−" sign. v_0 is the moving speed of the measured object (LO) relative to the medium. If the living object is moving towards the wave source, the symbol is a " + " sign. Otherwise, it is the "−" sign.

If substrate v with the speed of light c, then the frequency difference f_d can be obtained:

$$f_d = f^{'} - f = \left(\frac{v_s + v_0}{c - v_s}\right) \cdot f \tag{7.2}$$

For the living object detection system used for wireless power transfer systems, the sensors are regarded as integrated with the transmitter pad that is fixed. Hence, v_s should be 0, and the expression can be simplified as:

$$f_d = f^{'} - f = \frac{v_0}{c} \cdot f \tag{7.3}$$

In this formula, v_0 is regarded as positive when the living object moves towards the wave source which is the microwave sensor. As a result, this method can identify any moving objects, not only living ones. Relevant information in the detection region, such as the distance, speed, direction, and size of live things, may be determined, computed, detected, and outputted using a processor's analysis of the signal in the time and frequency domains. The wireless power carrier's low depth amplitude or phase modulation helps reduce signal system interference. At the same time, the microwave receiving device should be constructed to decrease the influence on the wireless charging system's charging efficiency by suppressing the formation of eddy currents in the wireless charging environment.

7.3 Introduction of Metal Object Detection

With the development of the times and the improvement of productivity, human life has undergone rapid changes, and these changes are largely due to the collection and utilization of fossil fuels. As an important fossil fuel, petroleum has always been one of the important channels of energy for human society. However, with the advancement of industry and the increase of the global population, fossil fuels have been over-exploited and increasingly depleted. The top priority of the society is to change the production mode of human society and prevent fossil fuels from becoming the shackles of scientific and technological progress and social development. In the automobile industry, traditional vehicles have social problems such as serious consumption of fossil fuels and a great impact on the environment. As a result, many types of new energy vehicles have appeared on the market. New energy vehicles essentially use non-traditional energy sources as power sources, while adopting different driving methods. According to different power sources and driving methods, new energy vehicles are divided into hybrid vehicles, battery electric vehicles, fuel cell vehicles, and other new energy vehicles. Because part of its power still comes from fossil fuels, hybrid electric vehicles have not fundamentally solved the problem. Fuel cell vehicles use fuel as an energy source, which is difficult to achieve and can be used as an important supplement to new energy vehicles.

Because electric energy is more convenient and has little impact on the environment, electric vehicles have gradually begun to occupy a broad market share and become an important part of people's daily lives. Among them, batteries are the core of new energy vehicles. Safe, effective, and convenient charging has become the primary problem that needs to be overcome with the rise of electric vehicles. The currently widely used charging method for electric vehicles is charging with cables. However, wired charging technology has some difficult problems, such as cumbersome plugging and unplugging of cables, risk of electric shock, and being severely affected by extreme weather like rain and snow. The concept of electromagnetic coupling resonant wireless energy transmission, that is, wireless charging technology, came into being. It can realize the battery charging function while perfectly solving the problems of wired charging technology. Currently, in the market, many companies, and laboratories such as Qualcomm, WiTricity, etc. have available wireless charging technologies. However, according to investigations, these products still have some problems, such as the lack of safety modules like metal detection modules, short power transmission distance, low charging efficiency, and poor anti-offset ability.

During the operation of the electric vehicle wireless charging system, a high-power strong magnetic field is excited between the primary and secondary side coils. If a metal object is mixed into the electromagnetic area (Huh et al. 2011), significant eddy current effects and magnetic effects will occur. Both the eddy current effect and the magnetic effect will have a certain impact on the electromagnetic parameters of the charging system, thereby reducing the efficiency of the entire charging system. In addition, the eddy current effect caused by metal will lead to eddy current loss and temperature rise, even severely damage the charging system (Jeong et al. 2018b). Therefore, the metal detection module is an indispensable part of a complete electric vehicle wireless charging system, an important guarantee for the safe operation of the electric vehicle wireless charging system, and a solid foundation for the popularization of electric vehicle wireless charging.

The metal object detection methods for wireless charging of electric vehicles can generally be divided into electrical detection and optical detection methods. Electrical detection methods include impedance change method, self-inductance-based metal detection method, induced voltage method, capacitance change method, efficiency loss detection method, etc. (Jeong et al. 2016b). Optical detection methods include temperature sensor detection method, infrared camera thermal imaging detection method, optical fiber induction light intensity detection method, radar ranging detection method, and refractive index detection method (Jeong et al. 2015).

7.3.1 Research State in China

In recent years, electric vehicles have received unprecedented attention in China. The market share occupied by electric vehicles has increased year by year. Therefore, the research on wireless charging metal object detection technology has also begun to bear fruit. In the article (Liu et al. 2019), the design and implementation of a multi-purpose TMR sensor matrix for electric vehicle wireless charging is proposed. TMR sensors are used to build a detection matrix to provide key information such as charging performance, coil positioning, and metal object position detection to achieve multi-purpose detection. Specifically, the magnetic field distribution between the primary and secondary side coils is measured by the above-mentioned TMR sensor matrix, and the matrix configuration is shown in Fig. 7.7. The charging performance indicators of wireless charging systems such as mutual inductance and charging power are derived from the measured magnetic flux density in the vertical direction. The horizontal position of the coil is determined by the symmetry coefficient of the time-varying magnetic field distribution within a range of 150 mm. Metal objects involved in the power magnetic field are detected by the measured magnetic field offset (Fukuda et al. 2012). Experimental verification in the article shows that the sensor matrix can monitor the real-time power of the wireless charging system,

Fig. 7.7 TMR sensor matrix configuration

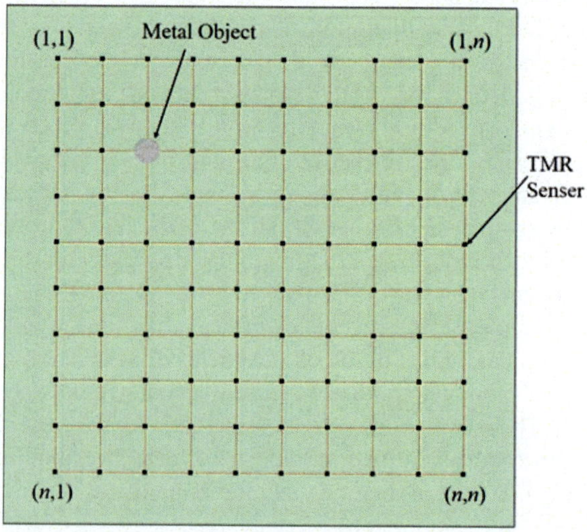

monitor the horizontal position of the coil, and realize the detection of small metal objects such as steel screws. However, the TMR sensor matrix still has some problems. This matrix requires the use of many expensive sensors. Moreover, the matrix can only be used when the electric vehicle is already charging, that is, it has no offline detection capability. If a metal object enters the magnetic field area before charging, this detection method has potential safety hazards.

In the article (Cai et al. 2020), a metal object detection method using auxiliary coils is proposed. This method uses only one auxiliary coil without external power supply excitation. The auxiliary coil is placed in the charging magnetic field. The metal object detection function is realized by measuring the change of the power factor of the auxiliary coil before and after the metal object enters the charging magnetic field. The method is not expensive, relatively easy to implement, but it is hard to detect metal objects under the working condition of the power magnetic field. Besides, the sensitivity of this method is not high and blind zones exist. The article (Zhang et al. 2017) proposed a method to detect metal by detecting the change in inverter output power caused by metal intervention when the wireless charging system is working. This method does not require additional sensors or auxiliary coils and is easy to operate. However, metal objects are usually relatively small. When the wireless charging system is working, the power loss caused by the metal eddy current is often less than 10 watts, and the working power of the charging system can reach several thousand watts or even higher. Because the sensitivity of this detection method is low, this method is unable to commercialize on a large scale.

In the article (Sun et al. 2020), a metal object detection method based on a balanced coil group is proposed. The coil configuration is a double-layer balanced coil group with metal detection and coil positioning functions, as shown in Fig. 7.8. In the two-layer coil, each group of coils is composed of two coils, and these two coils are symmetrical about the *Y* or *X* axis. Measure the mutual inductance between the two

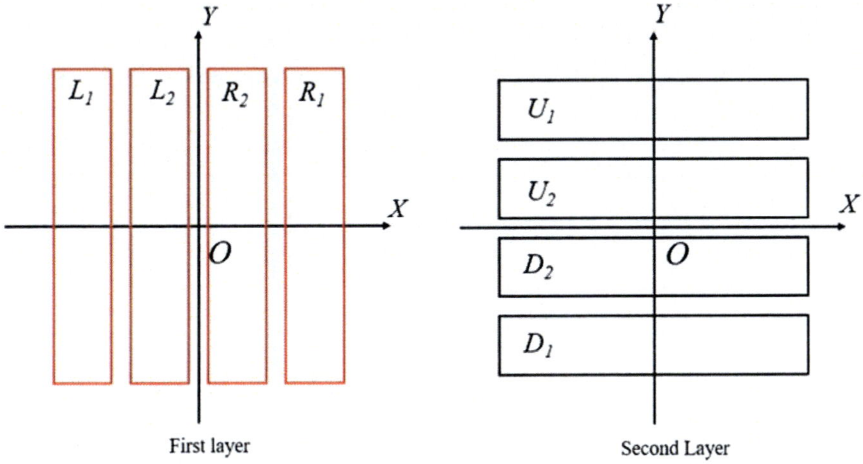

Fig. 7.8 Double-layer balanced coil configuration

detection coils and the primary side transmitting coil in each group and calculate the difference. If there is a spike in the difference of mutual inductance, it is judged that the metal foreign body exists and its position in the charging magnetic field is determined. Through finite element simulation, this article believes that the designed coil can accurately determine the presence of metal and locate its real-time position. However, the double-layer coil itself has a higher cost. Due to the large coil size design, sensitivity cannot be guaranteed. In addition, the center position of the coil design and the position of the symmetry axis of each group of coils are detection blind areas.

The article (Zhang et al. 2020) proposes a metal object detection system that can quickly and accurately detect the presence or absence of metal objects without interference from power magnetic fields. It includes a metal detection method based on the impedance change of the detection coil, a parallel resonant circuit, a bipolar detection coil magnetically decoupled from the wireless charging coil, and a multi-channel automatic tuning control method. First, the article analyzes two metal materials, non-ferromagnetic and ferromagnetic. The impedance vector diagram of the detection coil is shown in Fig. 7.9. The equivalent impedance change of the detection coil is used to sense and feedback the information of the metal object in the wireless charging area. It is proposed in the article that the smaller the size of the detection coil, the greater the rate of change of its impedance. For a fixed charging detection area, the smaller the size of the single detection coil, the more its quantity needs. But too many single detection coils mean longer detection time and more redundant detection circuits., Therefore, the improvement of detection sensitivity must consider the constraints of detection speed and hardware cost (Futatsumori et al. 2014; Kato et al. 2002). In addition, the detection coil must be magnetically decoupled from the charging coil to eliminate the magnetic coupling interference of the 85 kHz power magnetic field. This paper proposes a bipolar detection coil, which consists of two toroidal coils

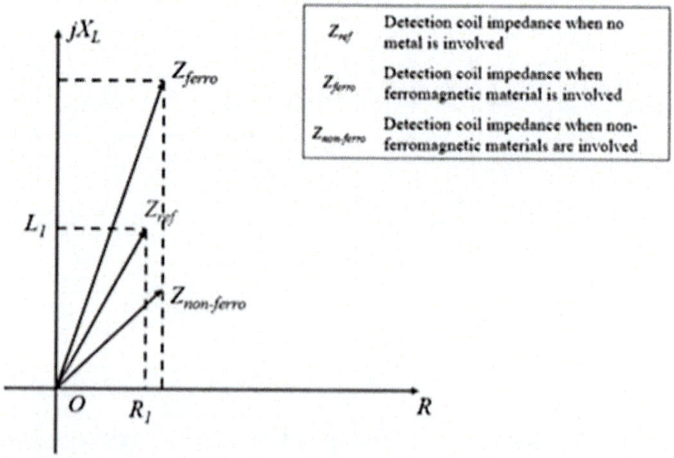

Fig. 7.9 Vector diagram of detection coil impedance

Fig. 7.10 Bipolar coil configuration

connected in series in reverse, as shown in Fig. 7.10. According to the finite element simulation, When the charging coil and the detection coil are completely aligned, the two are completely decoupled. In addition, the article also proposes a parallel resonant amplifier circuit and filter circuit, which can resonantly amplify the external excitation signal through the coil and the resonant capacitor, while filtering out the 85 kHz interference signal. The on and off of different coils are controlled by the multiplexer, and the best resonance point of each coil is found through the automatic tuning control algorithm so that the sensitivity of detecting metals achieves the best results. An experimental platform is built in the article, and the measured optimal excitation frequency is used to drive each single detection coil in turn. When one yuan coin, half yuan, and one dime coin are placed 3 mm above the single detection coil in turn, according to the output voltage, the intervention of the coin makes the output voltage corresponding to each single detection coil change significantly, that is, the detection of the coin can be realized. This method has high detection sensitivity and can perform metal detection in advance when the charging magnetic field is not working. But there is still a problem. Because the coil is a single-layer coil, there is a certain detection blind zone. When working in a high-power magnetic field, although the system has implemented some measures to reduce noise interference, it will still receive greater interference and even cause the coil to fail. There is room for improvement.

7.3.2 World Research Status

Foreign research on wireless charging metal object detection technology has been carried out for more than ten years, and many excellent results have been achieved. The article (Sonnenberg et al. 2019) proposed a combined metal and living foreign body detection system based on real-time thermal imaging. Based on real-time thermal imaging, this paper aims to propose a system that considers both metal and living object detection, so that it can detect metal objects and living objects of different sizes at the same time, ignoring non-threatening objects, and being unaffected by the system power magnetic field. The system proposed in the article uses a single sensor to collect the required image information, which is post-processed by a deep learning algorithm. In this paper, a neural network model is constructed. 500 normal images and 298 abnormal images are collected. They are divided into a training set and a test set. Among them, the training set contains 85% of the collected images, which are used to update the weights of the neural network. The test set consists of the remaining 15% of the images used to validate the model after training. Let the neural network model perform unsupervised learning to recognize the images acquired by the sensor and get better results (Kuyvenhoven et al. 2011). To add anomalous images when training the neural network, the Siamese network is used in the article, and appropriate sampling weights are set. After experimental verification, the system can detect human hands and other living bodies, screws, and other metal objects, and can eliminate the interference of plastics and other objects. However, this system requires an additional thermal imager, which is expensive, takes up a lot of space, and is not warranted. In addition, the system can only detect when the power magnetic field is working, which has certain limitations.

The article (Thai et al. 2020c) proposed a symmetrical induction coil design and applied it to the self-inductive metal object detection for wireless charging of electric vehicles, as shown in Fig. 7.11. The proposed detection coil is composed of multiple channels. Each channel includes two induction coils connected in series with opposite polarities to offset the induced voltage generated by the primary side transmitting

Fig. 7.11 Coil configuration

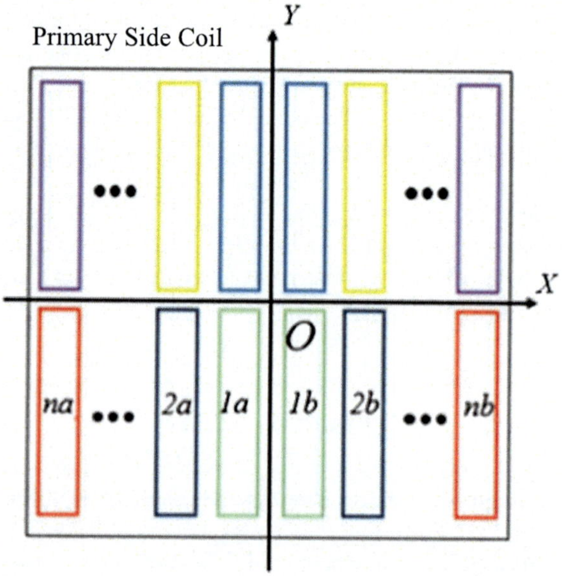

coil current. In the designed set of coils, two induction coils are arranged symmetrically about the y-axis of the Tx pad. This is because the z-direction component of the magnetic flux generated by the Tx coil current is symmetrical with respect to the *Y* axis of the Tx pad. The induction coils are arranged and connected in opposite polarities. The induced voltages can cancel each other out. After experimental verification, the coil can detect metal objects with a size of 20 mm × 20 mm. But obviously the coil design has a detection blind zone on the *Y* axis (symmetrical position).

In the article (Jang et al. 2016), a metal object detection circuit with non-overlapping coils is used to ensure the safe operation of wireless electric vehicle chargers. Non-overlapping coils include multiple pairs of copper wires. Each pair is composed of two copper wires installed in parallel. There is no overlap between the two copper wires. The non-overlapping structure can be manufactured on a single-layer printed circuit board, which reduces the manufacturing cost of the sensing coil. However, if the non-overlapping coils are arranged in a single layer, there must be a detection blind zone. Considering various extreme positions, when using a four-layer board to place the induction coil array, as shown in Fig. 7.12, no matter the position and shape of the metal, it can be detected in all areas. In addition, when placed horizontally and vertically, the position of the metal can also be accurately known. When the sensing coil is arranged horizontally and vertically, the position of the metal on the emitter pad can also be accurately known. The number of coils is determined by the size of the launchpad. In addition, a matching circuit is proposed in the article, which is composed of a variable resistor, a low-pass filter, and a precision rectifier with an amplification function for accurate data acquisition. One problem with non-overlapping coils is that the voltage difference between two non-overlapping coils is usually not exactly zero, even if there is no metal on the non-overlapping coils. Because the magnetic field is not uniformly distributed on the two wires, the induced voltage on each wire is not uniform (Kuyvenhoven et al. 2011). Use variable resistors or digital potentiometers to adjust the gain of each induced voltage to achieve zero voltage. The metal object enters the detection area so that the voltage difference of

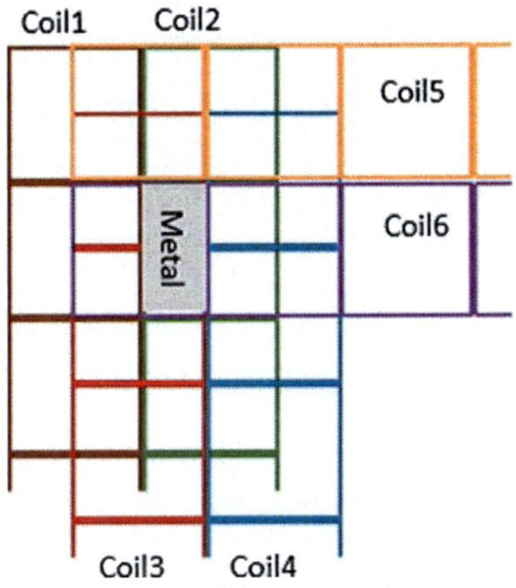

Fig. 7.12 Three-layer overlapping coil configuration diagram

a group of coils is not 0. After the matching circuit detects it, it transmits a signal to the inverter to stop the charging system. Experiments verify that the coil can detect metal foreign objects and their location. However, this coil is essentially composed of four layers of coils, with complex structure, high cost, and low detection accuracy.

The article (Jeong et al. 2018a) proposed a coil structure like the article (Kuyvenhoven et al. 2011), which can realize the metal detection function and charging coil positioning at the same time. Specifically, the presence of metal objects in the charging magnetic field is determined by the induced voltage difference of the non-overlapping coil groups. If the pressure difference is 0, there is no metal foreign object. If the pressure difference is not 0, the metal foreign object exists and is in the corresponding coil group. The position of the electric vehicle can be judged by the induced voltage of the non-overlapping coil group. The coil proposed in this paper is composed of two layers of coils arranged vertically, so the specific position of the secondary side coil, that is, the electric vehicle can be judged by the induced voltage of each group of coils. This detection method has a detection blind zone. The accuracy is not high, and it is difficult to detect small metal foreign objects.

In the article (Jeong et al. 2018b), a foreign object detection system based on a resonant circuit and using the self-inductance change of the detection coil is proposed. The coil board used in this article is composed of a plurality of toroidal coil groups, which are installed on the primary side transmitting coil of an electric vehicle. One of the toroidal coil groups is composed of two coils connected in series with opposite polarities to offset the induced voltage generated by the transmitting coil. According to the theoretical calculation in the article, the parallel resonant circuit is more sensitive than the series resonant circuit. The advantage of the parallel resonant circuit is that it is less sensitive to noise due to its high output voltage characteristics (Jeong et al. 2015; Ombach et al. 2013). Therefore, a parallel resonant circuit is used in the modem system proposed in the article. After the metal foreign body enters the charging magnetic field, the self-inductance change of the coil assembly is detected by the parallel resonance circuit, which is driven by a current source and works around 1 MHz. To improve the detection sensitivity of the proposed modem system, the parallel resonance circuit adopts- The non-resonant operating frequency near the 3 dB point, not the precise resonant frequency. After experimental verification, the coil can detect coin-sized metal foreign objects. However, the sensitivity of this detection method is related to the location of the metal foreign body, and the interference is relatively large under the high-power magnetic field.

In the article (Thai et al. 2020a), a simple and economical symmetrical induction coil set for wireless charging of electric vehicles is proposed. To solve the problem of blind spots and high costs, a single-layer symmetrical induction coil group is proposed in the article. As shown in Fig. 7.13, the induction coil group includes many rectangular induction coils, covering the entire area of the primary side transmitting coil. The coil group uses an even number of induction coils. Two coils form a set. Each set includes two induction coils arranged symmetrically about the y-axis of the primary side coil. When the number of sensing coils is 2n, the number of coil groups is n. Each group includes two symmetrical coils: group 1 includes coils 1a and 1b, group 2 includes coils 2a and 2b, and so on, group n includes coils na and nb.

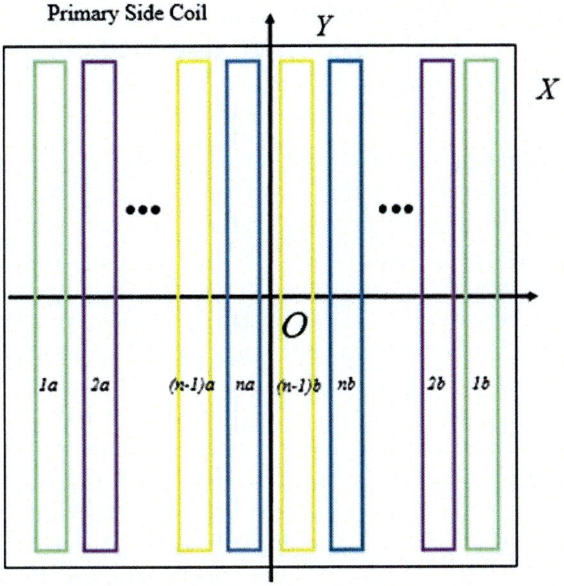

Fig. 7.13 Symmetrical coil design without blind zone

Because the z-direction component of the magnetic flux generated by the primary-side transmitting coil is symmetrical about the y-axis, the symmetrical coil group is designed in this way. No metal object enters the charging magnetic field. Due to the symmetrical design, the induced voltage difference between the two coils in each set of coils is 0; when a metal object enters the charging magnetic field and the voltage difference is not 0, it can be judged that the metal exists. But if there are n groups of coil groups, and each group is completely symmetric about the y-axis, there is a blind zone in the entire induction coil. In this paper, an innovative idea is adopted, and nb and (n−1)b are exchanged creatively. The charging magnetic field is simulated. The z-direction component of the magnetic field near the y-axis of the primary side coil is basically the same, that is, the magnetic field strength at the position of na and (n-1)a is almost the same, which also creates a premise for the coil exchange position. The asymmetric design of nb and (n−1)b eliminates the blind zone in the detection area. Experiments verify that the coil can detect metal objects with a size of 4 cm × 4 cm. Through testing at different positions, the coil eliminates the detection blind zone. However, it is difficult for coil detection to detect small-sized metal foreign objects, such as screws or coins.

7.4 Multi-channel Automatic Tuning Wireless Charging Metal Object Detection Method Based on Impedance Change

Aiming at the situation that the high-frequency and high-intensity electromagnetic field generated by the high-power magnetic induction wireless charging system can quickly heat the metal objects placed in it and cause danger, a metal detection method is proposed, which can quickly and accurately detect the presence or absence of metal objects without being interfered by the power magnetic field. First, by analyzing the influence of metal objects on the impedance characteristics of the coil, a metal detection method based on detecting changes in the impedance of the coil is proposed. A parallel resonant circuit is proposed, and the impedance change of the detection coil is amplified and mapped to the output voltage change of the resonant circuit, and the optimal excitation frequency of the detection circuit is obtained through theoretical calculation. Next, a bipolar detection coil that is magnetically decoupled from the wireless charging coil is proposed, which can eliminate the magnetic coupling interference of the power magnetic field, and its impedance can better reflect the metal intervention. Then, a multi-channel automatic tuning control method is proposed to improve the detection accuracy and response speed. Finally, the function of the system is verified through experiments, and the results show that the system can detect the intervention of metal coins.

7.4.1 Analysis of the Influence of Metal on the Impedance Characteristics of Detection Coils

When a metal object is placed in the alternating electromagnetic field generated by the detection coil, a toroidal eddy current will be induced inside the metal. At this time, the metal can be equivalent to a series circuit composed of an inductance and a resistance. The equivalent inductance of the metal is magnetically coupled with the detection coil. The equivalent circuit of the magnetic coupling between the detection coil and the metal is shown in Fig. 7.14a, b is a simplified circuit diagram of Fig. 7.14a.

Among them, L_1, L_{metal} are the equivalent inductance of the detection coil and the metal object respectively; R_1, R_{metal} are the parasitic resistance of the detection coil and the metal object respectively; U_{ac} is the excitation voltage of the detection coil; ω is the angular frequency of U_{ac}. According to Kirchhoff's voltage law, for the two closed-loop voltage equations in Fig. 7.14b:

$$\begin{bmatrix} R_1 & j\omega M \\ j\omega M & R_{metal} + j\omega M_{metal} \end{bmatrix} \begin{bmatrix} I_1 \\ I_2 \end{bmatrix} = \begin{bmatrix} U_{ac} \\ 0 \end{bmatrix} \tag{7.4}$$

Then the equivalent impedance of the detection coil is:

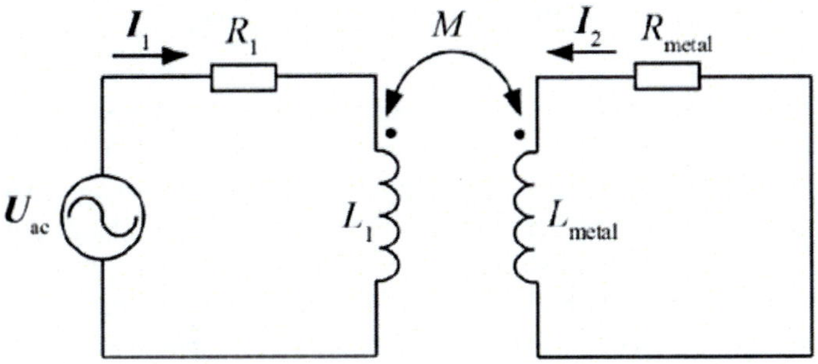

(a) Equivalent circuit of detecting coil and metal magnetic

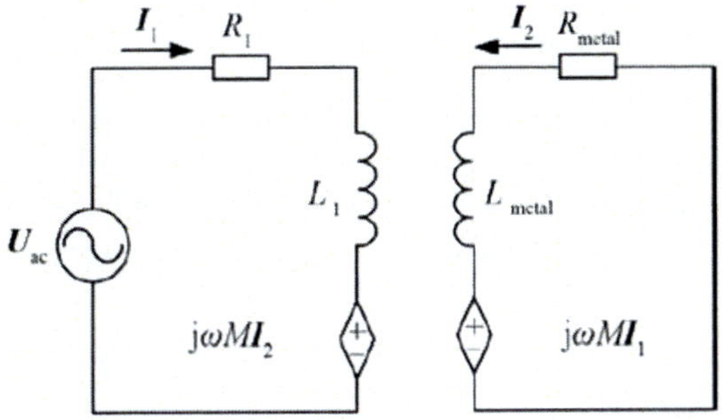

(b) Simplified circuit for magnetic coupling between detection coil and metal

Fig. 7.14 Model of detecting coil coupling with metal object (Ote et al. 2020)

$$Z_{non-ferro} = \frac{U_{ac}}{I_1} = \left(R_1 + \frac{\omega^2 M^2 R_{metal}}{R_{metal}^2 + \omega^2 L_{metal}^2}\right) + j\omega\left(L_1 - \frac{\omega^2 M^2 L_{metal}}{R_{metal}^2 + \omega^2 L_{metal}^2}\right) \tag{7.5}$$

From Eq. (7.5), when a metal object enters the detection area, the eddy current effect will reduce the equivalent inductance of the detection coil and increase the parasitic resistance. When the metal object is a ferromagnetic metal, its relative permeability μ_r is much greater than 1, the metal object will be magnetized by the electromagnetic field generated by the detection coil, and the equivalent inductance of the detection coil will increase significantly.

7.4.2 *Hardware Design of Metal Object Detection System*

To detect metal objects, a metal object detection system is proposed as shown accurately and effectively in Fig. 7.15, which includes a resonant circuit that can amplify and map the detection coil impedance change to the output voltage change, a filter circuit that can filter out power magnetic field interference, and a peak hold circuit that converts AC signal into DC analog signal, etc. Among them, the resonant circuit is composed of a resonant network and an amplifier circuit, and the resonant network is composed of a metal detection coil, a resonant capacitor, and a resistor in parallel. This section will focus on the research and analysis of the composition of the metal detection system.

7.4.2.1 Resonance Network Analysis

To measure the impedance change of the detection coil, a parallel resonant network of the detection coil is constructed, and the impedance change of the resonant network is amplified and mapped to the output voltage change through an amplifier circuit. The parallel resonant circuit composed of the parallel resonant network and the amplifier circuit is shown in Fig. 7.16.

Among them, C_p is the resonant capacitor of the parallel resonant network; R_p is the parasitic resistance of the resonant capacitor; R_a is the parallel resistance of the parallel resonant network; R_{11} and R_{12} are the two resistors that form the same direction amplifier circuit; U_AC and U_out1 are the input and output voltages of the parallel resonance circuit, respectively. According to resonance conditions:

$$C_p = \frac{1}{\omega_0^2 L_1} \tag{7.6}$$

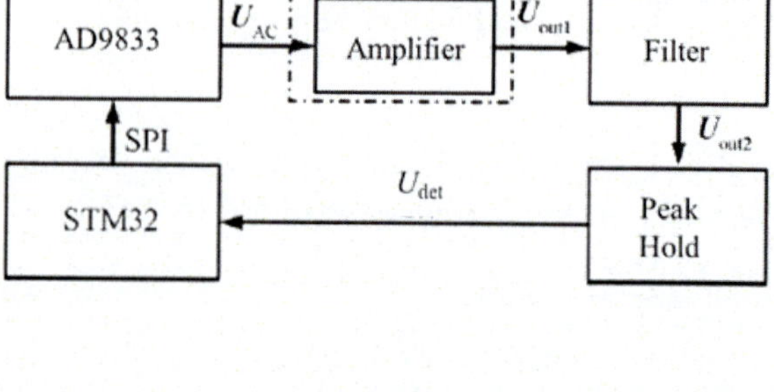

Fig. 7.15 Compositions of a metal object detection system

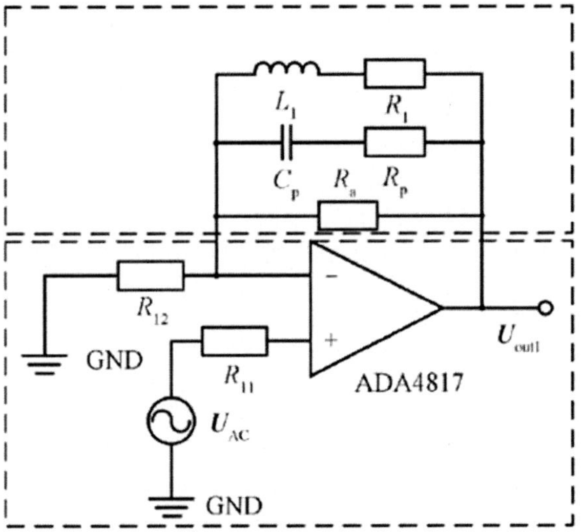

Fig. 7.16 Parallel resonant circuit

In the formula, ω_0 is the resonant angular frequency. From the characteristics of the amplifier circuit, the frequency response function of the parallel resonant circuit is:

$$H(L_1,\omega)=\frac{U_{out1}}{U_{AC}}=1+\frac{1}{R_{12}\left[\frac{1}{R_a}+\frac{1}{j\omega L_1+R_1}+\frac{1}{1/(j\omega C_p)+R_p}\right]} \tag{7.7}$$

Then the output amplitude gain is:

$$|H(L_1,\omega)|\approx\sqrt{\frac{b}{a}} \tag{7.8}$$

In the formula, the variables a and b satisfy the following relationship:

$$\begin{cases} a=m_1^2+\left(L_1^2R_{12}^2-2m_1^2f_p+m_1^2C_p^2p_2^2\right)\omega^2+m_1^2f_p^2\omega^4 \\ b=(m_1+p_1R_1)^2+f_p^2(m_1+m_3)^2\omega^4+\left[-2m_1^2f_p+L_1^2p_1^2+C_p^2(m_1p_2+m_2p_1)^2\right]\omega^2 \end{cases} \tag{7.9}$$

where $m_1=R_{12}R_a$;$m_2=R_1R_p$;$m_3=R_aR_p$;$p_1=R_{12}+R_a$;$p_2=R_1+R_p$; $f_p=C_pL_1$. When the value of L_1 changes, to obtain the maximum output voltage change, $H(L_1,\omega)$ should have the maximum rate of change at L_{ref} (L_{ref} is the equivalent inductance value when there is no metal foreign object near the detection coil L_1), that is, $\partial H(L_1,\omega)/\partial L_1$ takes the maximum at L_{ref}. $\partial H(L_1,\omega)/\partial L_1$ is as follows.

$$\frac{\partial H(L_1,\omega)}{\partial L_1}=\frac{aL_1p_1^2-am_1^2C_p-bL_1R_{12}^2+bm_1^2C_p}{\sqrt{a^3b}}\omega^2+\frac{aC_pf_p(m_1+m_3)^2-bC_pf_pm_1^2}{\sqrt{a^3b}}\omega^4 \tag{7.10}$$

Here we mainly discuss the numerical relationship between $\partial H(L_1,\omega)/\partial L_1$ and the excitation power frequency ω and the influence of the parasitic resistance R_p of the resonant capacitor on $\partial H(L_1,\omega)/\partial L_1$. To find the excitation power frequency ω_r at which $\partial H(L_1,\omega)/\partial L_1$ takes the maximum value, it can be obtained by solving Eq. (7.11):

$$\left.\frac{\partial^2|H(L_1,\omega)|}{\partial L_1\partial\omega}\right|_{\omega_r}=0 \tag{7.11}$$

Because the analytical solution of Eq. (7.11) is too complicated, this book uses numerical substitution to solve the process. Substitute the parameters of the bipolar detection coil $L_1=L_{ref}=17.5\mu H$, $R_1=2.2\Omega$, the resonance capacitance parameter $C_p=1.4nF$, $R_p=5\Omega$ and the amplifier circuit parameters $R_{12}=4.7k\Omega$, $R_a=20k\Omega$ into the formula (7.6) and formula (7.11):

$$\begin{cases}\omega_0=6.38877\times10^6\ \text{rad/s}\\ \omega_r=6.26768\times10^6\ \text{rad/s}\end{cases} \tag{7.12}$$

Equation (7.12) shows that the maximum point ω_r of $\partial H(L_1,\omega)/\partial L_1$ is slightly smaller than the maximum point ω_0 of $|H(L_1,\omega)|$, that is, when the excitation power frequency ω is slightly smaller than the resonance frequency ω_0, the output amplitude gain of the parallel resonance circuit is most sensitive to the change of the equivalent inductance of the detection coil. The curve of $\partial H(L_1,\omega)/\partial L_1$ with ω is shown in Fig. 7.17. When ω is 6.26768×10^6 rad/s, the curve corresponding to $R_p=5\Omega$ achieves a maximum value. In addition, the parasitic resistance R_p of the resonant capacitor has a greater impact on $\partial H(L_1,\omega)/\partial L_1$. As R_p increases, $\partial H(L_1,\omega)/\partial L_1$ decreases rapidly, and the extremes large value point ω_r of $\partial H(L_1,\omega)/\partial L_1$ also gradually decreases. Therefore, in the parallel resonant circuit, a resonant capacitor with a smaller loss angle factor is a better choice.

7.4.2.2 Metal Detection Coil Design

When a metal object enters the detection area, the impedance characteristics of the detection coil will change. For a metal object with a specific material, a specific size, and a specific distance from the detection coil, the greater the impedance change rate of the detection coil, the higher the sensitivity of metal detection. This book defines the impedance change rate as follows:

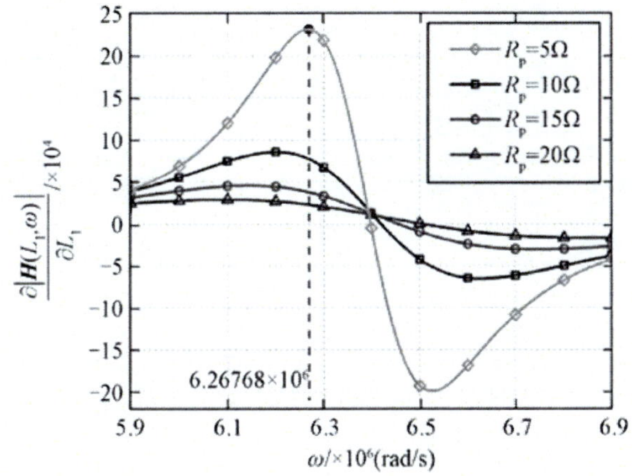

Fig. 7.17 Variation of $\partial H(L_1, \omega)/\partial L_1$ with ω

$$\varphi = \left| \frac{\Delta Z_{\text{detection}}}{Z_{\text{detection}}} \right| \times 100\% \quad (7.13)$$

In the formula, $Z_{\text{detection}}$ is the impedance mode of the detection coil; $\Delta Z_{\text{detection}}$ is the change of the impedance modulus of the detection coil; φ is the impedance change rate of the detection coil. The size of φ is related to the size, shape, and number of turns of the detection coil. Literature (Jeong et al. 2016b) draws the conclusion that the smaller the detection coil size, the larger φ through simulation. For a fixed area charging detection area, the smaller the size of the single detection coil, the more its quantity needs. Too many single detection coils mean longer detection time and more redundant detection circuits, therefore, the improvement of detection sensitivity must consider the constraints of detection speed and hardware cost. In addition, the detection coil must be magnetically decoupled from the charging coil to eliminate the magnetic coupling interference of the 85 kHz power magnetic field. Aiming at the DD–DD wireless charging coil used in this book, a bipolar detection coil is proposed, which is composed of two toroidal coils in reverse series. The assembly structure is shown in Fig. 7.18.

To verify the basic magnetic decoupling of the proposed bipolar detection coil and the charging coil, this paper simulates the spatial magnetic field distribution of the DD–DD wireless charging coil in the ANSYS MAXWELL simulation environment. The simulation results are shown in Fig. 7.19.

When the charging coil is perfectly aligned, the magnetic flux generated by the power magnetic field penetrating into and out of the detecting coil is basically the same, that is, the charging coil and the detecting coil are basically magnetically decoupled. The mutual inductance values of some single detection coils and charging coils measured by simulation are shown in Table 7.1. It can be seen that the mutual inductances between the detection coils and the DD-type coils at the transmitting and receiving ends are very small. The mutual inductance can be further reduced by

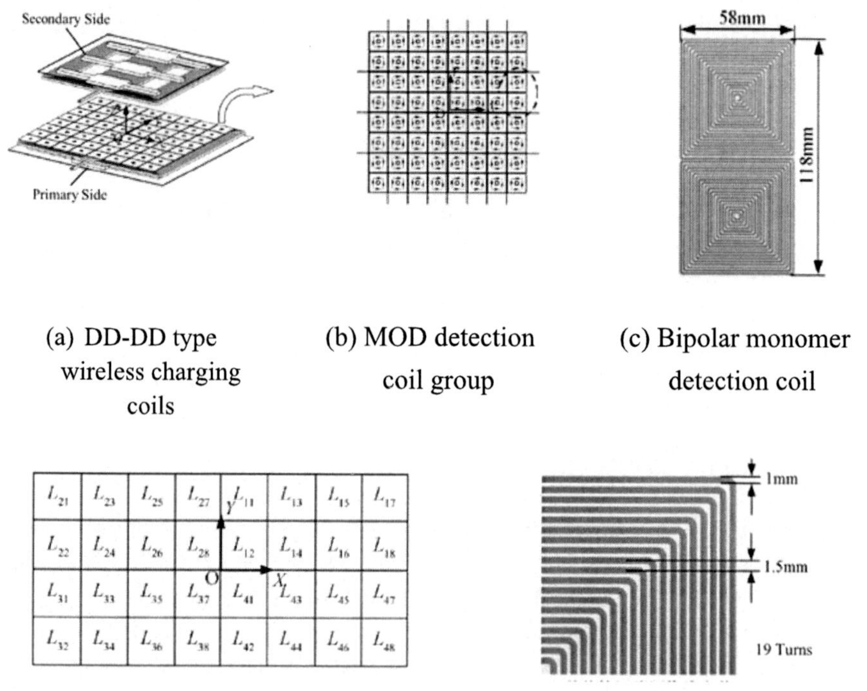

(a) DD-DD type wireless charging coils (b) MOD detection coil group (c) Bipolar monomer detection coil

(d) Naming rules (e) Detection coil parameter

Fig. 7.18 Assembly diagram of detecting coils

Fig. 7.19 Magnetic field distribution of DD–DD coupler

slightly adjusting the number of turns of the detection coil. Meet the requirements of magnetic decoupling.

In the ANSYS MAXWELL simulation environment, the equivalent inductance change of the detection coil caused by metal objects is measured. The aluminum sheet with an outline size of 40 mm × 40 mm × 1 mm and 3 mm from the upper surface of the detection coil is selected as a metal object. When the aluminum sheet

Table 7.1 Mutual inductance between detecting coils and charging coils

Detection coil	*M* with primary side coil	*M* with secondary side coil
L_{31}	526.37	183.99
L_{32}	1013.31	182.26
L_{33}	1413.64	274.55
L_{34}	880.29	124.17
L_{35}	1413.05	229.38
L_{36}	724.45	10.92
L_{37}	37.28	119.65
L_{38}	276.64	256.19

moves in two directions, the equivalent inductance of the detection coil is changed. The moving direction of the aluminum sheet is shown in Fig. 7.20. The simulation results are shown in Fig. 7.21. When the aluminum sheet appears above the single test coil, the equivalent inductance of the single test coil changes significantly, and the rate of change exceeds 20%. Since the detection coil group covers the entire charging area, the metal objects in the entire charging area can be identified by sequentially detecting the change in the equivalent inductance of each individual coil.

When the DD coil at the receiving end is not completely aligned with the DD coil at the transmitting end, the equivalent inductance of the detecting coil will also change to a certain extent. This book analyzes the change of the equivalent inductance of the detection coil when the DD coil at the receiving end shifts along the X-axis and Y-axis. The simulation result is shown in Fig. 7.22. The change rate of the equivalent

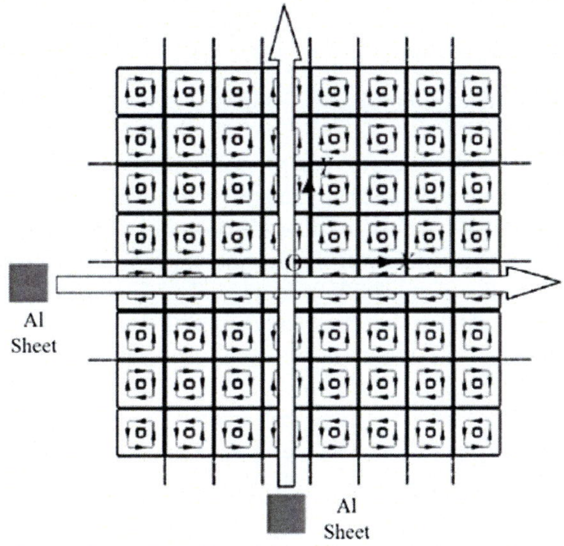

Fig. 7.20 Two moving directions of Al sheet

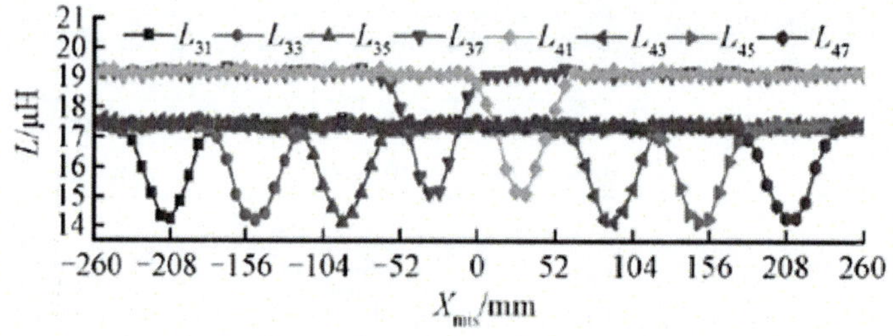

(a) The equivalent inductance of the detection coil changes with the *X*-axis coordinate of the aluminum sheet

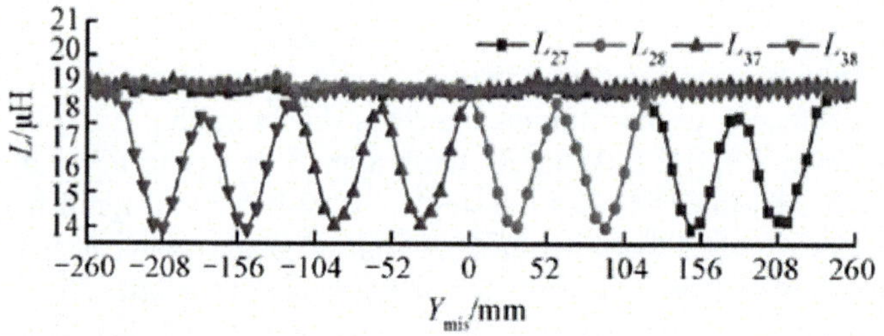

(b) The equivalent inductance of the detection coil changes with the *Y*-axis coordinate of the aluminum sheet

Fig. 7.21 Variation of detecting coils' equivalent inductance with the coordinate of Al sheet

inductance of the detection coil does not exceed 1.5%, much less than the influence of metal objects.

7.4.2.3 Filter Circuit Design

The bipolar detection coil proposed in the previous section is magnetically decoupled from the wireless charging coil, that is, theoretically, the 85 kHz power electromagnetic field excited by the wireless charging coil will not induce a voltage in the detection coil. In fact, the assembly position of the detection coil often inevitably has deviations, and the 85 kHz AC voltage interference that cannot be ignored is often induced in the detection coil. In addition, the AC excitation power supply often has a certain DC voltage component, which will eventually affect the detection accuracy. Therefore, it is necessary to design a filter circuit at the back end of the resonant circuit. On the one hand, it amplifies the detection signal of 1 MHz, and on the other hand, it eliminates the interference of 85 kHz induced voltage and DC voltage. The fourth-order Butterworth filter used in this system is shown in Fig. 7.23, and U_{out2} is the output voltage of the filter circuit.

The transfer function of the filter circuit is:

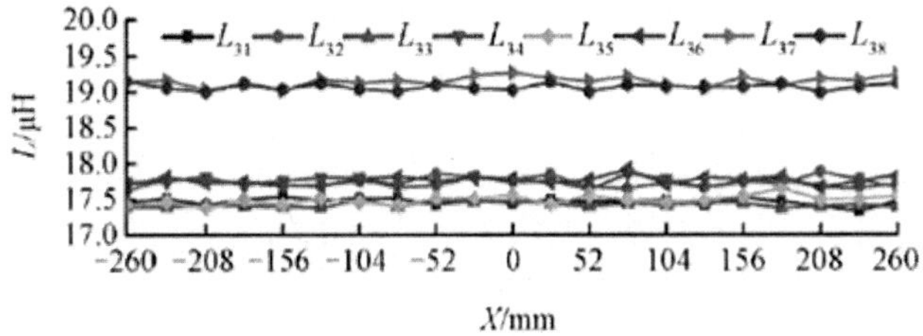

(a) The equivalent inductance of the detection coil varies with the *X*-axis offset of the DD coil at the receiving end

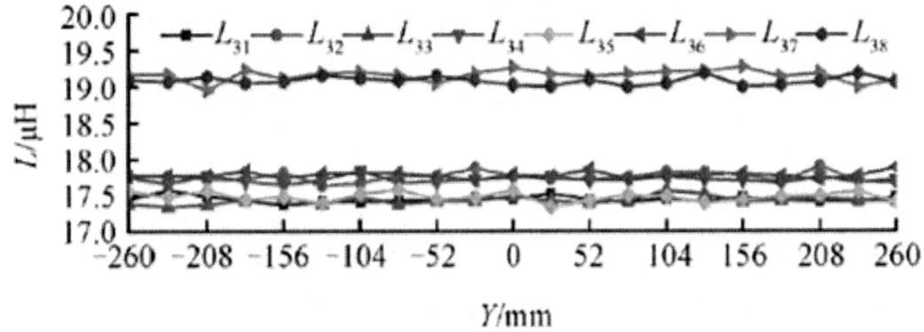

(b) The equivalent inductance of the detection coil varies with the *Y*-axis offset of the DD coil at the receiving end

Fig. 7.22 Variation of detecting coils' equivalent inductance with alignment state of charging coils

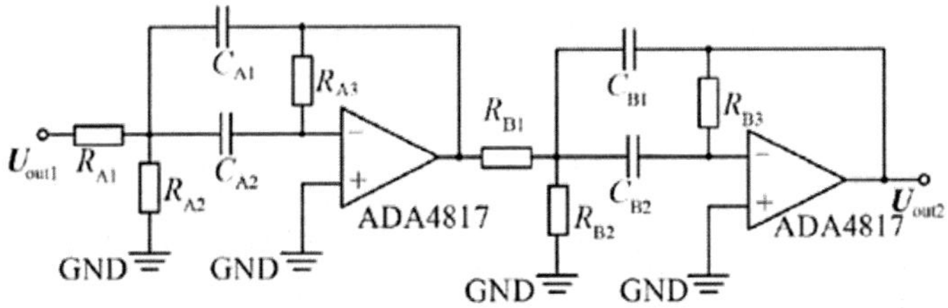

Fig. 7.23 Filter circuit

$$\frac{U_{out2}(s)}{U_{out1}(s)} = \frac{\chi_s^2}{\left(\alpha_1 s^2 + \beta_1 s + \gamma_1\right)\left(\alpha_2 s^2 + \beta_2 s + \gamma_2\right)} \tag{7.14}$$

where:

$$\begin{cases} \chi = C_{A2}C_{B2}R_{A2}R_{A3}R_{B2}R_{B3} \\ \alpha_1 = C_{A1}C_{A2}R_{A1}R_{A2}R_{A3} \\ \beta_1 = C_{A1}R_{A1}R_{A2} + C_{A2}R_{A1}R_{A2} \\ \gamma_1 = R_{A1} + R_{A2} \\ \alpha_2 = C_{B1}C_{B2}R_{B1}R_{B2}R_{B3} \\ \beta_2 = C_{B1}R_{B1}R_{B2} + C_{B2}R_{B1}R_{B2} \\ \gamma_2 = R_{B1} + R_{B2} \end{cases} \tag{7.15}$$

Obviously, the filter circuit can amplify the amplitude of the 1 MHz detection signal by about 9.7 dB, while suppressing the 85 kHz interference signal, so that its amplitude gain is lower than −55 dB. In addition, for DC or other low-frequency and ultra-high frequency signals, the filter circuit also has a strong blocking effect.

7.4.3 Multi-channel Metal Object Detection System and Automatic Tuning Method

7.4.3.1 Multi-channel Metal Detection System Composition

For the detection coil group composed of 32 single detection coils, a multi-channel metal detection system needs to be designed, which can cyclically drive each detection coil to extend the detection range to the entire charging area.

Under the control of the STM32F103 chip, the AD9833 module and the digital potentiometer can generate a standard sinusoidal signal with adjustable amplitude and frequency. This signal can be used as the input signal of the parallel resonant circuit to obtain an output signal reflecting the change in impedance of the detection coil. Pass the output signal of the parallel resonant circuit into the filter circuit and the peak detection circuit in turn to obtain the DC analog signal $U_{\text{detection}}$ readable by the STM32F103 chip. For 32 single-cell detection coils, each single-cell detection coil is cyclically driven in the driving sequence. When the STM32F103 chip detects the entry of metal foreign objects, it sends a stop signal to the main control chip of the wireless charging system TMS320F28335, and the main control chip controls the inverter bridge to shut down, terminates the charging process, and lights up the indicator light to indicate that the foreign object has detected a fault.

7.4.3.2 Automatic Tuning Method Realization

Figure 7.22 shows that the equivalent inductance of each individual detection coil is different, and the resonant capacitance is fixed, the maximum point ω_r of each individual detection coil $\partial H(L_1, \omega)/\partial L_1$ is different. To determine the ω_r value of each single detection coil, the STM32F103 chip controls the AD9833 module to generate a series of standard sinusoidal signals with different frequencies and the same amplitude before the detection, and the STM32F103 chip reads and records the output voltage value of the peak hold circuit. Among them, the excitation power frequency corresponding to the maximum output voltage is the resonance frequency ω_0, slightly adjust the value of ω_0 downward (generally reduce by 5–10 kHz) to obtain the approximate value of ω_r. Since the equivalent inductances of the individual detection coils are similar, their ω_r values are also distributed within a certain range (the distribution range of ω_r /(2π) is 950–1050 kHz). Through this process, the individual detection coils can be obtained quickly This method is called the automatic tuning method in this article. The automatic tuning method can effectively improve the robustness of the metal detection system. In the wireless charging application scenario of electric vehicles, the receiving end coil installed under the vehicle often cannot be completely aligned with the transmitting end coil, the equivalent inductance of each detection coil will change to a certain extent. The automatic tuning method can record the optimal excitation frequency of each detection coil at this position after the car has stopped, thereby eliminating the interference of alignment errors.

7.4.4 Experiment Verification

The wireless charging metal object detection experimental platform built in this paper is shown in Fig. 7.24. Among them, the wireless charging coil is DD–DD type, the size of the transmitting end coil is 500 mm × 500 mm, the metal detection coil group is composed of 32 bipolar single detection coils, and the overall size is 480 mm × 480 mm. Set the charging power of the wireless charging system to 2 kW. To verify the magnetic decoupling of the detection coil and the wireless charging coil, a digital oscilloscope is used to measure the induced voltage $U_{\text{induction}}$ in the single detection coil L_{31}, the waveform is shown in Fig. 7.25. Due to the assembly error of

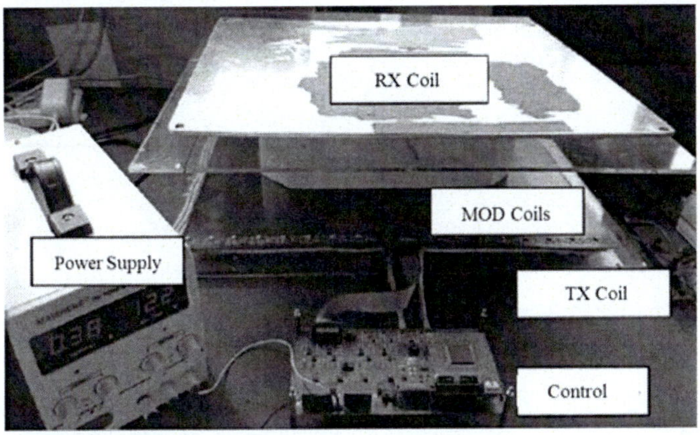

Fig. 7.24 MOD experiment for wireless charging

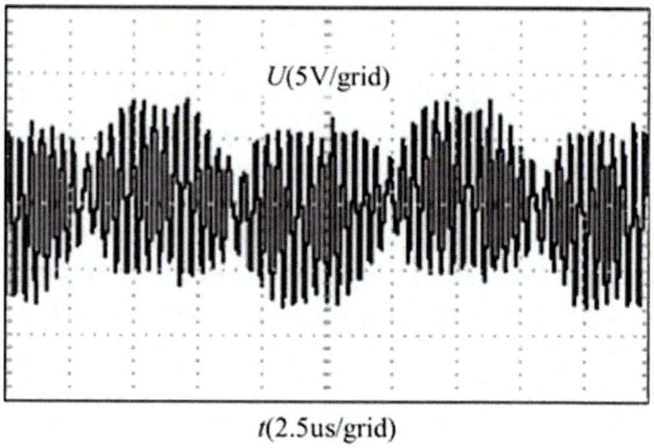

Fig. 7.25 Inductive voltage waveform in detecting coil L_{31}

the detection coil and the winding error of the charging coil, a certain interference signal is still inevitably induced in the detection coil, but the peak value of the signal does not exceed 8 V, which can be blocked and filtered by the back-end filter circuit.

The measurement shows that the input signal of the filter circuit is mixed with many clutters of different frequencies, while the output signal of the filter circuit is a more standard sine wave, so the filter circuit realizes the blocking of the clutter.

Finally, to verify the realization of the metal object detection function, the "Special Requirements for Electric Vehicle Wireless Charging System" (draft for comments) limits the objects of the foreign object detection temperature rise test and mentions that the temperature rise test of a round, five-corner, and one-corner simulated coin is required. So this book chooses one circle, five corners, and one dime as the detection object. Set the charging power of the wireless charging system to 2 kW. When the coin distribution position is shown in Fig. 7.26, and the vertical height from the upper surface of the detection coil is 3 mm, the detection signal waveform is shown in Fig. 7.27. When a round coin enters the detection area, the corresponding output voltage of the detection coil L_{34} decreases by more than 1.1 V; when a penny coin

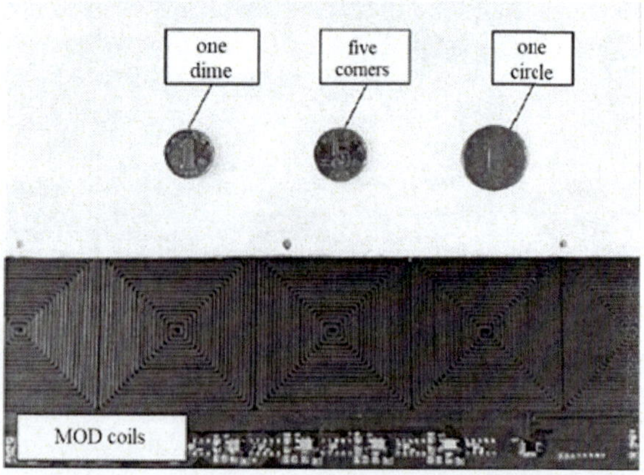

Fig. 7.26 Positions of target coins

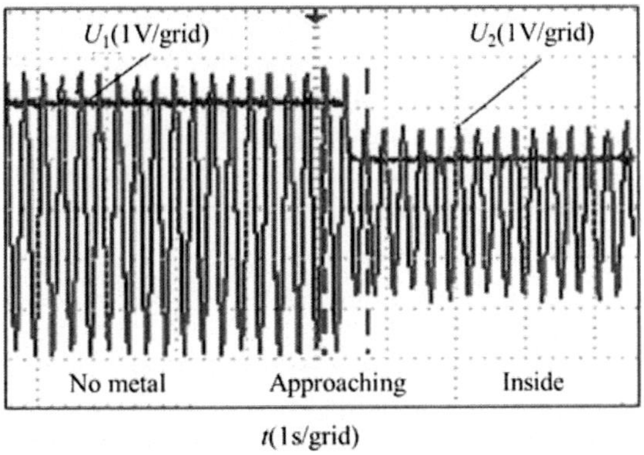

(a) Waveform diagram of detection when a circle coin is intervened

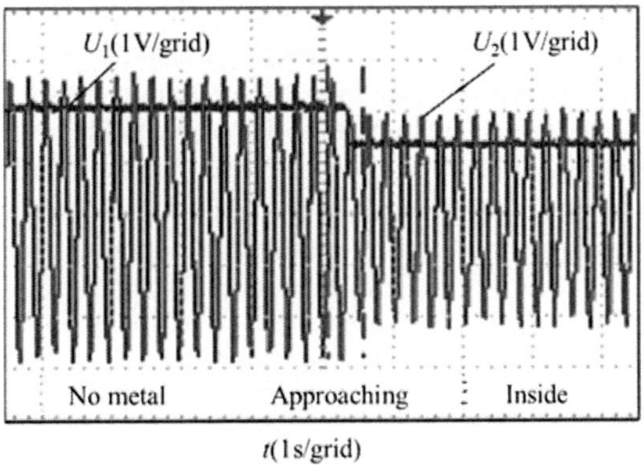

(b) Waveform diagram of detection when five corners coin is intervened

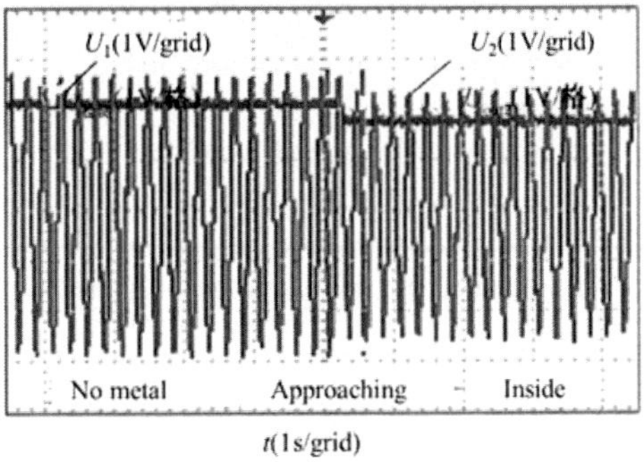

(c) Waveform diagram of detection when a dime coin is intervened

Fig. 7.27 Experimental waveform of coins detection

enters the detection area, the corresponding output voltage of the detection coil L_{36} decreases by more than 0.7 V; when a dime enters the detection area, the detection 4 V. Coil L_{38} corresponding to the output voltage drop exceeds 0.4 V. By setting the threshold value of the output voltage change, the sensitivity of the metal detection system can be adjusted. When the threshold value of the output voltage change is 0.3 V, the system can detect the intervention of a dime.

7.4.5 Chapter Summary

This section proposes a multi-channel automatic tuning wireless charging metal object detection system based on impedance detection, and studies and analyzes its component modules and working principles. Through the equivalent circuit method, the influence of metal on the impedance of the detection coil is theoretically analyzed, a parallel resonant circuit is proposed, and the method for determining the optimal excitation frequency is obtained through theoretical calculation. For the DD–DD wireless charging coil, a bipolar detection coil is proposed, and the basic magnetic decoupling of the two is proved. Next, the amplitude-frequency characteristics of the filter circuit are analyzed by calculating the transfer function, and the automatic tuning method for determining the optimal excitation frequency of each single detection coil by sweeping frequency is explained through the flowchart. Finally, experiments verify the feasibility and efficiency of the metal object detection function.

The advantages of the proposed metal object detection method are:

(1) High sensitivity: It can detect the intervention of metal foreign objects as small as a dime.
(2) Strong anti-interference: The bipolar detection coil and the DD–DD wireless charging coil are basically magnetically decoupled; the filter circuit can filter out the 85 kHz interference in the detection circuit and amplify the amplitude of the detection signal.
(3) Fast detection speed and high stability: Multiple detection coils are driven in parallel, and the detection speed is fast; the automatic tuning method can actively obtain the optimal excitation frequency of each single detection coil and eliminate the influence of the alignment deviation of the charging coil on the detection accuracy.

7.5 Summary and Outlook

Section 7.3 explains that electric vehicle wireless charging metal object detection technology has achieved some results at home and abroad. Section 7.4 introduces a resonant metal object detection method, but there is no reliable and popular metal object detection method. exist. Optical detection methods such as infrared imaging detection method, camera detection method, radar detection method, etc., have problems such as high cost, difficulty in maintaineince, large environmental constraints, and large space occupation, and are not widely accepted in the industry. However, electrical detection methods generally have problems such as low sensitivity, difficulty in detecting small-sized metal objects, detection blind spots, and poor anti-interference ability in high-power magnetic fields. However, metal object detection systems are crucial in the process of wireless charging of electric vehicles. The wireless charging system for electric vehicles without metal object detection poses great safety risks. Practitioners and researchers related to electric vehicles still need to work hard to design a reliable, high-sensitivity, anti-interference ability, no blind zone, economical and simple metal object detection system to avoid the operation of the electric vehicle wireless charging system and to promote the marketization of wireless charging for electric vehicles.

References

Cai W, Yi F, Jiang L (2020) Fast Online foreign object detection method using auxiliary coils in wireless charging systems. In: 2020 IEEE transportation electrification conference & Expo (ITEC), Chicago, IL, USA, pp 979–984

Dimke B, Stamenic M (2019) Wireless charging transmitter with foreign object and living object detection systems. U.S. Patent 16,292,792, Jun. 9 Apr 2019.

Fukuda S, Nakano H, Murayama Y, Murakami T, Kozakai O, Fujimaki K (2012) A novel metal detector using the quality factor of the secondary coil for wireless power transfer systems. In: 2012 IEEE MTT-S international microwave workshop series on innovative wireless power transmission: technologies, systems, and applications, Kyoto, pp 241–244

Futatsumori S, Morioka K, Kohmura A, Yonemoto N (2014) Design and measurement of W-band offset stepped parabolic reflector antennas for airport surface foreign object debris detection radar systems. In: 2014 international workshop on antenna technology: small antennas, novel EM structures and materials, and applications (iWAT), Sydney, NSW, pp 51–52

Hoffman PF, Boyer RJ, Henderson RA (2015) Foreign object detection system and method suitable for source resonator of wireless energy transfer system. U.S. Patent 9,304,042 B2, 5Apr 2016.

Huh J, Lee SW, Lee WY, Cho GH, Rim CT (2011) Narrow width inductive power transfer system for online electrical vehicles. IEEE Trans Power Electron 26(12):3666–3679

Jafari H, Moghaddami M, Sarwat A (2019) Foreign object detection in inductive charging systems based on primary side measurements. IEEE Trans Ind Appl 55(6):6466–6475

Jang GC, Jeong SY, Kwak HG, Rim CT (2016) Metal object detection circuit with non-overlapped coils for wireless EV chargers. In: 2016 IEEE 2nd annual southern power electronics conference (SPEC), Auckland, pp 1–6

Jeong J, Ryu S, Lee B, Kim H (2015) Tech tree study on foreign object detection technology in wireless charging system for electric vehicles. In: 2015 IEEE international telecommunications energy conference (INTELEC), Osaka, pp 1–4

Jeong SY, Kwak HG, Jang GC et al (2016a) Living object detection system based on comb pattern capacitive sensor for wireless EV chargers. In: IEEE 2nd annual southern power electronics conference (SPEC), pp 1–6 (in proc)

Jeong SY, Kwak HG, Jang GC, Rim CT (2016b) Living object detection system based on comb pattern capacitive sensor for wireless EV chargers. In: 2016 IEEE 2nd annual southern power electronics conference (SPEC), Auckland, pp 1–6

Jeong SY, Kwak HG, Jang GC, Choi SY, Rim CT (2018a) Dual-purpose nonoverlapping coil sets as metal object and vehicle position detections for wireless stationary EV chargers. IEEE Trans Power Electron 33(9):7387–7397

Jeong SY, Thai VX, Park JH, Rim CT (2018b) Self-inductance-based metal object detection with mistuned resonant circuits and nullifying induced voltage for wireless EV chargers. IEEE Trans Power Elect 34(1):748–758

Kato T, Ninomiya Y, Masaki I (2002) An obstacle detection method by fusion of radar and motion stereo. IEEE Trans Intell Transp Syst 3(3):182–188

Katz N, Roy AM, Jonas JR et al (2015) Object detection for wireless energy transfer systems. U.S. Patent 15,0233,988 A1, 20 Aug 2015.

Kuyvenhoven N, Dean C, Melton J, Schwannecke J, Umenei AE (2011) Development of a foreign object detection and analysis method for wireless power systems. In: 2011 IEEE symposium on product compliance engineering proceedings, San Diego, CA, pp 1–6

Liu X, Liu C, Han W, Pong PWT (2019) Design and implementation of a multi-purpose TMR sensor matrix for wireless electric vehicle charging. IEEE Sens J 19(5):1683–1692

Lu J, Zhu G, Mi C (2021) Foreign object detection in wireless power transfer systems. IEEE Trans Ind Appl (accepted; in press)

Moghaddami M, Sarwat A (2018) A sensorless conductive foreign object detection for inductive electric vehicle charging systems based on resonance frequency deviation. IEEE Ind Appl Soc Ann Meet (IAS) 1–6

Ombach G (2013) Design considerations for wireless charging system for electric and plug-in hybrid vehicles. In: IET hybrid and electric vehicles conference 2013 (HEVC 2013), London, 2013, pp 1–4

Ombach G (2014) Design and safety considerations of interoperable wireless charging system for automotive. In: International conference on ecological vehicles and renewable energies (EVER), pp 1–4

Ote M, Jeong S, Tentezeris MM (2020) Foreign object detection for wireless power transfer based on machine learning. 2020 IEEE Wireless Power Transfer Conference (WPTC)

Poguntke T, Schumann P, Ochs K (2017) Radar-based living object protection for inductive charging of electric vehicles using two dimensional signal processing. Wireless Power Transfer 4(2):88–97

Sieber L, Chopra S (2018) Systems, methods, and apparatus for living object protection in wireless power transfer applications. U.S. Patent 14,801,559, 29 May 2018

Sonnenberg T, Stevens A, Dayerizadeh A, Lukic S (2018) Combined foreign object detection and live object protection in wireless power transfer systems via real-time thermal camera analysis. In: 2019 IEEE applied power electronics conference and exposition (APEC), Anaheim, CA, USA, 2019, pp 1547–1552

Stillig J, Edviken M, Parspour N (2020) Overview and aspects of foreign object detection in wireless power transfer applications. 2020 IEEE Wireless Power Transfer Conference (WPTC)

Sun T, QI C (2020) Design of metal object detection coil set with positioning function for WPT system. In: IECON 2020 the 46th annual conference of the IEEE industrial electronics society, Singapore, Singapore, pp 3884–3888

Thai VX, Park JH, Jeong SY et al (2018) Multiple comb pattern-based living object detection with enhanced resolution design for wireless electric vehicle chargers. In: International exhibition and conference for power electronics, intelligent motion, renewable energy and energy management, Nuremberg, Germany, pp 1–6 (in Proc)

Thai VX, Jang GC, Jeong SY, Park JH, Kim Y-S, Rim CT (2020a) Symmetric sensing coil design for the blind-zone free metal object detection of a stationary wireless electric vehicles charger. IEEE Trans Power Electron 35(4):3466–3477

Thai VX, Jang GC, Jeong SY et al (2020b) Symmetric sensing coil design for the blind-zone free metal object detection of a stationary wireless electric vehicles charger. IEEE Trans Power Electron 35(4):3466–3477

Thai VX, Park JH, Jeong SY, Rim CT, Kim Y-S (2020c) Equivalent-circuit-based design of symmetric sensing coil for self-inductance-based metal object detection. IEEE Access 8:94190–94203

Von Novak WH, Monat P, Kallal E (2017) Wireless power system with capacitive proximity sensing. U.S. Patent 14,875,449, 28 Feb 2017

Wang Q, Lee JR, Huttere M (2019) Methods and apparatus utilizing time division access of multiple radar modules in living object detection for wireless power transfer applications. U.S. Patent 14,935,098, 15 Oct 2019

Widmer H, Sieber L, Bittner M (2016) System and method for detecting the presence of a moving object below a vehicle. U.S. Patent 14,059,299, 14 Jun 2016

Zhang X, Jin Y, Yang Q, Yuan Z, Meng H, Wang Z (2017) Detection of metal obstacles in wireless charging system of electric vehicle. In: 2017 IEEE PELS workshop on emerging technologies: wireless power transfer (WoW), Chongqing, pp 89–92

Zhang Y, Yan Z, Zhu J et al (2019) A review of foreign object detection (FOD) for inductive power transfer systems. eTransportation

Zhang Y, Zhang X, Zhang Z (2020) Impedance detection based multi-channel automatic tuning metal object detection system for wireless charging device. Adv Tech Electr Eng Energy 39(202)(04):56–66 (in Chinese)

Chapter 8
Communication System

8.1 Introduction

In the era of the Internet of Things (IoT), vehicles, as an important link in the IoT, carry a huge quantity of traffic. In the system of the Internet of Vehicles, the general overall topology of charging is shown in Fig. 8.1. In this topology, all modules are connected by the Internet or 3G/4G/5G signals, and the network communication becomes the link among these modules. Generally speaking, the on-board charger (OBC) communicates with the charging pile via the CAN bus. The charging pile transmits the information of the charging pile status, vehicle status, and fault diagnosis to the cloud via a router or a wireless module. The end-user can apply for the corresponding services using mobile phone APPs, web pages, clients, etc. The operations of the user will be transmitted to the cloud by the Internet or 3G/4G/5G signal, and the current status of the user can be seen in real time. The cloud management platform is regarded as the brain of this network, which is responsible for monitoring the status of each charging pile, handling the needs of users, fault diagnosis, and traffic monitoring to ensure smooth communication between each module.

The wireless charging system replaces the traditional charging pile in this communication network and is responsible for two-way communication with the cloud management platform and the electric vehicle:

(1) The communication between the wireless charging system and the cloud management platform mainly includes the transmission of users' operation instructions and the running status of the system. This will serve as an important basis for the cloud to manage the entire wireless charging platform and scheduling strategy. The wireless charging system transmitter controller is responsible for this communication. However, as far as the current situation is concerned, wireless charging is not yet popular. It's difficult to build a complete communication network platform like a wired charging pile. Nevertheless, the behavior of the wireless charging system in this respect should be the same as

X. Zhang et al., *Wireless Power Transfer Technologies for Electric Vehicles*,
Key Technologies on New Energy Vehicles,
https://doi.org/10.1007/978-981-16-8348-0_8

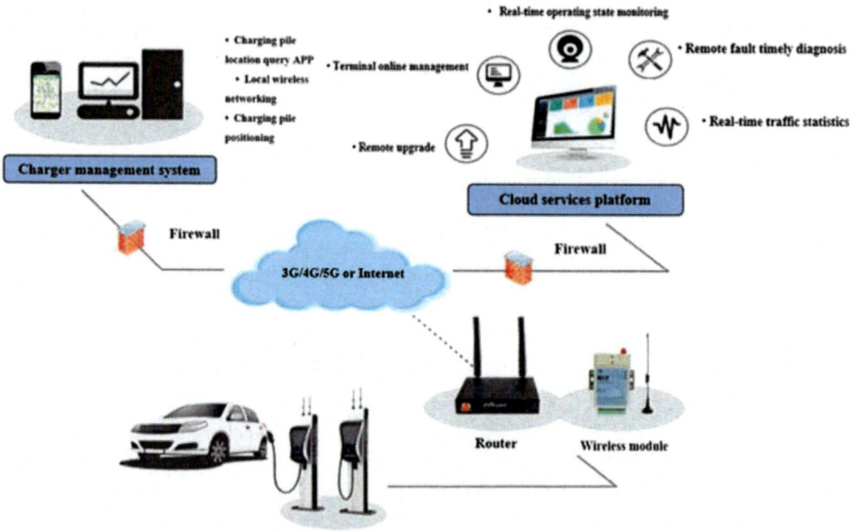

Fig. 8.1 Topology of the electric vehicle charging network

that of the wired charging pile, and the cloud communication network can be built on the experience of building the cloud communication network of the wired charging pile.

(2) The communication between the wireless charging system and the electric vehicle mainly includes the data information of the specific charging process and the corresponding fault diagnosis information. The on-board controller is mainly responsible for this function. Due to the difference between wireless charging and wired charging, the charging processes of these two can be completely different. Hence, it is necessary to design a set of the wireless charging process.

At present, the most widely used field bus in the vehicle is the CAN bus, and the data communication between the wireless charging system and the electric vehicle also mainly depends on the CAN bus. The CAN bus was first developed by BOSCH. Due to its high performance and reliability, it eventually became the ISO international standardized serial communication protocol and was widely used in the field of industrial automation, especially in internal node communication of the vehicle. The structure of CAN bus is simple, the number of nodes in the network is theoretically unlimited, and the nodes within the network can communicate freely. These features make many controller nodes in the vehicle able to achieve data interaction via two simple lines (CANH and CANL). The vehicle-mounted wireless charging system communicates with the nodes related to battery charging via the CAN bus. These data ensure the reliability and completeness of the charging process. The following section will give a detailed introduction to the wireless charging process and CAN bus protocol.

8.2 Process of Wireless Charging

Electric vehicles, as a widely used transportation, must guarantee the safety of passengers. Therefore, safety is the foremost goal to be satisfied in the charging scenario of electric vehicles. To make the wireless charging process as safe as possible, the reliability of the wireless charging device itself and the security of the entire charging process should be taken into consideration. The reliability of the wireless charging device can be achieved by designing various corresponding protection circuits, such as the open-circuit protection measures of the bilateral integrated LCC compensation topology. The safety of the charging process is reflected in the fact that the wireless charging system can avoid the emergence of various emergencies as much as possible so that the charging device can operate under the working conditions. Moreover, in an emergency, there should be corresponding solutions to ensure the safety of the charging process.

At the same time, considering that the current mainstream charging method of electric vehicles is wired charging. Wired charging is a relatively mature charging method, and many corresponding standards have been promoted for wired charging. Therefore. the charging process of wireless charging systems needs to be compatible with the charging process of wired charging. Moreover, this compatibility can show these advantages:

(1) Wireless charging, as a charging method complementary to wired charging, shows similarities in the charging process with wired charging. Meanwhile, for an electric vehicle that supports both wired charging and wireless charging, the compatibility of the two charging methods is extremely convenient for simplifying design and maintenance.
(2) At the current stage, most electric vehicles only support wired charging. If wireless charging device is regarded as an expanded component to be installed on the vehicle to achieve two types of charging, the wireless charging device should make full use of the vehicle's existing devices, which requires that the charging process of wireless charging is compatible with wired charging.

The power-on process and power-off process of the wireless charging system are shown in Figs. 8.2 and 8.3, respectively. When the user wants to start charging, the start charging signal is sent via the start button of the vehicle or the mobile phone APP. When the wireless charging controller of the vehicle receives this signal, it firstly judges the current position of the vehicle. Excessive parking offset leads to a decrease in the transmission power and transmission efficiency of the wireless charging system. Therefore, only when the offset between the transmitting coil and the receiving coil is within a specified value, can the charging be started. When the vehicle is parked in a correct position, the wireless charging system will be initialized, and the on-board controller will activate the corresponding network nodes inside the vehicle via the CAN bus and give the charging CC signal.

Charging CC (Connection Confirm) signal is a kind of connection signal commonly used in wired charging. This signal can be used to determine whether the

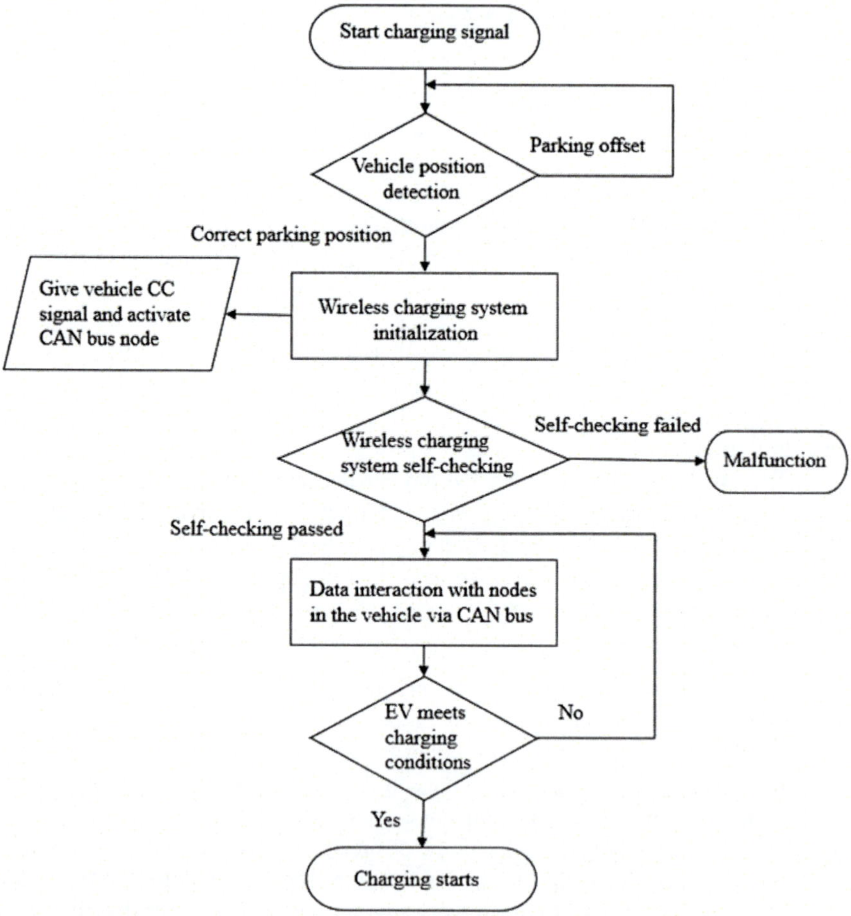

Fig. 8.2 Power-on flowchart for wireless charging system

charging pile is plugged in, which can avoid accidents caused by forcible charging when the power plug is not inserted properly. This signal is achieved by detecting the level of the point. The simplified schematic diagram is shown in Fig. 8.4, where R_1 ~ R_5 are all 1 k ohm resistors, and S is a normally closed switch. For specific implementation, see reference. To be compatible with the wired charging communication process, the wireless charging system retains this signal. The CC signal generally has two detection points: detection point 1 and detection point 2, which are in the ground controller and the vehicle controller, respectively. In a typical vehicle, the voltage represented by Vcc is 12 V, so detection point 1 and detection point 2 respectively show levels of 6 V and 12 V when the vehicle interface is disconnected. The controller judges that the vehicle is not charging currently according to this level. When the driver tries to insert the gun for charging, the button on the charging gun will be pressed. At this time, the normally closed switch S will be opened, and the

Fig. 8.3 Power-off flowchart for wireless charging system

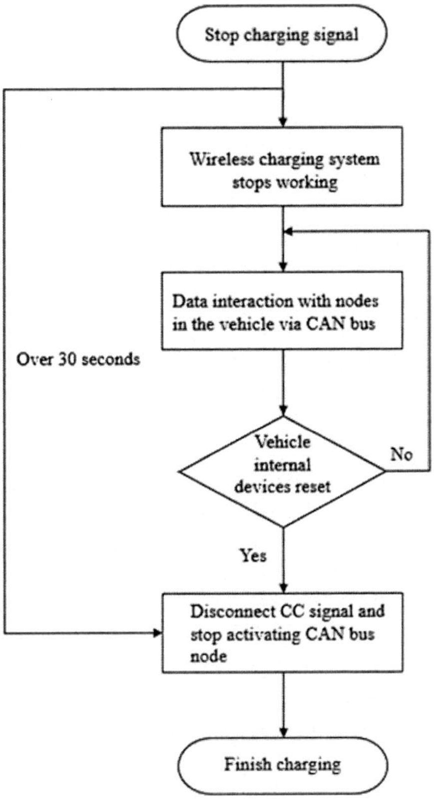

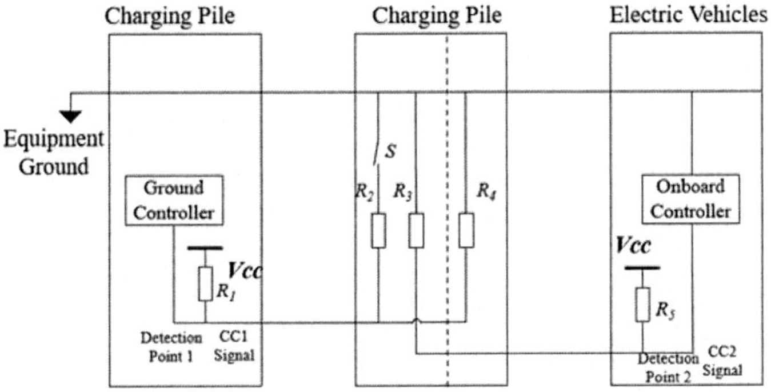

Fig. 8.4 Simplified structure of the CC signal

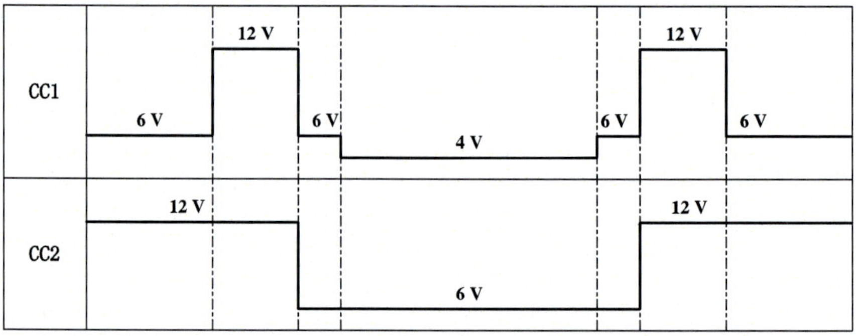

Fig. 8.5 CC1 and CC2 signal waveform

voltage at the detection point 1 will temporarily rise to 12 V. When the charging gun is plugged in and the button is not released, the voltage will temporarily become 6 V, and then when the button is released, the voltage will eventually become 4 V. For the detection point 2, when the vehicle is not charged, the level at CC2 should be 12 V. When the charging interface is connected, the voltage at detection point 2 will become 6 V, as shown in Fig. 8.5. The ground controller and the vehicle-mounted controller will judge the vehicle state and the driver's intention at this time through the change of the levels of the detection point 1 and 2. For wireless charging, it is necessary to consider that wireless charging has no ground charging piles and wireless charging need being compatible with wired charging piles. So, the voltage at detection point 2 should be ensured that it will jump from 12 to 6 V before and after the charging command is issued. The on-board controller can be used to control whether the resistor R_3 is connected to the detection point 2 by a relay to achieve this function.

After the CC signal is given and the vehicle CAN bus wakes up, the wireless charging system will perform a self-test. If there is a problem with the self-test, the system will issue a fault signal for fault diagnosis. If the self-test passes, the controller of the wirelessly charged vehicles will interact with the internal controller of the vehicle according to the charging protocol. When the vehicle meets the charging conditions, the wireless charging can be carried out.

When the driver voluntarily issues a command to stop charging, or the charging is completed, or when the wireless charging system fails during operation, the on-board controller will send a signal to stop charging. At this time, the wireless charging system will stop charging, and then the wireless charging vehicle controller will interact with the vehicle's internal controller through the CAN bus, so that the relevant devices in the vehicle, such as the vehicle's operating status, charging permission signal, slow charging relay status, etc. to restore the state before charging. Then the wireless charging vehicle controller will release the CC signal and stop the bus wakeup, and the wireless charging is over. If the blockage of a certain step in this process makes it difficult to proceed at a certain step, the onboard controller will directly release the CC signal and stop the bus wakeup after 30 s, so that the vehicle can end charging.

8.3 Introduction of Communication Protocol

The wireless charging vehicle-mounted controller and the vehicle's internal controller exchange data through the CAN bus and the data to be transmitted is sent to the CAN bus according to the predetermined coding method. The corresponding nodes receive the data and follow the predetermined methods to analyze the data to make corresponding decisions. The communication protocol is such a pre-defined encoding and decoding method.

In the CAN bus, the data field generally corresponds to 0 ~ 8 bytes. Each byte contains 8 bits. The most significant bit (MSB) of the byte is sent first, and the other bits of the byte are sent subsequently. Under normal circumstances, when defining the CAN bus protocol, OEMs will clearly define the order in which bytes are sent in the data field. Usually, the order is Byte 0 (MSB), Byte 1, …, Byte 7 (LSB). Thus, the bit/byte order table of the communication matrix as shown in Table 8.1 can be obtained. In this matrix, the sequence of transmitted data streams should be: 7, 6, 5, 4, 3, 2, 1, 0, 15, 14, 13, 12, 11, 11, 10, 9, 8, 23, … 40, 55, 54, 53, 52, 51, 50, 49, 48, 63, 62, 61, 60, 59, 58, 57, 56.

There are two ways to encode the CAN bus signal, namely Intel format and Motorola format. In the wireless charging system studied in this book, the Motorola format encoding method is adopted. In Motorola format coding, when the length of a signal does not exceed 8 bits and is implemented in a byte, the most significant bit of the signal (S_msb) is stored in the high bit of the byte, and the least significant bit of the signal (S_lsb) is stored in the low bit of the byte, as shown in (a) and (b) of Fig. 8.6. In some cases, the capacity of one byte is difficult to store certain signals. For example, when representing information such as battery voltage and charging

Table 8.1 Bit/byte sequence order of the communication matrix

	Bit order	MSB	→ → → → → → → → →					LSB
Byte order		Bit 7	Bit 6	Bit 5	Bit 4	Bit 3	Bit 2	Bit 1	Bit 0
MSB	Byte 0	7	6	5	4	3	2	1	0
↓	Byte 1	15	14	13	12	11	10	9	8
↓	Byte 2	23	22	21	20	19	18	17	16
↓	Byte 3	31	30	29	28	27	26	25	24
↓	Byte 4	39	38	37	36	35	34	33	32
↓	Byte 5	47	46	45	44	43	42	41	40
↓	Byte 6	55	54	53	52	51	50	49	48
LSB	Byte 7	63	62	61	60	59	58	57	56

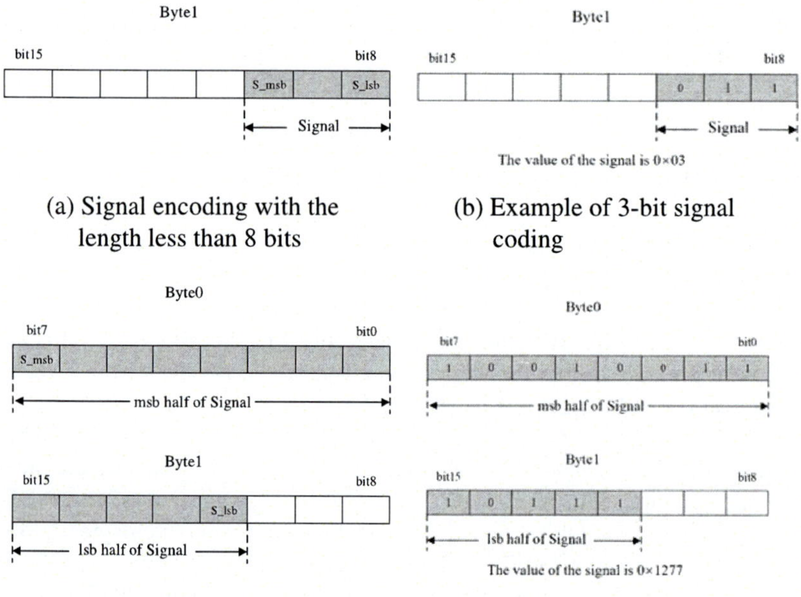

(a) Signal encoding with the length less than 8 bits

(b) Example of 3-bit signal coding

(c) Signal encoding with the length greater than 8 bits

(d) Example of 13-bit signal coding

Fig. 8.6 Demonstration for the encoding of Motorola

current, the capacity of 8 characters is difficult to ensure the accuracy and amplitude of data. So, it is often necessary to store this signal in multiple bytes. In the case where the length of such signal data exceeds 1 byte, Motorola encoding requires that S_msb is placed in the high bit of MSB, and S_lsb will be placed in the low bit of LSB, as shown in (c) and (d) of Fig. 8.6.

According to Motorola's coding rules, information such as control signals, component status, battery voltage, and charging current that is used in the wireless charging process can be encoded into the data field of CAN messages. After parsing the message according to predetermined rules, information interaction between the wireless charging onboard controller and the in-vehicle controller can be achieved. Besides, a check bit needs to be added to the message. The common method is a parity bit to ensure the validity of the data.

8.4 Summary

This section mainly introduces the communication function of the onboard wireless charging system. In the era of the IoT, the charging device is in a network environment constructed by the cloud, the user, and the facility. Every section of the network handles a large quantity of data. For the wireless charging of electric vehicles, the wireless charging system replaces the wired charging pile, and it needs to communicate with the cloud, the user, and the vehicle. Besides, to complement wired charging, the communication process needs to be compatible with wired charging as much as possible, and the communication between the wireless charging system and the user side, and the communication between the wireless charging system and the vehicle side are introduced in this chapter.

In the communication process, the power-on and power-off processes are designed. At the same time, considering the abnormal conditions that may occur in the system, the process of troubleshooting is added. To be compatible with wired charging, the CC signal processing method in the wired charging process is used in the charging process. The wireless charging onboard controller communicates with the internal control node of the vehicle via the CAN bus to ensure the stability of the charging process.

This section also gives a brief description of the wireless charging communication protocol, including the sequence order and coding method of the data stream. The wireless charging system studied in this chapter adopts Motorola coding method. For signals that are longer than 8 bits, the high bit (S_msb) of the signal is stored in the high bit of the byte, and the low bit (S_lsb) of the signal is stored in the low bit of the byte. According to this encoding method, the number of signals, device status, battery status and fault information that need to be considered in the wireless charging process can be encoded and decoded to achieve the data interaction and ensure the safe and effective wireless charging process.

Chapter 9
EMF Problems and Evaluation

9.1 Introduction

Due to the environmental friendliness of electric vehicles (EVs), governments around the world have issued policies to support the research, manufacturing, and promotion of EVs. As a new type of charging method for EVs, wireless power transfer (WPT) uses the electromagnetic field as the medium to realize power transmission (Wei et al. 2009). During the whole charging process, the ground assembly has no contact with the vehicle assembly. Thus, the WPT technology enjoys a simple charging operation, a high space utilization rate, no mechanical wear, resistance to bad weather, etc. The research of this paper is based on the inductive power transfer (IPT) technology. However, compared with the traditional plug-in method, the WPT method is faced with some problems that hinder its marketing. Two major problems are the efficiency of the WPT system and the public's concerns about the electromagnetic exposure (EME). To promote the development of WPT, the optimal design of the WPT system and the electromagnetic safety (EMS) evaluation are indispensable.

In terms of the optimal design of the WPT system, the research team led by Professor John T. Boys from The University of Auckland has started the research on the WPT system based on the IPT technology for trams, implantable medical equipment and EVs since the 1990s, and has made remarkable achievements. In (Budhia et al. 2011a), a pair of circular coils with a diameter of 700 mm, an air gap of 200 mm, a maximum allowable horizontal misalignment of 130 mm, and a rated transmission power of 2 kW was designed. (Wu et al. 2012) presented a circular WPT prototype with a diameter of 813 mm, an air gap of 200 mm, a maximum allowable horizontal misalignment of 170 mm and a rated transmission power of 5 kW. Reference (Budhia et al. 2011b) proposed a pair of bipolar double D (DD) coils and simulation results showed that the maximum allowable misalignment of DD coils is much larger than that of circular coils. In (Zaheer et al. 2012), the double D quadrature (DDQ) pad was designed and because the DDQ pad could generate the polarized and non-polarized field, the flexibility of the WPT system was enhanced. To

X. Zhang et al., *Wireless Power Transfer Technologies for Electric Vehicles*, Key Technologies on New Energy Vehicles,
https://doi.org/10.1007/978-981-16-8348-0_9

expand the charging zone, this paper adopts the DD-DDQ system for the transmitting coil and the receiving coil.

The research team led by Professor Chris Mi from the University of Michigan has also made great progress in the WPT technology for EVs. In (Li et al. 2015), the characteristics of the double-sided LCC compensation topology in WPT were analyzed in detail by the equivalent circuit method. With this compensation topology, the resonant frequency would not be affected by the load condition as well as the coupling coefficient between the coils. This demonstrates that the double-sided LCC compensation topology is ideal for WPT of EVs. In (Kan et al. 2017), an optimization design method of coils and system parameters is proposed to simplify the design process. To obtain a constant switching frequency, this paper adopts the double-sided LCC compensation topology.

As for the EMS assessment, because the transmission power of the WPT system for EVs is usually large (above 3 kW), it is necessary to evaluate the human exposure to the electromagnetic environment. According to ICNIRP 1998 (ICNIRP 1998), exposure to an electromagnetic field beyond a certain limit may cause adverse health effects such as shocks and burns, elevated tissue temperatures, and brain dysfunction which impairs the visual processing (a perception of faint flickering light in the periphery of the visual field) or motor coordination (stimulation of muscles). Besides, the central and peripheral nerves may be affected if the electromagnetic field exceeds the basic restrictions. More importantly, even lower than the limit, the electromagnetic field may interfere with medical electronic equipment and devices such as cardiac pacemakers. At present, there is no clear and unified test plan at home and abroad for human exposure to electromagnetic environment. There are many kinds of test positions and methods, and more measurement indexes need to be explored to determine the EMS. The trade standard "SAE TIRJ2954" issued by Society of Automotive Engineers (SAE) sets the standard frequency of wireless charging for light-duty EVs as 85 kHz, and the frequency range is 81.38–90.00 kHz. At present, among all the electromagnetic safety standards and guidelines formulated at home and abroad, ICNIRP 1998 by International Commission on Non-Ionizing Radiation Protection and IEEE C 95.1a-1998 by Institute of Electrical and Electronics Engineers are representative (Wang and Peng 2013). Compared with other standards and guidelines, ICNIRP 1998 more comprehensively considers the adverse effects of general public exposure and occupational exposure. Hence, this paper takes ICNIRP 1998 as the reference (ICNIRP 1998).

In this paper, a 22 kW WPT system with the DD-DDQ structure and double-sided LCC compensation topology for EVs is developed, which enjoys a high efficiency. Firstly, the structure of the magnetic coupler is optimized using Maxwell to have the largest coupling coefficient between the primary and secondary coils. Secondly, the parameters of the LCC compensation circuit are optimized with Pspice to yield the highest transmission efficiency of the whole WPT system.

After the 22 kW WPT system model at the operating frequency of 85 kHz is established, the EV model and human models with vital organs are built to evaluate the EMS of the human body in different postures and positions. The safety of the human body exposed to the electromagnetic environment is analyzed by comparing the simulation results with the reference levels in ICNIRP 2010. This will provide a basis for the establishment of electromagnetic standards of the high-power WPT systems for EVs.

This paper is organized as follows. Section 9.2 optimizes the 22 kW WPT system from two aspects which are the structure of the magnetic coupler and the compensation parameters. In Sect. 9.3, we construct the human body models and the EV model. Section 9.4 presents the EMS assessment for different scenarios. Finally, Sect. 9.5 gives the conclusion.

9.2 Design and Optimization of the 22 kW WPT System

In this section, first we use ANSYS Maxwell to optimize and determine the structure of the main coupling coils. Then, the efficiency formula is derived to calculate the compensation parameters with MATLAB. Finally, we use Pspice to verify the optimization.

9.2.1 Optimization of the Main Coupling Coils

According to Li and Mi (2015), we have

$$\eta_{max}=\frac{k^2 Q_1 Q_2}{\left(1+\sqrt{1+k^2 Q_1 Q_2}\right)^2} \tag{9.1}$$

where η_{max} is the maximum efficiency of the coupling coils, k is the coupling coefficient between two coils, Q_1 and Q_2 are the quality factors of the primary and secondary coils, respectively. It can be seen that the transmission efficiency of the coupling coils is dependent on the coupling coefficient between two coils and the quality factor of each coil. The relationship is plotted as shown in Fig. 9.1.

The larger the coupling coefficient and the higher the quality factor, the higher the transmission efficiency. Generally, the quality factor of coils can reach more than 200, so the coupling coefficient between two coils is the main factor that affects the transmission efficiency of the coil. The primary purpose of coil optimization is to improve the coupling coefficient.

The size of both coils is set as 600 * 600 mm, and the air gap between two coils is set as 150 mm. The Litz wire has a diameter of 5.5 mm. The heights of the ferrite and the aluminum shielding are set as 9.5 and 2 mm, respectively. The overall three-dimensional configuration of the DD-DDQ system is shown in Fig. 9.2.

The number of turns for the primary coil N and the number of turns for the secondary coil n are set as variables. The size and shape of the ferrite change with the number of turns. The rest of the parameters remain the same. Both coils are

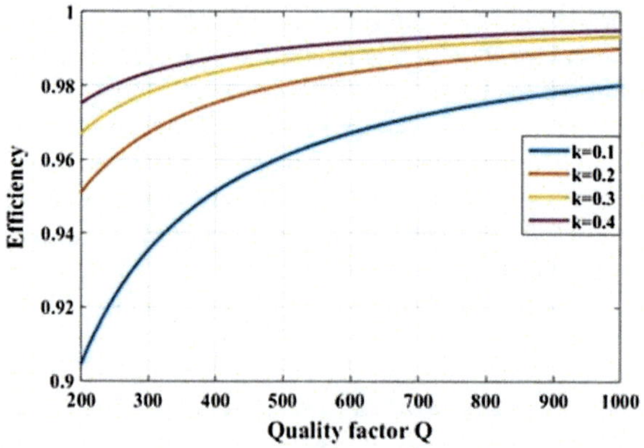

Fig. 9.1 Maximum transmission efficiency VS coupling coefficient and Quality factor of coils

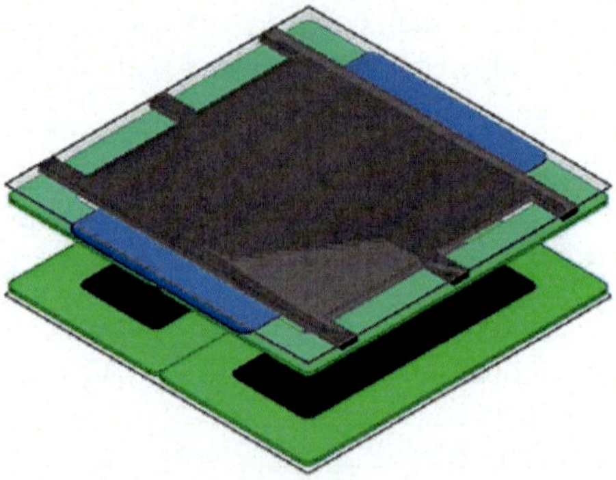

Fig. 9.2 3D model of the DD-DDQ system

aligned at the initial state. Through simulation, the relationship between the coupling coefficient and turn numbers of the two coils can be obtained as shown in Fig. 9.3.

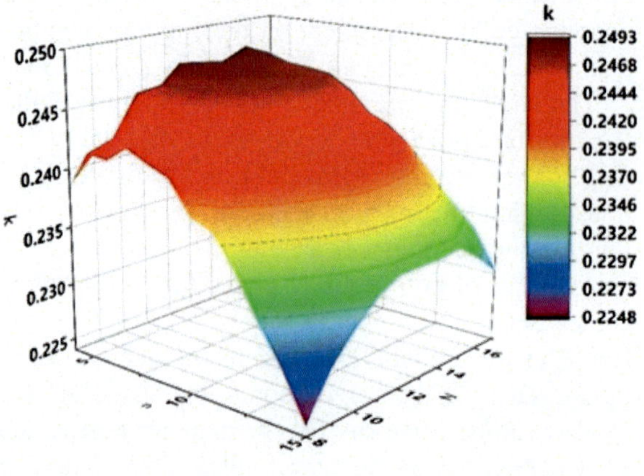

Fig. 9.3 Relationship between the coupling coefficient and turn number of DDQ coils

According to Fig. 9.3, the comprehensive coupling coefficient increases first and then decreases with the increase of turn numbers. When the turn numbers for the primary and secondary coils are 13 and 12 respectively, the maximum coupling coefficient of is achieved. After the turn numbers for both sides are determined, the structural parameters of the DD-DDQ system are displayed in Fig. 9.4 and Table 9.1.

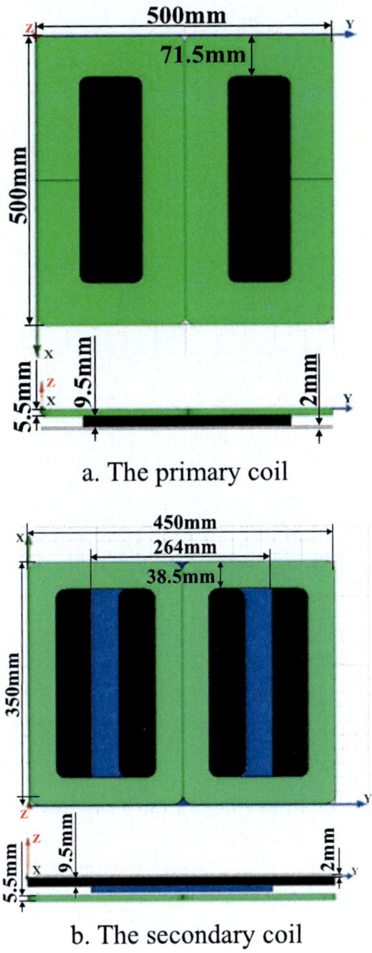

Fig. 9.4 Configuration of the DD-DDQ system

Table 9.1 Parameters of the DD-DDQ system

Parameter	Self-inductance of the primary coil L_1	Self-inductance of the secondary coil L_2	Coupling coefficient k
Value	506.22μH	566.38μH	0.378157

9.2.2 Optimization of the LCC Compensation Topology

Considering the parasitic resistance of the circuit components, the equivalent circuit of the double-sided LCC compensation topology is shown in Fig. 9.5. R_{Lf1}, R_{Cf1}, R_{C1}, R_1, R_2, R_{C2}, R_{Cf2}, and R_{Lf2} are the parasitic resistances of L_{f1}, C_{f1}, C_1, L_1, L_2, C_2, C_{f2}, and L_{f2}. The rectifier and battery at the rear end can be treated as the equivalent resistance R_{eq}. According to SAE (2017), we have

$$R_{eq}=\frac{8}{\pi^2}\frac{U_{ab}^2}{P_{out}} \tag{9.2}$$

where U_{ab} is the output voltage and P_{out} is the output power.

Then the resistances of compensation coils and capacitors can be determined by:

$$R_1=\frac{2\pi f L_1}{Q_1},\ R_2=\frac{2\pi f L_2}{Q_2},\ R_{Lf1}=\frac{2\pi f L_{f1}}{Q_{f1}},\ R_{Lf2}=\frac{2\pi f L_{f2}}{Q_{f2}} \tag{9.3}$$

$$R_{C1}=\frac{D}{2\pi f C_1},\ R_{C2}=\frac{D}{2\pi f C_2},\ R_{Cf1}=\frac{D}{2\pi f C_{f1}},\ R_{Cf2}=\frac{D}{2\pi f C_{f2}} \tag{9.4}$$

where f denotes the operating frequency of the system. Q_1, Q_2, Q_{f1}, and Q_{f2} are the quality factors of the primary-side mail coil, the secondary-side main coil, the primary-side compensation coil, and the secondary-side compensation coil, respectively. D is the dissipation factor of the capacitor.

To facilitate the calculation of the transmission efficiency for the equivalent circuit shown in Fig. 9.5, the circuit is divided into six blocks.η_1, η_2, η_3, η_4, η_5, and η_6 are the efficiency of Block 1, Block 2, Block 3, Block 4, Block 5, and Block 6, respectively. Then we have

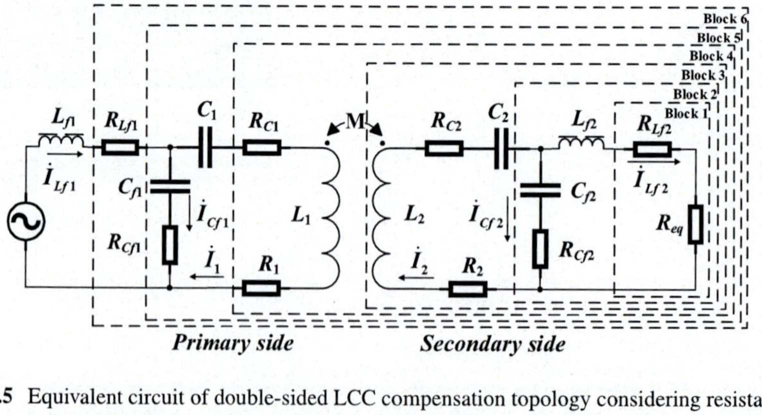

Fig. 9.5 Equivalent circuit of double-sided LCC compensation topology considering resistance

$$\eta_1 = \frac{R_{eq}}{R_{eq} + R_{f2}} \tag{9.5}$$

$$\eta_2 = \frac{\left|R_{Cf2} + \frac{1}{j2\pi f C_{f2}}\right|^2 \left(R_{eq} + R_{Lf2}\right)}{\left|R_{Cf2} + \frac{1}{j2\pi f C_{f2}}\right|^2 \left(R_{eq} + R_{Lf2}\right) + \left|R_{eq} + R_{Lf2} + j2\pi f L_2\right|^2 R_{Cf2}} \tag{9.6}$$

$$\eta_3 = \frac{Re(Z_{block2})}{Re(Z_{block2}) + R_2 + R_{C2}} \tag{9.7}$$

$$\eta_4 = \frac{Re\left(Z_{refl}\right)}{Re\left(Z_{refl}\right) + R_{C1} + R_1} \tag{9.8}$$

$$\eta_5 = \frac{\left|R_{Cf1} + \frac{1}{j2\pi f C_{f1}}\right|^2 \left[R_{C1} + R_1 + Re\left(Z_{refl}\right)\right]}{\left|R_{Cf1} + \frac{1}{j2\pi f C_{f1}}\right|^2 \left[R_{C1} + R_1 + Re\left(Z_{refl}\right)\right] + \left|R_1 + R_{C1} + j2\pi f L_1 + \frac{1}{j2\pi C_1} + Z_{refl}\right|^2 R_{Cf1}} \tag{9.9}$$

$$\eta_6 = \frac{Re(Z_{block5})}{Re(Z_{block5}) + R_{Lf1}} \tag{9.10}$$

where Z_{block2} is the impedance of Block 2:

$$Z_{block2} = \frac{\left(R_{Cf2} + \frac{1}{j2\pi f C_{f2}}\right) * \left(j2\pi f L_{f2} + R_{Lf2} + R_{eq}\right)}{R_{Cf2} + \frac{1}{j2\pi f C_{f2}} + j2\pi f L_{f2} + R_{Lf2} + R_{eq}} \tag{9.11}$$

Z_{refl} denotes the reflected impedance of the secondary-side circuit in the primary-side circuit:

$$Z_{refl} = \frac{\left(2\pi f k\sqrt{L_1 L_2}\right)^2 R_{eq}}{\left(2\pi f L_{f2}\right)^2} \tag{9.12}$$

Z_{block5} is the impedance of Block 5:

$$Z_{block5} = \frac{\left(R_{C1} + R_1 + \frac{1}{j2\pi f C_1} + j2\pi f L_1 + Z_{refl}\right)\left(R_{Cf1} + \frac{1}{j2\pi f C_{f1}}\right)}{R_{C1} + R_1 + \frac{1}{j2\pi f C_1} + j2\pi f L_1 + Z_{refl} + R_{Cf1} + \frac{1}{j2\pi f C_{f1}}} \tag{9.13}$$

Thus, the overall efficiency of the equivalent circuit shown in Fig. 9.5 is:

$$\eta = \eta_1 \cdot \eta_2 \cdot \eta_3 \cdot \eta_4 \cdot \eta_5 \cdot \eta_6 \tag{9.14}$$

Table 9.2 Parameters of two-sided main coils and the system

Parameter	L_1	R_1	L_2	R_1	k	f	U_{AB}	U_{ab}	P_{out}
Value	216 μH	0.254 Ω	201.7 μH	0.204 Ω	0.25	85 kHz	650 V	600 V	22 kW

It can be seen that the overall efficiency η is related to many parameters. When the structure of the main coils is optimized, we can determine the inductances L_1 and L_2, their parasitic resistances R_1 and R_2, and the coupling coefficient k between the main coils. Besides, the input voltage U_{AB}, the output voltage U_{ab}, the operating frequency $\omega(\omega{=}2\pi f)$, and the output power P_{out} are predetermined according to the actual application. These parameters are listed in Table 9.2.

In addition, the output power of the double-sided LCC compensation topology satisfies

$$P_{out} = \mathrm{Re}\left(\dot{U}_{ab}\dot{I}_2^*\right) = \frac{k\sqrt{L_1L_2}U_{AB}U_{ab}}{\omega L_{f1}L_{f2}} \tag{9.15}$$

To guarantee that the input and output voltages do not change with the mutual inductance of main coils, we have

$$\omega^2 L_{f1}C_{f1} - 1 = 0 \tag{9.16}$$

$$\omega^2 L_{f2}C_{f2} - 1 = 0 \tag{9.17}$$

Due to the characteristic that the input voltage and the input current are in phase, we have

$$1 - \omega^2\left(L_1 - L_{f1}\right)C_1 = 0 \tag{9.18}$$

$$1 - \omega^2\left(L_2 - L_{f2}\right)C_2 = 0 \tag{9.19}$$

By combining (2–3), (2–4), and (2–14)-(2–19), we can derive that the overall efficiency is only related to $L_{f2.}$ The curve of the transfer efficiency for the double-sided LCC compensation topology with respect to the inductance of the secondary side inductor can be plotted by using MATLAB, as shown in Fig. 9.6. The optimal value of L_{f2} is obtained when η reaches the peak. After L_{f2} is determined, the rest of the circuit parameters can be determined based on the above equations. The calculated values of parameters are listed in Table 9.3.

The simulation is carried out with Pspice under optimized parameters and the output power and efficiency are 22.104 kW and 95.4415% respectively at the steady state.

According to the above procedures, the design flow of the double-sided LCC compensation topology is shown in Fig. 9.7.

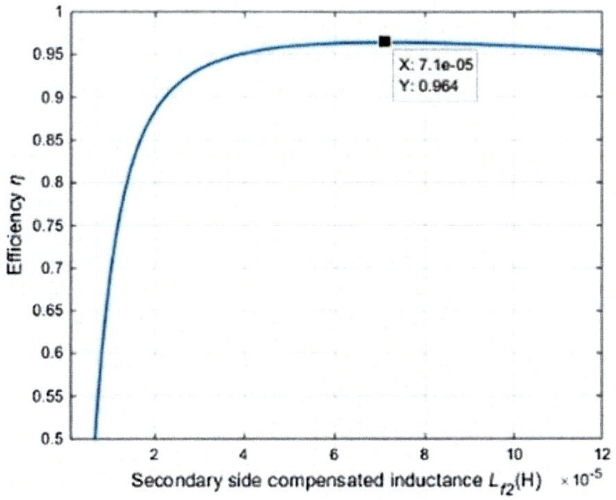

Fig. 9.6 Transfer efficiency of double-sided LCC compensation topology VS inductance of secondary side inductor

Table 9.3 Parameters of the DD-DDQ coils

Primary side		DD coil	Secondary side		DD coil	Q coil
Series compensation capacitor	C_1(nF)	12.515	Series compensation capacitor	C_2(nF)	82	162
	R_{C1}(Ω)	0.030		R_{C2}(Ω)	0.006	0.003
Parallel compensating capacitor	C_{f1}(nF)	49.947	Parallel compensating capacitor	C_{f2}(nF)	106.24	206.23
	R_{Cf1}(Ω)	0.075		R_{Cf2}(Ω)	0.035	0.018
Compensation inductor	L_{f1}(μH)	70.192	Compensation inductor	L_{f2}(μH)	33	17
	R_{f1}(Ω)	0.093		R_{f2}(Ω)	0.043	0.023

9.3 Modelling

During the operation of the WPT system, the electromagnetic field (EMF) will be generated between the coils on both sides. If the magnetic field is not effectively shielded, the magnetic field radiated to the environment will have various effects on the surrounding living beings, especially the human body. Thus it is important to conduct the safety evaluation of the human body in the WPT environment. The previous section establishes the 22 kW WPT system. This section will construct the EV model and the human body models.

9.3.1 The EV Model

In practice, the transmitter might be mounted above the ground surface, flush with the ground surface or buried below the ground surface (Zhai et al. 2012), while the receiver is integrated with the vehicle chassis. During the operation of the WPT

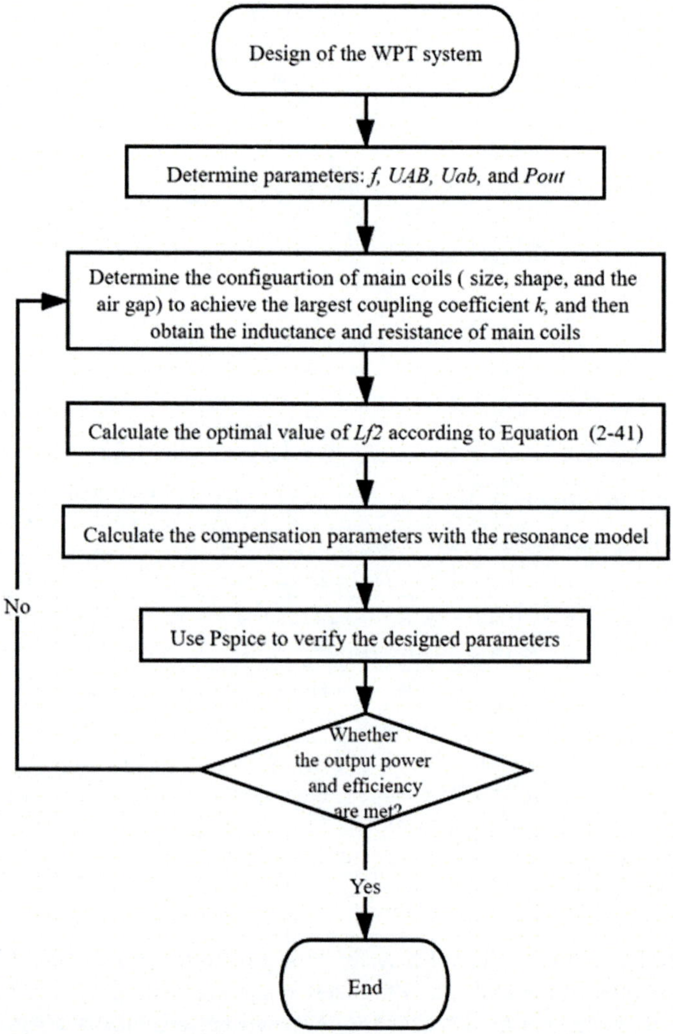

Fig. 9.7 Parameter optimization flow chart of two-sided LCC compensation topology

system, the EMF around the system is affected by not only the coupling of the magnetic coupler itself, but also the vehicle body, the chassis, the coupling with the external environment, and other factors (Zhang et al. 2016, 2015; Huang et al. 2012; Gabriel et al. 1996). Therefore, to simulate the actual environment, a simple EV model with a size of 4460 × 1800 × 1580 mm is built by SolidWorks, and then converted into the stp format. Because of its good compatibility with the finite element electromagnetic simulation software, ANSYS Maxwell, the stp file can be imported into Maxwell. Next, we set the material for the EV body as stainless steel. Besides, we construct the ground and respectively set the relative dielectric constant and conductivity of the ground as 100 and 0.02 S/m to include the impact of the ground. Figure 9.8 displays the whole EV model and the WPT system. In addition, to reduce the simulation time and improve the calculation accuracy, a proper mesh unit size is selected for different parts of the whole model.

An experiment was carried out to validate the simulation. The experimental set up is as shown. The probe Narda-EHP-200 is used to measure the magnetic field strength (MFS) on two lines in the ZY plane: the line from (0, 0, 0) to (0, 900, 0) and the line from (0, 0, 75) to (0, 900, 75). The experimental current waveforms and the comparison of the measured and simulated values are shown in Fig. 9.9. It can be seen that the simulation results agree well with the experimental results, which demonstrates the correctness of the simulation model.

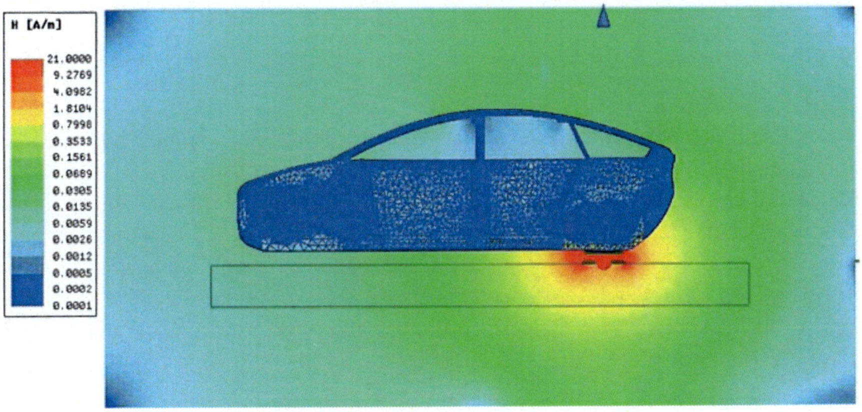

Fig. 9.8 Simulation model of the EV and the WPT system

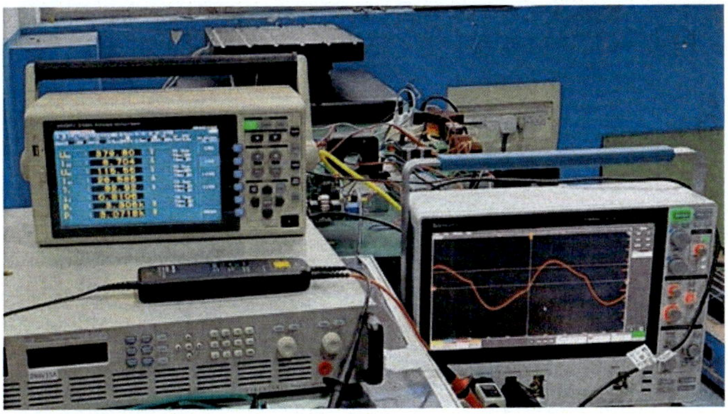

Fig. 9.9 The experimental waveforms, output power, and efficiency

Table 9.4 Dielectric parameters of the human model at the operating frequency (85 kHz)

Tissues(Organs)	Relative permittivity	Relative permeability	Conductivity (S/m)	Mass Density (Kg/m^3)
Cerebrum	2804	1	0.129	1030
Lung	5011.87	1	0.158	563
Kidney	7126.4	1	0.179	1147
Liver	11,294.6	1	0.113	1151
Heart	8500	1	0.316	1059
Spleen	10,000	1	0.113	1010
Stomach	2988.34	1	0.535418	1126
2/3 Muscle	5664.43	1	0.23919	1033.3

9.3.2 Human Body Models

After the EV model and the WPT system are completed, human body models including adult models in the standing, lying, and sitting postures as well as a child model are derived according to the standard GB 10,000—88 Human Dimensions of Chinese Adults. Then, to investigate effects of the EMF on key tissues and organs, simplified models of various organs and tissues, such as brains and hearts, are established, which we find to be sufficiently accurate to represent the details of the human body. Table 9.4 presents the dielectric properties of seven tissues and organs at 85 kHz based on (Gabriel et al. 1996). Human body models are shown in Fig. 9.10.

9.4 Numerical Results and Analysis

To investigate the influence of the EMF on the human, and to assess the compliance of the WPT system to human exposure limits suggested by ICNIRP, seven numerical simulations are carried out. That is, an adult standing behind an EV; an adult sitting in the left-front, right-front, left-rear, right-rear positions inside an EV; an adult lying behind an EV; and a child lying behind an EV, which is the worst case scenario.

According to ICNIRP Guidelines (ICNIRP 1998), reference levels and basic restrictions for the occupational and general public EME are listed in Tables 9.5 and 9.6.

Basic restrictions refer to restrictions on EME "that are based directly on established health effects" (ICNIRP 1998). For ease of assessment, Reference levels are used to practically evaluate "whether the basic restrictions are likely to be exceeded". If the measured or calculated values are below the reference levels, the relevant basic restrictions are satisfied. However, it is important to note that the incompliance with the reference level does not mean the incompliance with the basic restriction. In this

Fig. 9.10 Human body models

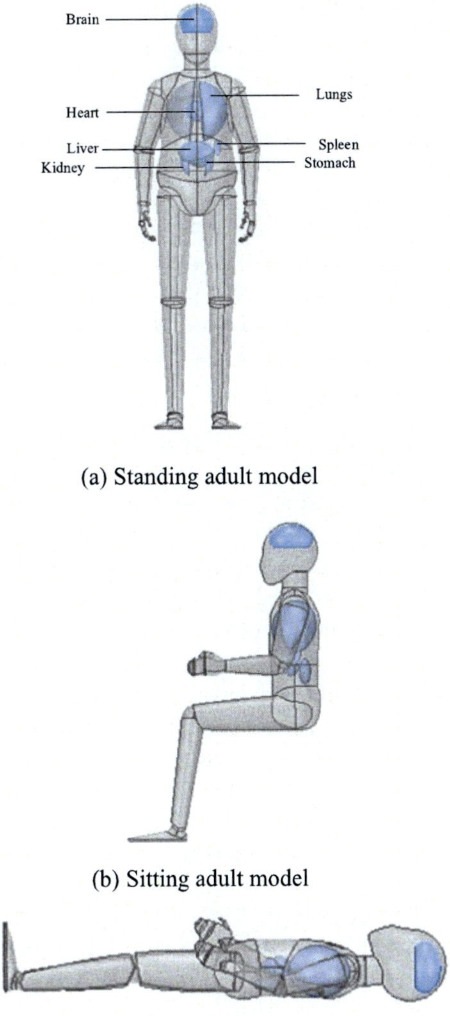

(a) Standing adult model

(b) Sitting adult model

(c) Lying child model

Table 9.5 Reference levels for occupational and general public EME at 85 kHz

Exposure characteristics	Frequency range	Electric field strength *E* (V/m)	Magnetic field strength *H* (A/m)	Magnetic flux density *B* (μT)
Occupational exposure	3kH-400 kHz	$\left(1.7\text{x}10^2\right)\text{V/m}$	80A/m	$\left(1\text{x}10^2\right)\mu\text{T}$
General public exposure	3 kHz-400 kHz	83V/m	21A/m	$27\mu\text{T}$

Table 9.6 Basic restrictions for occupational and general public EME at 85 kHz

Exposure characteristics	Exposed areas	Frequency range	Internal electric field (IEF) (V/m)	Internal electric field (IEF) at 85 kHz (V/m)
Occupational exposure	CNS tissue of the head	3 kHz-10 MHz	$(2.7\text{x}10^{-4}f)V/m$	22.95 V/m
	All tissues of head and body	3 kHz –10 MHz	$(2.7\text{x}10^{-4}f)V/m$	22.95 V/m
General public exposure	CNS tissue of the head	3 kHz -10 MHz	$(1.35\text{x}10^{-4}f)V/m$	11.475 V/m
	All tissues of head and body	3 kHz –10 MHz	$(1.35\text{x}10^{-4}f)V/m$	11.475 V/m

case, further analysis is required to determine whether additional protective measures are necessary.

We first simulated the adult standing behind the EV, which is a general exposure scenario. The distribution of the MFS and IEF on the body is as shown in Fig. 9.11.It can be seen the MFS gradually decreases from the feet to the head, which is predictable because MFS is in reverse proportion to the distance from the magnetic field source, i.e. the WPT system. However, the IEF distribution is quite non-uniform. The reason is that the IEF values are determined by the MFS and the direction of the magnetic flux together. It is obvious that the chest of the body has the highest IEF value 1.36 V/m, which is still beyond the limit 11.475 V/m. This localized IEF is caused by the direct contact with the vehicle body. The IEF values for the rest of the body are far below the limit. Therefore, it is important for people to avoid contact with the vehicle body during charging. As for the MFS, the highest MFS value in the adult body reaches 65.6454A/m, and the red area indicates that the area is exposed to a magnetic field exceeding 21A/m. Although the MFS exceeds the reference level, the IEF value is within the basic restriction, which means compliance with the safety limit. Thus, the IEF value determines whether the scenario is safe for people. In the following scenarios, we only consider the IEF value as the safety index.

Figure 9.12 a shows an adult respectively sitting in four different positions inside an EV. And Fig. 9.12b shows the IEF distribution for the adult model in the driver's position, i.e. the left-front position. Table 9.7 lists the maximum IEF values and their corresponding locations for the adult model in different sitting positions. It can be inferred from Table 9.7 that the maximum IEF values appear at locations near the stainless steel vehicle shell. For example, when the adult model is in the driver's position, the maximum IEF values are generated on the head and the left arm, which are close to the shell. So, when people are sitting inside an EV, the dominant factor that influences the IEF distribution on the human body is the stainless steel vehicle shell instead of the distance from the WPT system. In addition, due to the shielding effect of the shell, the highest IEF value is 0.7135 V/m, which is within the safety limit.

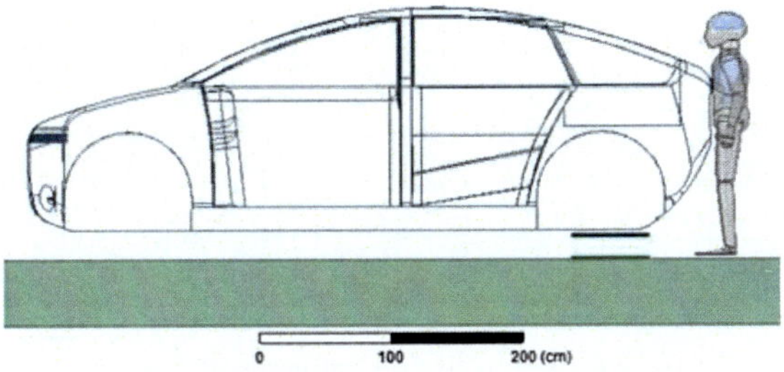

(a) Standing adult behind the charging vehicle

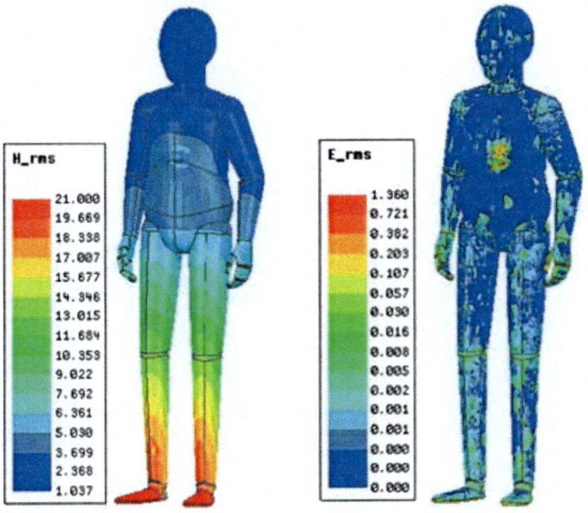

(b) Root mean square value of MFS (c) Root mean square value of IEF

Fig. 9.11 Distribution of the MFS and IEF for the standing adult

To obtain the safety estimation of the WPT system, we have to consider the worst-case scenario, that is, lying on the ground near the WPT system. The adult model is simulated first. The position of the lying adult is as shown in Fig. 9.13a. The IEF distribution of the adult model is displayed in Fig. 9.13b. The right arm close to the WPT system is exposed to an IEF greater than 11.475 V/m, with the highest IEF value being 30.42 V/m, which is dangerous. Likewise, the child model is simulated next. Because the child model has a smaller size, it is closer to the WPT system. Therefore, the child model is faced with a much more severe electromagnetic exposure compared with the adult model and the highest IEF value is 41.66 V/m, as shown in Fig. 9.13c. Then, the IEF distribution on the cutting surface through the child model is presented in Fig. 9.13d. The IEF inside the body is less intense compared with that outside the body, which indicates the damping effect of the human skin.

From the cutting surface, we can also observe that the IEF distribution is influenced by the different dielectric properties of the tissues and organs. So, next we will investigate the maximum IEF values of the tissues and organs for the adult and child models in the aforementioned exposure scenarios. To facilitate the evaluation, we

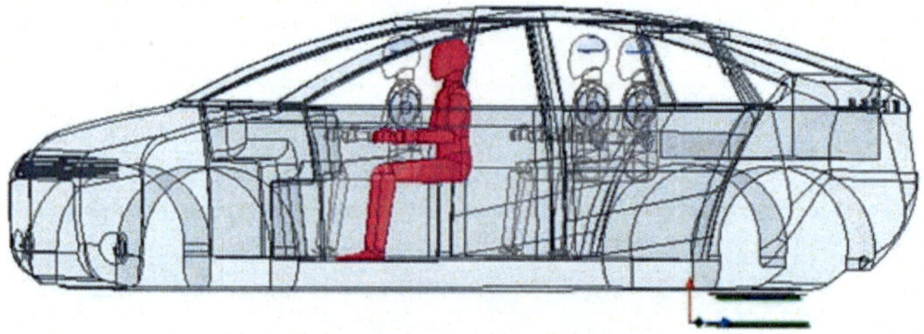

(a) adults in four sitting positions

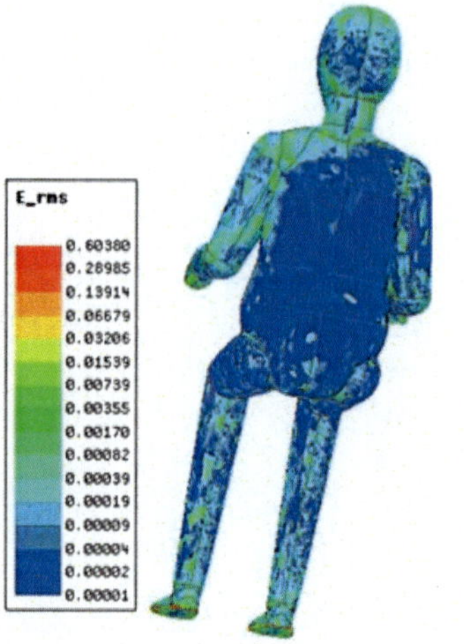

(b) the IEF distribution for the sitting adult in the left-front position

Fig. 9.12 Simulation results for the sitting adults inside the charging EV

Table 9.7 Maximum IEF values and the corresponding locations for the four sitting adults

Positions inside the EV	Maximum IEF values (V/m)	Locations of maximum IEF values
Left-front	0.6038	Head and the left arm
Right-front	0.4905	Head and the right arm
Left-rear	0.7135	Head and the left arm
Right-rear	0.45164	Head and the right arm

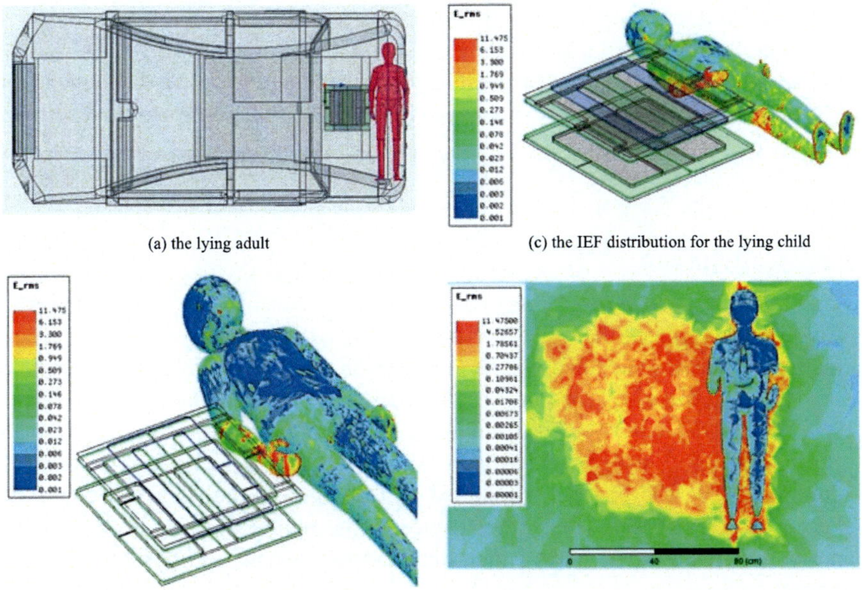

(a) the lying adult

(c) the IEF distribution for the lying child

(b) the IEF distribution for the lying adult

(d) the cross-sectional view for the lying child

Fig. 9.13 Simulation results for the lying adult and child behind the charging EV

define a safety coefficient S_{IEF} as $S_{IEF} = \ln \frac{E_{simulation}}{E_{guideline}}$. Figure 9.14 displays the safety coefficients for important organs in the aforementioned scenarios. It shows that when the child is lying behind the vehicle, both the heart and the lungs are exposed to an IEF that exceeds the safety limit. As for the adult lying case, only the heart will be endangered by the IEF. When the models are sitting inside or standing behind the charging vehicle, they are safe. These results demonstrate that when the

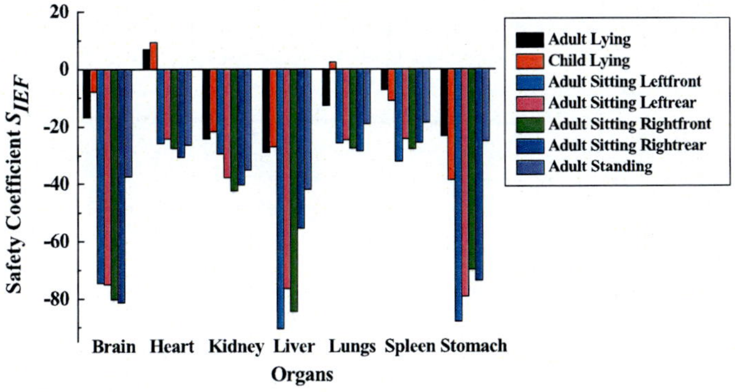

Fig. 9.14 Safety coefficients for important organs of the child and adult models in different postures and positions

WPT system is working, to protect themselves from an excessive electromagnetic exposure, people should avoid getting close to the Litz wires, which is the case of the lying adult and child. Apart from this, designers should work on the living object detection and power control strategies in case of people falling down on the ground close to the WPT system.

9.5 Conclusion

This paper completes the WPT system optimal design considering the magnetic coupler and the double-sided LCC compensation topology to achieve the specified output power and the highest transmission efficiency. The simulation results by Pspice show that the designed WPT system achieves the output power of 22.104 kW and the efficiency of 95.4415%, which proves the correctness of the simulation model. In addition, a 3 kW WPT system is designed with the same optimization procedure. The experimental results show that the system can achieve the output power of 3.306 kW, and the efficiency of 92.92%, which demonstrates the feasibility of the practical application.

To conduct the EMS evaluation of the human body in the electromagnetic field, the EV model, and the human models with vital organs are integrated with the 22 kW WPT system model. Simulation results show that the heart and lungs of the lying child and the heart of the lying adult exceeds the basic restriction of the IEF value. The scenarios of sitting inside the vehicle and standing behind the vehicle are rather safe for people. This provides a basis for designers to work on living object detection and power control strategies to protect people.

References

Budhia M, Covic GA, Boys JT (2011a) Design and optimization of circular magnetic structures for lumped inductive power transfer systems IEEE Trans. Power Electron 26:3096–3108

Budhia M, Covic GA, Boys JT, Huang H (2011) Development and evaluation of single sided flux couplers for contactless electric vehicle charging. IEEE Energy Conversion Congress Exposition

Gabriel S, Lau RW, Gabriel C (1996) The dielectric properties of biological tissues: II. measurements in the frequency range 10 to 20 GHz Phys. Med Biol 41:2251–2269

Huang X, Ji Q, Cao W, Tan L (2012) The magnetic field simulation and measurement of resonator in wireless power transmission based on magnetic resonant coupling. Trans China Electrotech Soc 28:105-109

ICNIRP (1998) Guidelines for limiting exposure to time-varying electric, magnetic and electromagnetic fields (Up to 300GHz) Health Phys 75:494–522

Kan T, Nguyen TD, White JC, Malhan RK, Mi CC (2017) A new integration method for an electric vehicle wireless charging system using LCC compensation topology: analysis and design. IEEE Trans Power Electron 32:1638–1650

Li S, Mi CC (2015) Wireless power transfer for electric vehicle applications emerging and selected topics in power. Electronics 3:4–17

Li S, Li W, Deng J, Nguyen TD (2015) A double-sided LCC compensation network and its tuning method for wireless power transfer IEEE Trans. Veh Technol 64:2261–2273
SAE Wireless power transfer for light-duty plug-in/electric vehicles and alignment methodology (2017). https://doi.org/10.4271/J2954_201711
Wang J, Peng X (2013) Electromagnetic field management related laws and regulations and standards at home and abroad. J Environ Health 30:162–166
Wei H, Wang J, Liu R (2009) Technical analysis of wireless power transmission in the electric power system. J Southwest Univer 31:163-167
Wu H, Gilchrist A, Sealy K, Bronson D (2012) A 90 percent efficient 5 kW inductive charger for EVs. IEEE Energy Conversion Congress Exposition
Zaheer A, Kacprzak D, Covic GA (2012) A bipolar receiver pad in a lumped IPT system for electric vehicle charging applications IEEE Energy Convers Congress Exposition
Zhai Y, Sun Y, Dai X, Su Y, Wang Z (2012) Modeling and analysis of magnetic resonance wireless power transmission systems. Proceedings of the CSEE 32:155–160
Zhang P, Yang Q, Zhang X, Yuan Z (2015) Optimization and characteristic analysis of wireless power driven vehicle featuring magnetic coupled resonance. Trans China Electrotech Soc S1:286-291
Zhang X, Zhang P, Yang Q, Yuan Z, Su H (2016) Magnetic shielding design and analysis for wireless charging coupler of electric vehicles based on finite element method. Trans China Electrotech Soc 31:71–79

Chapter 10
Optimal Charging Control

10.1 Introduction

Because of its better energy density and longer lifespan compared to lead-acid and nickel-metal hydride alternatives, lithium-ion (Li-ion) batteries have been marketed for electric vehicles (EVs) (Xiong et al. 2018, 2017; Wei et al. 2018). In comparison to a fuel-powered internal combustion engine, the battery charging process is far more difficult, owing to its sluggish charging pace and the unknown effects of charging techniques on battery performance. The charging speed of lithium-ion batteries has become a hurdle in the popularization of EVs (Zhang et al. 2017; Sun et al. 2016). The US Department of Energy (DOE) has set a rapid charging objective of 10 miles per minute (Chandrasekaran 2014). In other words, the DOE's objective for an EV with a 100-mile range (24-kWh battery pack) is to charge it completely in 10 min (6C rate). However, merely raising the charging rates might result in a significant temperature rise and hasten adverse reactions (Keyser et al. 2017; Meintz et al. 2017; Waldmann et al. 2018). The trade-off between rapid charging and battery health should be considered concurrently (Parvini et al. 2018; Amanor-Boadu et al. 2018). As a result, the battery optimum charging method has received a lot of interest in the EV/PHEV research field (Liu et al. 2017; Khan and Choi 2018; Jannesar et al. 2018). An optimum charging methodology is desirable to increase charging efficiency, limit performance degradation, and ensure the safe operation of a LIB system. Many studies have been conducted throughout the years in an attempt to determine the best pricing method.

Based on internal mathematical models, the various Li-ion battery charging strategies may be classified into three groups. The first kind of methodology is model-free, and it includes constant-current (CC), constant-voltage (CCCV), multi-stage CCCV, and pulse charging strategies (Abdel-Monem et al. 2017; Anseán et al. 2016; Hussein and Batarseh 2011). These methods are distinguished by their predefined charging profiles with fixed current, voltage, and/or power limitations while disregarding battery dynamics. Given the operability of model-free techniques, the

X. Zhang et al., *Wireless Power Transfer Technologies for Electric Vehicles*,
Key Technologies on New Energy Vehicles,
https://doi.org/10.1007/978-981-16-8348-0_10

related programs are referred to as heuristic. As a result, it is necessary to investigate improved charging techniques to fulfill quick charging requirements while minimizing the impact on battery state-of-health (SOH).

The second type of charging strategy makes use of empirical models such as comparable circuit-based models and neural network models (Jokar et al. 2016). Using previous experimental data, these models anticipate battery states and compute electrical elements. Kalman-type filters (Campestrini et al. 2017), recursive least squares (Wei et al. 2018) sliding mode observers (Lu et al. 2013), and moving horizon estimations (Hu et al. 2018) were used to predict battery states using various circuit models. Meanwhile, to enhance charging performance, frequency optimization (Chen et al. 2013), multi-objective optimization (Fadaee and Radzi 2012), fuzzy control (Hsieh et al. 2011), linear quadratic control (Fang et al. May 2017), and model predictive control (Kujundžić et al. 2017) were developed. The empirical models are computationally quick and simple, but they cannot account for physics-based factors or battery aging (Kujundžić et al. 2017). As a result, an empirical model-based charging control protocol may fail to function effectively after a certain number of cycles (Docimo and Fathy ; Patnaik et al. 2019).

More complex charging algorithms are based on electrochemical models guided by kinetics and transport equations (Zou et al. 2016). Closed-loop optimization problems may be written to reduce charging time while accounting for model uncertainty and disruptions (Zou et al. 2018). Furthermore, temperature variations may be anticipated using thermal-related relationships. As a result, when employed as a state observer, the electrochemistry-based control algorithm is very near to the actual battery mechanism. However, the intractable computing complexity associated with full-order nonlinear partial differential equations (PDEs) precludes future applicability to a real-time charging controller (Zou et al. 2017).

As a result, presenting an acceptable optimum charging method for a commercial Li-ion battery is a hard problem. This objective indicates that the battery should be charged as quickly as feasible while the temperature increase and aging effects are kept within the permissible range. Recently, the variety and number of current research dealing with optimal pricing techniques have provided a wealth of information.

In this chapter, we are going to provide the most recent findings on various battery optimum charging techniques. The first component discussed here is passive charging techniques such as constant-current (CC), constant-voltage (CCCV), multi-stage CCCV, and pulse charging. Their attributes are summarized and compared. The generalized structure of the active optimal charging procedure follows. Furthermore, the examined optimal charging procedures in the book are summarized, including their data, findings, studied battery type, and charging techniques. Their advantages and disadvantages were contrasted and analyzed based on the information. The suggestions and challenges for battery optimal charging strategies are proposed in the end.

10.2 Passive Charging Profiles

The passive charging techniques are distinguished by charging the battery according to pre-programmed instructions, as shown in Fig. 10.1. When the battery reaches the terminal condition, the charging process is terminated. Although the passive charging technique is simple to use, feedbacks from battery states and health-related optimization restrictions are not taken into account throughout the charging process, which may reduce battery longevity (Vo et al. 2015; Chen et al. 2018; Notten et al. 2005).

10.2.1 Constant-Current Constant Voltage Charging Methods

Because of its simplicity and ease of implementation, the constant current constant voltage (CC/CV) charging method is extensively used in charging Li-ion batteries.

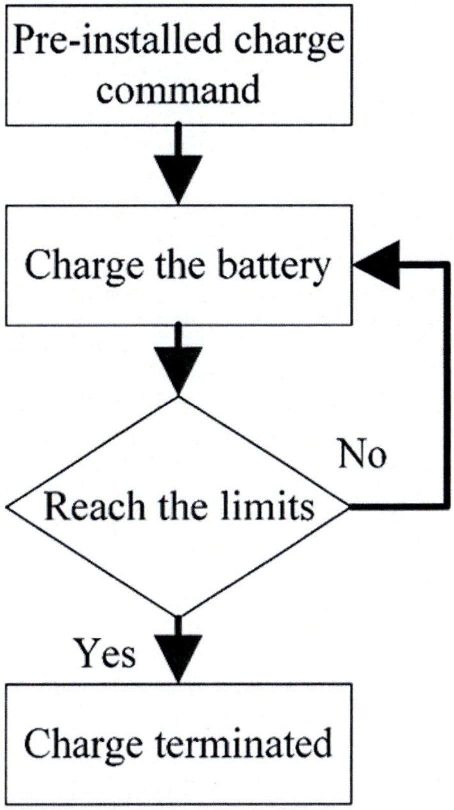

Fig. 10.1 Passive charging structure

The CC/CV algorithm charges the battery with constant current until the battery voltage reaches a predetermined maximum charging voltage, then the charging voltage is maintained constant until the current is decreased to a preset minimum value (Serhan and Ahmed 2018). The charging curve of the CC/CV is shown in Fig. 10.2.

Several variations of the CC/CV charging technique have been developed. A suggested multistage rapid charging profile is divided into three stages, illustrated in Fig. 10.3, referred to as CC-I, CC-II, and CV-I. The algorithm is based on how internal

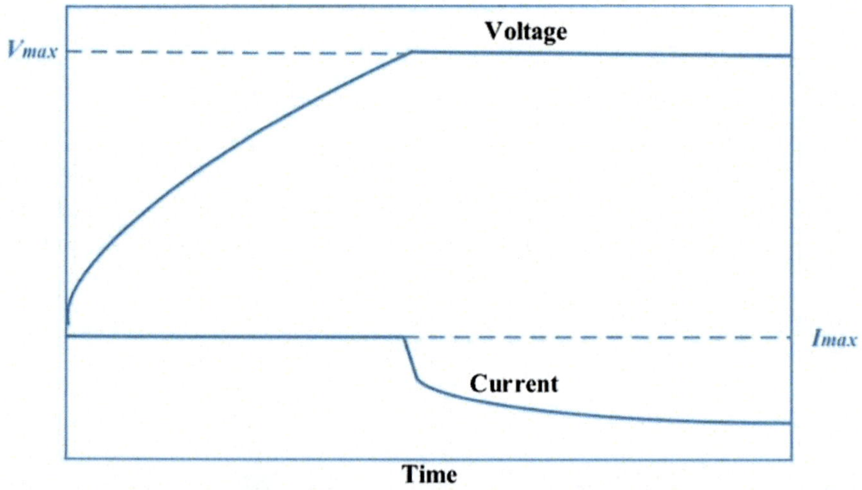

Fig. 10.2 Constant current-constant voltage curve (Gao et al. 2019)

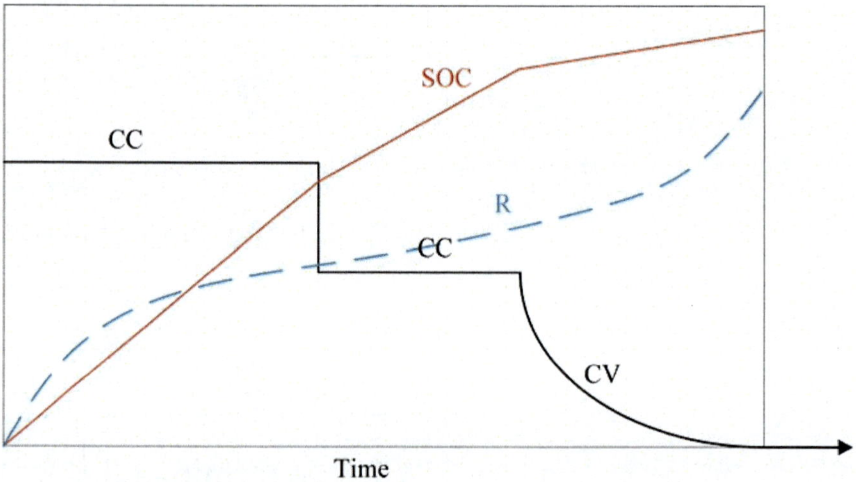

Fig. 10.3 Current, state of charge, resistance profiles of the fast charging technique

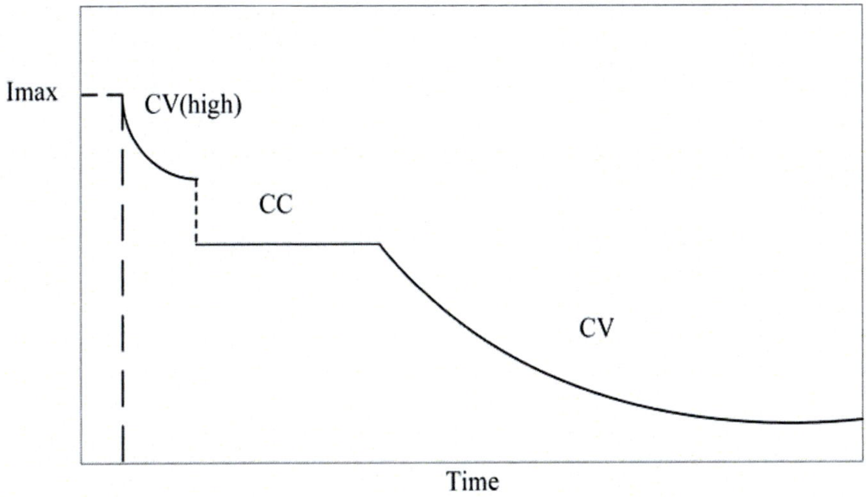

Fig. 10.4 Current profiles of boost-charging technique

resistance changes during charging (Anseán et al. 2016, 2013). The greatest current is applied due to the cell's reduced resistance in the lower SOC range. As the cell's internal resistance grows fast, the last two phases are employed. The experimental findings show that the technique is effective for preventing a significant temperature rise and increasing the cycle life.

In Fig. 10.4, another method is presented. A Vmax voltage is used to charge the battery in mode CV over a very brief period before a conventional CCCV process is taken (Notten et al. 2005). This boost-charging method assures that a totally drained battery may be recharged to one-third of its rated capacity without causing additional deterioration.

Some research also focuses on applying Taguchi-based approaches to optimal charging (Amanor-Boadu et al. 2018; Liu et al. 2011; Liu and Luo 2010). As shown in Fig. 10.5 (Liu et al. 2011), the entire charging process is composed of five stages of the constant current process. Compared to the conventional CC-CV method, this pattern can provide 57% more cycles and reduce 11% charging time.

10.2.2 Pulse Charge Based Charging Curve

Pulse charging has been touted as a quick and efficient charging method for lithium-ion batteries. (Smith et al. 2010; Smith and Wang 2006; Chen 2007).

Purushothaman et al. concluded that the side reactions produced by lithium saturation at the particle interface may be avoided by properly selecting the current waveform parameters (Purushothaman et al. 2005). In less than 3/4 h, a nonlinearly declining current density that corresponds to the mass transfer coefficient fluctuation

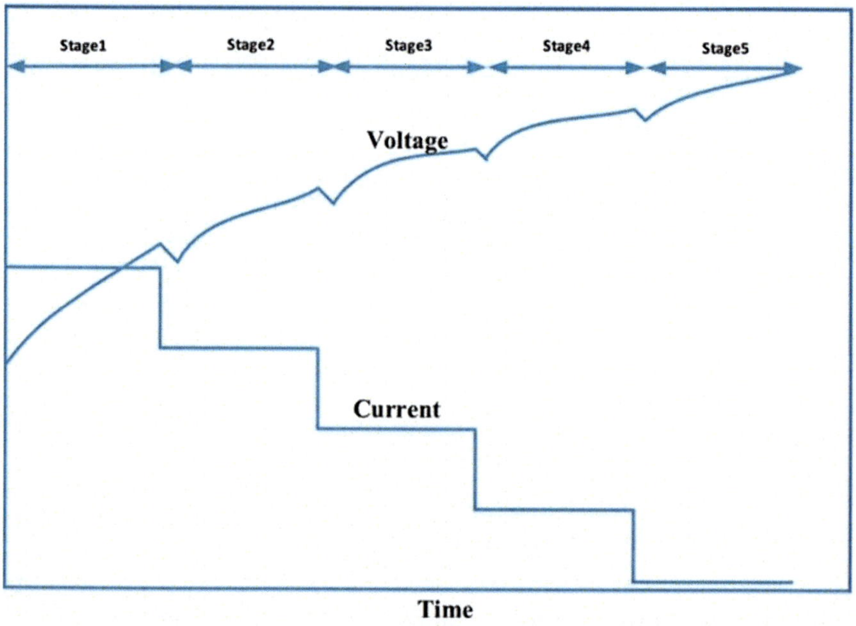

Fig. 10.5 Five-step constant current charging algorithm (Gao et al. 2019)

may enable complete charge (Purushothaman and Landau 2006). This approach, however, was based on computer analyses, and no comparable tests were conducted. Furthermore, following the fluctuation of the diffusion coefficient in real time proved computationally challenging. A state of charge (SOC) controlled rapid charging approach was utilized to charge the battery in a similar way, with the goal of minimizing parasitic reactions (Naznin 2013). The first charging step involves gradually increasing current pulses to compensate for the increased impedance of the battery at lower SOC levels. The charging amplitude gradually lowers at the end of the charging process to compensate for the battery's reduced charge uptake at higher SOC levels. The disadvantage of this method is that there isn't enough theory to back it up, and the pulse amplitude and breadth are picked at random. Based on the pulsed currents charging technique, Lu et al. showed the design of charging strategies for lithium-ion batteries in 2016, taking into account the balance between diffusion generated stress and overall charge time (Lu et al. 2016). The galvanostatic action is utilized first, followed by a potentiostatic operation in the two-stage charge techniques. In the three-stage charge techniques, two connected galvanostatic stages with differing currents were added, followed by a potentiostatic stage (Fig. 10.6).

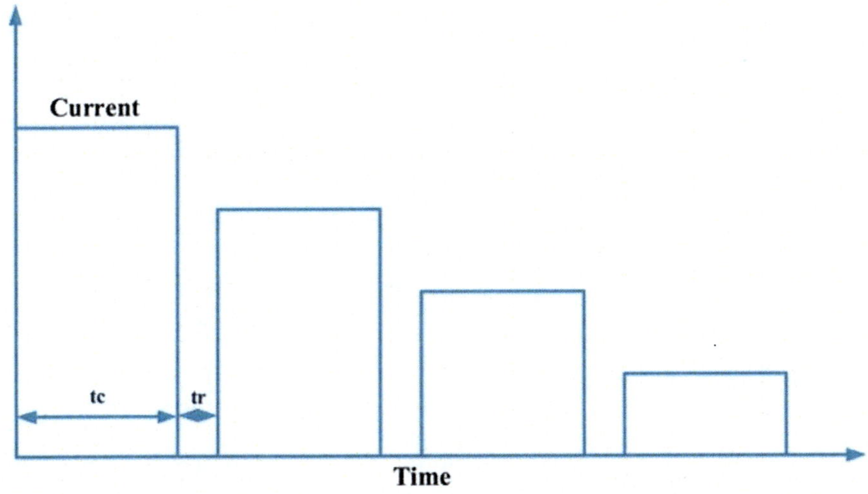

Fig. 10.6 Nonlinearly decreasing pulse charge technique (Gao et al. 2019)

10.2.3 Summary of the Reviewed Passive Optimal Charging

Table 10.1 summarizes the data and findings of the passive optimum charging protocols discussed in this work, including the battery type, charging methods, and parameters that are required to be improved. Their advantages and disadvantages were also discussed. Most research used commercial cylindrical batteries as the examined battery type for passive optimum charging techniques ranging from LMO to LFP, as shown in Table 10.1. The energy density of this cell is low, but the power density is great. The prismatic LP battery was also utilized for validation in Ref (Chen et al. 2018).

The variations of the conventional CC-CV approach are resistance evolution-based, diffusion stress-based, Taguchi-based, SOC controlled, and boost charging methods. The charging switching terms are the difference between them. The charging procedure changes with the internal resistance of the cell in the resistance-based technique. The ion concentration at the surface is seen as a trigger to initiate the next charging stage in the diffusion stress-based process. The charging process is divided into five CC phases using the Taguchi Orthogonal Arrays technique, and the current was tuned at each stage. The multi-stage charging technique guided by SOC attempted to determine the charging current based on SOC changes while charging. To attain the voltage threshold, a short burst of high current was given to the battery during boost charging. For the multi-stage charging method, the number of stages, current magnitude in each stage, and stage length were three key factors that needed to be optimized. Because of their model-free nature, these approaches were shown to have superior performance in terms of charging speed, cycle life, and power capability when compared to CC-CV charging. However, battery health indicators such as battery deterioration and temperature were not taken into account or closely

Table 10.1 Comparison of the reviewed passive charging strategies (Gao et al. 2019)

	Resistance evolution based three-stage CC/CV charging	Diffusion stress based charging	Taguchi-based five step charging	SOC governed multi-stage fast charging	Boost charging	Pulse charging based
Charging method	Step1:CC Step2:CC Step3:CV	Switching the CC mode to CV when $C_{surf} = 0$	Five steps of CC	Step 1: Multistage CC Step 2: Multistage CC-CV Step 3: Multistage CC Step 4: CV	High pulse currents followed by CC-CV mode	Nonlinearly decreasing charging current tracking the mass transfer coefficient evolution
Charging result	100% SOC within 20 min ($V_{max} =$ 3.6V $I_{max} = 4$ C)	100% SOC within 18 min $I_{max} = 8$ C	95% SOC within 130.7 min $I_{max} =$ 1.4 C	80% SOC in 41.6 min $I_{max} = 2$ C	30% SOC within 5 min $I_{max} = 4.5$ C	100% SOC in 43.2 min $I_{max} = 8$ C
Parameter need to be optimized	1. Number of the stage 2. Current magnitude 3. Voltage threshold	1. Minimum value of Li + concentration at particle surface 2. Number of stage	1.Number of stage 2. Current magnitude 3. Charging duration at each stage	1.Switching point of each charging stage	1. Voltage threshold at the initial pulse charging stage	1. Initial charging current 2. Current decreasing rate 3. Pulse frequency, magnitude and duty ratio
Advantages	1. Easy implementation 2. Extended cycle life 3. Invariable power capability	1. Accelerate the charging process and lower the peak stress	1.57% more cycles, 11.2% less charging time and improved charging efficiency by 1.02% compared to CCCV	1. Reducing the charging time and providing optimum battery performance and thermal management	1. Easy implementation 2. No significant impact on cycle life	1. The reactant concentration buildup at the electrode and the concentration over potential are minimized

(continued)

Table 10.1 (continued)

	Resistance evolution based three-stage CC/CV charging	Diffusion stress based charging	Taguchi-based five step charging	SOC governed multi-stage fast charging	Boost charging	Pulse charging based
Disadvantages	1. Charge rate is not optimized 2. Charge termination is based only on the cell voltage measurement 3. Temperature is not controlled	1. Battery degradation is not considered 2. Temperature is not controlled	1. Require very stable current and temperature, currents must be low	1. SOC during charging process needs to be estimated precisely	1. Charge rate is not optimized 2. Charge termination is based only on the cell voltage measurement 3. Temperature is not controlled	1. Hard to choose the proper parameters for pulse sequence 2. No corresponding experiment was carried out

monitored. The charging current waveform varies depending on the pulse frequency, duty ratio, and current amplitude in a nonlinear decreasing pulse charging profile. According to Table 10.1, the pulse-based charging technique necessitates online optimization of five parameters to attain optimal performance. As a result, among the studied passive optimum charging techniques, the calculation load on the controller is the highest. As a result, the implementation of a pulse-based charging method in real-world applications may be difficult. Existing passive charging approaches, to the best of our knowledge, are unable to meet the overall optimal charging aim in terms of implementation, charging time, and health-conscious criteria, necessitating the creation of an active charging algorithm.

10.3 Generalized Active Optimal Charging Structure

The most typical active battery charging management structure is made up of three key components: the battery model, state estimator, and model-based controller. A closed-loop control structure for the battery optimal charging method is constructed, taking process noise into account, as shown in Fig. 10.7.

In general, the battery model has been deliberately developed to be lower-order and simple to apply for excellent controllability (Amanor-Boadu et al. 2018; Lu et al. 2016; Zou et al. Aug. 2018). The goal of creating such a battery model is to replicate the real-world dynamics of a battery system under a certain loading current profile. The observer receives the model's output variables along with the noise vector as input. Given the battery's various state variables, many of which are unmeasurable, such as concentration and overpotential, to monitor the internal states of the battery system, a robust and effective model-based estimator is required (Zou et al. 2018). An active charging method is developed using an optimum control algorithm and a

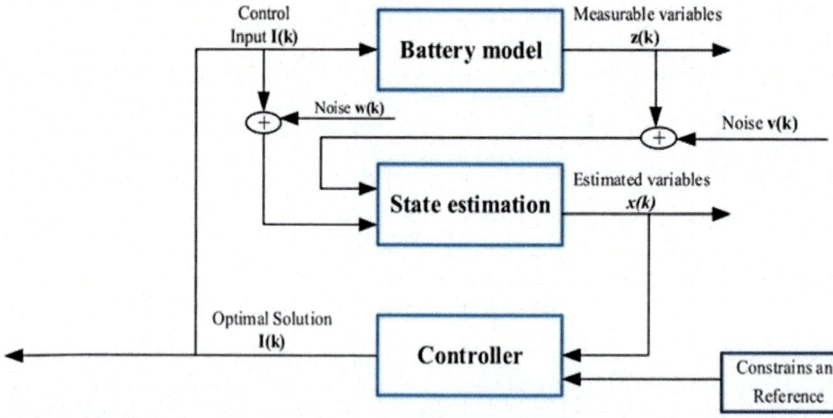

Fig. 10.7 Active optimal charging structure (Gao et al. 2019)

reduced-order battery model and a state estimator. Meanwhile, restrictions of battery health-related factors (temperature increase, side reaction rate, and so on) must be regarded as outputs references for improved charging performances.

10.3.1 Commonly-Used Battery Model

It is critical to rely on a quick and accurate real-time simulation on BMS for regulating and estimating battery conditions online (Cheng et al. 2011; Zou et al. 2015; Vašak and Kujundžić 2018). The broad classification of control-oriented Li-ion battery models is shown in Fig. 10.8.

Due to the benefit of rapid calculation, comparable circuit models have recently been popular in BMSs (Waag and Sauer 2013; Barillas et al. 2015; Perez et al. 2017) (Fig. 10.9).

Based on the first-order RC model, Hu et al. proposed a dual-objective optimum charging method for LiNMC and $LiFePO_4$ batteries (Hu et al. 2013). For the two types of batteries, the effects of the charging voltage threshold, temperature, and

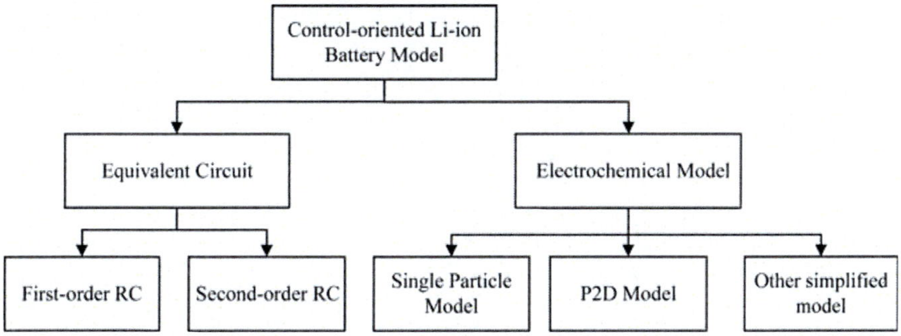

Fig. 10.8 Classification of commonly-used battery models

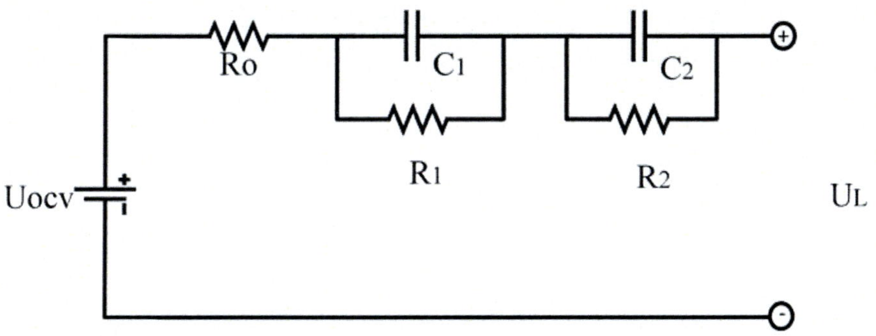

Fig. 10.9 Equivalent circuit model

health condition on the charging outcomes were investigated.

$$\frac{dz(t)}{dt} = \frac{\eta I(t)}{3600C_n} \tag{10.1}$$

$$\frac{dU(t)}{dt} = -\frac{U(t)}{\tau_1} + \frac{R_1}{\tau_1}I(t) \tag{10.2}$$

$$\frac{dh(t)}{dt} = -|\kappa I(t)|h(t) + |\kappa I(t)|H \tag{10.3}$$

$$V(t) = V_{oc}(z(t)) + R_0 I(t) + U(t) + h(t) \tag{10.4}$$

where I, U, Cn, z, and h are the current, output voltage, nominal capacity, SOC, and hysteresis voltage for $LiFePO_4$ cell, respectively.

Model-based charging management was proposed in 2017 (Zou et al. 2017) using an equivalent circuit consisting of an ideal voltage source, an internal resistor, and two resistor–capacitor (RC) pairs, as shown in Fig. 10.10. The average cell temperature is roughly comparable to the radial average temperature in battery thermal modeling:

$$T_a = \frac{1}{2}(T_s + T_c) \tag{10.5}$$

where T_s is the surface temperature, T_c is the core temperature and T_a is the average temperature. The governing equations for T_s and T_a are expressed as:

$$\frac{dT_s(t)}{dt} = \frac{T_f - T_s(t)}{R_u C_s} - \frac{2(T_s(t) - T_a(t))}{R_s C_s} \tag{10.6}$$

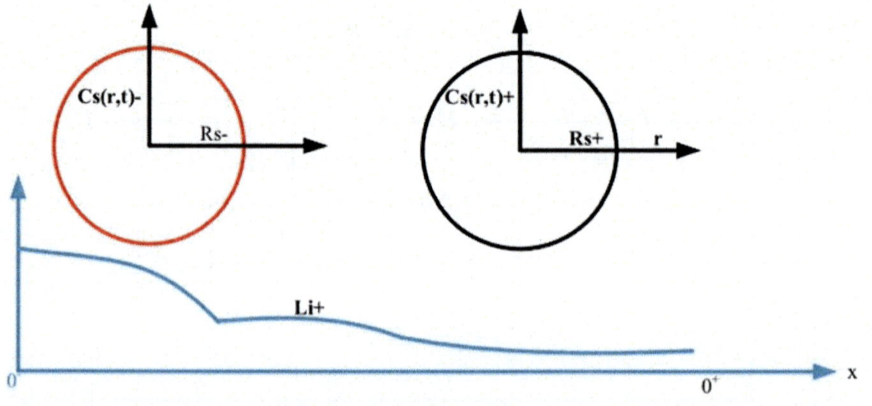

Fig. 10.10 Single particle model (Gao et al. 2019)

$$\frac{dT_a(t)}{dt} = \frac{T_f - T_s(t)}{2R_uC_s} + \frac{T_s(C_s - C_c) - T_a(C_s - C_c)}{R_sC_sC_c} + \frac{Q(t)}{2C_c} \tag{10.7}$$

where T_f is the ambient temperature, and R_c and R_u are the heat conduction and convection resistances, respectively. The surface and core heat capacities are denoted by C_s and C_c. However, its model parameters must be set first based on experimental results, and battery aging effects are not taken into account.

Because of their physical basis equations, these electrochemical-based models offer considerable benefits over equivalent circuit models (Zou et al. 2018; Lee et al. 2012; Han et al. 2015). The Partial-Two-Dimensional (P2D) model is undeniably accurate and rigorous (Ashwin et al. 2017; Ashwin et al. 2017). The Single-Particle-Model (SPM) is good for rapid reactions, but it's not good for modeling high (dis)charge rates (Jokar et al. 2016; Moura et al. 2017; Guo et al. 2011). Due to the shortcomings of the SPM and P2D models, simpler versions of the P2D model have been developed for use in battery charging control. Perez et al. used a linked single-particle model with electrolyte and thermal dynamics to establish the best rapid charging technique (Perez et al. 2017). The anode and cathode solid concentration dynamics were represented in the linked model using two PDE single-particle subsystems. A three-PDE electrolyte subsystem was used to measure the electrolyte content in three distinct domains (anode, separator cathode). The voltage output and solid/electrolyte dynamics were both affected by the temperature. The optimization issue is extremely nonlinear due to the combined electrochemical-thermal dynamics. Zou et al. have put a lot of effort into developing physics-based low-order battery models that can be used to simulate charging techniques (Zou et al. 2017). To model the battery dynamics as described in Eq. 8.11, a PDE-based SPM with electrolyte states was proposed in 2018. Three assumptions were used to decrease the model order (Cheng et al. 2011).

$$\dot{\bar{C}}_s^-(t) = -\frac{3}{FR_p^-a_s^-L^-}I(t) \tag{10.8}$$

$$\dot{\bar{C}}_s^+(t) = \left[n_s - L^-\varepsilon_s^-\bar{C}_s^-(t)\right]/\left(L^+\varepsilon_s^+\right) \tag{10.9}$$

$$C_{ss}^{\pm}(t) = \overline{C}_s^{\pm}(t) + \lambda_1\omega_1^{\pm}(t) + \lambda_2\omega_2^{\pm}(t) \tag{10.10}$$

$$\frac{dC_e^{\pm}(0,t)}{dt} = \frac{D_e^{\pm}\mu^{\pm}}{L^{\pm,2}\varepsilon_e^{\pm}}\left[C_e^{\pm}(0,t) - C_{e0}^{\pm}\right] \mp \frac{1 - t_c^0}{FL^{\pm}\varepsilon_e^{\pm}}I(t) \tag{10.11}$$

In order to account for aging effects, the electrolyte concentration C_e and the anode over-potential of side reactions η_s^- were incorporated in the output vector. The battery model's state-space function is as follows:

$$u(t) := I(t) \tag{10.12}$$

$$x(t) := \left[\overline{C_s^-}(t), \omega_1^-(t), \omega_2^-(t), \omega_1^+(t), \omega_2^+(t), C_e^-(0,t) - C_{e0}^-, C_e^+(0,t) - C_{e0}^+, T\right]^T \tag{10.13}$$

$$y(t) := \left[SOC(t), T(t), C_e^-(0,t), \eta_s^-(t)\right]^T \tag{10.14}$$

$$z(t) = [V(t), T(t)]^T \tag{10.15}$$

the system input vector is $u(t)$, $x(t)$ is the state vector, $z(t)$ is the measurable system output vector and controllable output vector is $y(t)$.

10.3.2 Classification of Optimal Charging Strategies

The temperature increase and aging rate are two characteristics of battery health-aware performance that are frequently linked to optimal charging objectives. As a result, the best charging strategy is devised by taking into account both battery performance and charging speed.

The battery aging characteristics are chosen from a subset of the battery's internal states. Side effects include the overpotential of side reactions at the cathode or anode electrodes, as well as the development of the solid electrolyte interface (SEI) δ_{SEI} (Choe et al. 2013; Subramanian et al. 2005; Zhou et al. 2017). C_e, the electrolyte concentration, is also thought to be a motivator for adverse effects (Zou et al. 2016; Yang et al. 2017; Fu et al. 2015). The lithium concentration at the surface of particles C_s is restricted to zero when considering the particle mechanical stress caused by diffusion (Lu et al. 2016; Perez et al. 2017). The state of health (SOH) is chosen as a deterioration representation based on empirical relationships between capacity loss and total discharge Ah throughout (Perez et al. 2017). The rapid temperature rise might cause thermal runaway in the battery. As a result, in most situations, battery temperature management is combined with aging prevention. (Zou et al. 2017, 2018).

To tackle the optimal charging problem, an effective charging strategy based on the battery model must be developed.

To improve charging profiles, Zou et al. used a linear time-varying (LTV) model predictive controller (MPC) (Zou et al. 2018). The rapid charging process was defined as a restricted LTV-MPC problem by employing a fast-moving horizon estimate (MHE) to detect battery internal states in the face of model mismatch, noise, and disturbances. Meanwhile, health-related constraints were factored into the solution. The following is the cost function for monitoring a reference trajectory SOC^r with input current $u(k)$ and associated total optimization current:

$$u^*(k) = \arg\ \min_{u(k),s} \sum_{i=0}^{N} \left\| y_1(k+i) - y_1^r(k+i) \right\|_Q^2 + \|\Delta u(k+i)\|_R^2 + \|s\|_p^2 \tag{10.16}$$

$$J(\boldsymbol{x}(k),\boldsymbol{u}(k)) = \sum_{i=0}^{N} \left\| y_1(k+i) - y_1^r(k+i) \right\|_Q^2 + \left\| \Delta u(k+i) \right\|_R^2 \tag{10.17}$$

$$\begin{aligned} & s.t. \forall i \in \{0, \cdots, N\}, \\ & \hat{x}(k+i+1) = A_k \hat{x}(k+i) + B_k u(k+i) + \hat{\varphi}(k) \\ & \hat{y}_1(k+i) = C\hat{\boldsymbol{x}}(\mathrm{k}+\mathrm{i}) + \mathrm{d} \\ & \hat{\boldsymbol{x}}(k) = \hat{\boldsymbol{x}}_{k|k} \\ & Hu(k+i) \leq L \\ & \varepsilon_k \hat{x}(k+i) + \boldsymbol{F}_k u(k+i) \leq S_k + s \\ & s \geq 0 \end{aligned} \tag{10.18}$$

MPC-based charging methods outperform their CCCV equivalents in terms of computational efficiency and charging rate. However, the recommended charging strategy first charged the battery at a maximum current of 15 C, and the cell voltage surpassed the top bound of 4.2 V during charging. By setting the weight factor to zero in Ref (Zou et al. 2017), the possible quickest charging mode of the MPC-based algorithm is examined. It is discovered that charging the examined battery from 10% SOC to the stated capacity level takes no less than 788 s.

Liu et al. used generalized predictive control (GPC) in conjunction with a suitable battery model to keep the internal temperature of the battery within a specific range during the charging process (Liu et al. 2017). As an online self-tuning prediction model for a GPC controller, a controlled auto-regressive integrated moving average (CARIMA) model was utilized. By minimizing a multistage cost function that included both battery charging time and energy loss, the predictive control sequence was created:

$$\begin{aligned} J = (1 - a_1) * t_f + a_1 * \int_{t=0}^{t=t_f} & i(t) * (V(t) - U_{ocv}(t)) \\ & + i(t) * T_{in}(t) * dU_{ocv}(t)/dT_{in}(t)dt \end{aligned} \tag{10.19}$$

The moment when the battery achieves its ultimate SOC level is denoted by t_f. The weighting factor for balancing the two objectives (charging time and energy loss) is $0 \leq a_1 \leq 1$. The primary disadvantage of the GPC-based control method is that numerous tuning parameters must be carefully set in advance, and the polynomial matrices are large, which complicates real-time implementation. Using the dynamic programming (DP) approach, Lin et al. established better charge schemes (Lin et al. 2018). Both charging time and battery deterioration were optimized and traded off. In the SOC domain, the cost function took the charging time t_f, SEI development δ_{SEI}, and lithium plating $\delta_{plating}$ into account:

$$min(I) \int_{SOClo}^{SOChi} \left(\alpha \cdot t_{char} + \beta \cdot \delta_{SEI} + \gamma \cdot \delta_{plating}\right) \tag{10.20}$$

The minimal time approach and health-conscious rapid charging method were studied by setting different weight factors (α, β, γ) to specified values. The charging time climbed to 29 min 7 s for the health-conscious method, while the minimal time strategy took 19 min 18 s to attain a goal SOC of 61% from 1.7% SOC. Unfortunately, the weight parameters for the recyclable lithium used in SEI development and lithium plating were randomly assigned equal weights.

To address the resultant nonlinear multi-state optimum charging issue, the Legendre–Gauss-Radau (LGR) pseudo-spectral approach with adaptive multi-mesh-interval collocation was used in Perez et al. (2017). The objective function J is defined as follows:

$$min_{I(t),x(t),t_f} \int_{t_0}^{t_f} 1 \cdot dt \tag{10.21}$$

where $(t_f - t_0)$ is the charging time required to achieve a given SOC. The input current $I(t)$ and final time t_f are the optimization variables, with state variables:

$$x(t) = \left[c_s^+(r,t), c_s^-(r,t), c_e^+(x,t), \times c_e^{sep}(x,t), c_e^-(x,t), T_c(t), T_s(t)\right]^T \tag{10.22}$$

Fang et al. suggested two charge techniques in 2017 (Fang et al. 2017) utilizing the linear quadratic control theory. One is based on linear quadratic control with a set charging rate at the terminal. The other is created by following a charging route that is used as a reference. The following is a linear quadratic state-feedback formulation for monitoring a reference trajectory r_N:

$$\begin{aligned} & min_{u_0,u_1,\cdots,u_{N-1}} \frac{1}{2}(x_N - r_N)^T S_N (x_N - r_N) \\ & + \frac{1}{2}\sum_{k=0}^{N-1}\left[(x_k - r_k)^T Q (x_k - r_k) + u_k^T R u_k\right] \\ & s.t. x_{k+1} = A x_k + B u_k, x_0 \end{aligned} \tag{10.23}$$

Trippe et al. (2014) published another variant of the quadratic constraint program in 2014, which reduced overall charging cost, charging energy cost, and battery aging cost. The linear objective function of the optimization problem is described by Eq. 10.18, where the variables to be improved are charging power P and battery aging cost c_{aging}. The electricity price is reflected by pr and Δt is the time step. n_{park} and t are indices for the number of parking events an automobile has and the time it takes to charge. The primary flaw of Annette's method is that it needs a massive amount of data on vehicle usage history (Trippe et al. 2014).

$$argmin \sum_{n_{park}} \sum_{t} \left(P\left(n_{park}, t\right) \cdot pr(t) \cdot \Delta t + c_{aging}\left(n_{park}, t\right)\right) \tag{10.24}$$

Pramanik and Anwar (Pramanik and Anwar 2016) proposed an optimum charging method based on the Pontryagins principle with both state and input constraints in 2016. When compared to normal CCCV charging, the suggested charging algorithm is capable of reducing the charging time while maintaining the temperature restriction. The performance index was established as follows:

$$P.I. = \int_{0}^{T_f} \left[\alpha(I_{max} - I(t))^2 + \beta(T_{max} - T(t))^2 + \delta I^2(t)\right]dt \tag{10.25}$$

The performance index was designed to save time and effort by keeping the current and bulk cell temperatures near their maximum values. The Hamiltonian function was created by combining the battery model with it:

$$\begin{aligned} H(C_s, T, \lambda_1, \lambda_2, t) &= \alpha(I_{max} - I(t))^2 + \beta(T_{max} - T(t))^2 \\ &+ \delta I^2 + \lambda_1\left(-\frac{6i_0}{RF} sinh\left(\frac{\alpha F}{RT}\eta(t)\right)\right) \\ &+ \lambda_2\left(\frac{1}{\rho^{avg} c_p}\right. \\ &\left.\left[h_{cell}(T_{max} - T(t)) + I(t)V(t) - \sum_{i=1}^{n}\left[\int_{0^-}^{0^+} \frac{3\varepsilon}{R} F J(t)\left(U_i(t) - T(t)\frac{\partial U}{\partial T}\right)dx\right]\right]\right) \end{aligned} \tag{10.26}$$

$$\lambda_1^* = \frac{\partial H(\cdot)}{\partial C_s} = 0 \tag{10.27}$$

$$\begin{aligned} \lambda_2^* = \frac{\partial H(\cdot)}{\partial T} &= 2\beta(T_{max} - T) + \lambda_2^* \frac{h}{\rho^{avg} c_p} - \lambda_2^* \frac{h}{\rho^{avg} c_p} \\ &\times \left[\sum_{i=0}^{n}\left[\int_{0^-}^{0^+} \frac{3\varepsilon}{R} F J(t)\left(U_i(t) - T(t)\frac{\partial U}{\partial T}\right)dx\right]\right] + \lambda_1^* \frac{\alpha F}{RT^2}\eta\left[\frac{6i_0}{RF} cosh\left(\frac{\alpha F}{RT}\eta\right)\right] \end{aligned} \tag{10.28}$$

in which α, β are tunable parameters that give the flexibility to tune the charging performance based on charging current and maximum rated limit. The ideal current trajectory at each time step was found by solving the equation:

$$I(t) = I_{max} - \lambda_2 \frac{V(t)}{\rho^{avg} c_p} \times \frac{1}{2\alpha} \tag{10.29}$$

10.3.3 Summary and Comparison

To better understand the links and differences between the evaluated optimal charging techniques, we'll look at them from two perspectives: the internal battery model and optimal control algorithms.

The commonly-used battery types are $LiFePO_4$, $LiCoO_2$, and NMC. To assess the internal states of batteries, a variety of battery models are used. Due to its simple circuitry description and ease of configuration and identification of parameters, the first-order RC (Liu et al. 2017) and second-order RC (Fang et al. 2017) models have been utilized in control and optimum applications. In battery optimum charging, the SPM is the most commonly used electrochemical model. Both are frequently associated with thermal and aging dynamics, which are critical for controlling and optimizing charging curves. For similar circuit models, the thermal impact is usually integrated into the controller using a two-stage approximation of the radially distributed thermal model (Perez et al. 2017). However, it is thought that the energy wasted by the electrode represents the Li-ion contribution and has an influence on overall heat output (Ye et al. 2012; Xia et al. 2017; Edouard et al. 2016). For the analogous second-order RC model, an empirical equation between input current and capacity fade is used to show the aging effects (Perez et al. 2017). To prevent the battery from degrading, the Li-ion concentration in both the solid and electrolyte phases, as well as the overpotential of side reactions, are regulated within a restricted range in SPM. In terms of SOC estimate, the equivalent circuit model uses coulomb counting to compute it, whereas the SPM uses the Li-ion concentration distribution and integration to forecast it (Perez et al. 2017). Methods such as electrochemical models and equivalent models work well, but they can't be applied to other batteries directly (technology, design, materials) (Barré et al. 2013; de las Casas and Li 2012; Etacheri et al. 2011). Furthermore, these two methods are ineffective in modeling all deterioration mechanisms that occur during the battery's lifetime (Li et al. 2014; Verma et al. 2010; Castaing et al. 2014).

Every technique described tries to address the optimal charging problem in a different way, but each has its own set of drawbacks. Most studies charge the battery with the maximum current (up to 6C) (Lin et al. 2018), which puts a lot of strain on the power electronics (Appiah et al. 2016; Agubra and Fergus 2013; Baghdadi et al. 2016). Furthermore, several studies have used accelerated life tests to confirm the efficiency of optimum charging (Trippe et al. 2014), but this technique has one major flaw. A test bench is typically used for an accelerated life test. As a result, the influence of all environmental factors that occur in real-life situations is not considered, resulting in certain inaccuracies. Each of the algorithms given is capable of finding a balanced solution for the two competing objectives of charging time and battery health. It is clear that the MPC-based control approach is the most often used method (Zou et al. 2017, 2018; Ito et al. 2018; Liu et al. 2017). The popularity of MPC methods can be attributed to two factors. For starters, its ability to handle constraints and nonlinearities, as well as model-based optimum or suboptimal control, makes it relevant to a wide range of industrial issues (Vazquez et al. 2017). Second, it has a high level of resilience and stability in the face of a wide range of sounds (Mayne 2014; Vazquez et al. 2014). However, it has two limitations: (1) no hardware-in-the-loop experiment was performed to test the method; and (2) the suggested MPC framework's sensitivity to parameter uncertainties, such as weighting matrices and starting values, was not investigated. Because of the problem's strong nonlinearity, the electrolyte dynamics approximation, SEI, and plating static map were created to allow DP-based charging optimization (Lin et al. 2018). The problem here is that the DP process has made it difficult to manage large amounts of data in the controller. This infinite-dimensional optimum control issue was transcribed into a finite-dimensional optimization problem with algebraic constraints at the discretized nodes using the Legendre–Gauss-Radau (LGR) pseudo-spectral based technique (Perez et al. 2017), but the convexity and convergence were not guaranteed. The algorithm calculates the states and co-state values to provide the rectified input current at each iteration for Pontryagins minimal principle-based approach (Pramanik and Anwar 2016). As a result, exact beginning values for state variables were necessary. Because of their remarkable simplicity, linear optimization-based techniques (Fang et al. 2017) have higher computational appeal in terms of time and space complexity. The selection of the gain matrix, on the other hand, is a multifaceted problem since it must account for both battery health protection and charging speed, as well as the economic cost and user happiness more widely.

10.4 Suggestions and Challenges

Based on the foregoing assessment of the available literature on optimum battery charging techniques, it appears that closed-loop and model-based charging is the most promising choice for health-aware optimal charging. For this, you'll need two things: 1. A battery model that is both simple and controllable. 2. A computationally efficient and well-fitting optimization technique for the battery model. As a result, the following is a list of concerns related to the two elements that should be addressed in future studies.

10.4.1 Suggestions for the Control-Oriented Battery Models

1. The electrochemistry-based battery model can reproduce battery dynamics with great fidelity, which is crucial in health-conscious charging protocols. Most research used side reaction-related criteria as a part of the cost function for

battery aging prevention (Zou et al. 2018; Chaturvedi et al. 2010). For example, to prevent deterioration, the overpotential of side reactions is kept below zero (Zou et al. 2018). To the authors' knowledge, however, the aging process of battery charging has not been fully explored. What are the ramifications of choosing higher or lower bonds? Which parameter, other than SEI growth, should be computed and taken into account for a more optimum curve?

2. As a battery ages, its characteristics alter (Pop et al. 2007; Börner et al. 2017; Tanim and Rahn 2015; Cheng et al. 2015). As a result, depending on the input historical data such as current, voltage, and temperature, the real-time controller should adjust battery settings. Finally, the performance of online monitoring, battery control, and optimization will be improved.

10.4.2 Suggestions for the Charging Optimization Algorithms

1. The majority of the optimum control techniques discussed in this work are implemented by addressing the optimal problem using a set of new performance indices and restricted linear matrix inequalities. The expense of hardware implementation, on the other hand, is rarely considered. Indeed, improving charging performance necessitates a considerable increase in computing overhead (Andre et al. 2013; Cai and White 2010; Hu et al. 2012). As a result, developing a simple and quick optimum method that can be successfully replicated in BMSs is critical.
2. The evaluated best charging technique may not be directly applicable to other batteries (technology, design, materials) due to the great diversity in battery kinds and sizes due to the wide variety of applications (Bahiraei et al. 2017; Dubarry et al. 2011; Jin et al. 2017). Developing universal optimum charging techniques is thus important. This feature may be accomplished by using a large data-driven method or by using the controller to autodetect the battery chemical responses when they are skewed to an external signal (Nuhic et al. 2013).

10.4.3 Charging Strategies at Low Temperatures

1. Due to Li-ion batteries' poor low-temperature performance (Jaguemont et al. 2016), fast charging at subzero temperatures is a difficult operation. At low temperatures, Li plating is thought to be the main undesirable side reaction (Tippmann et al. 2014). Though some efforts have been made to address these issues. To reduce battery deterioration due to Li plating, a two-phase charging technique with constant current (CC) and constant voltage (CV) phases is given (Remmlinger et al. May 2014). The cell is charged above a temperature that prevents Li plating by adding a fast internal heating step before charging (Yang et al. 2018). The research on the topic is still lacking. It's critical to look at

the electrochemical model properties and aging mechanisms of batteries at low temperatures to come up with the best charging techniques.

10.4.4 Conclusion

In recent years, there has been a lot of study and development on battery charging techniques. This chapter provided an in-depth analysis of the most recent optimum charging techniques for commercial lithium-ion batteries. Passive and active controllers are two types of controllers that are widely classified based on their mathematical model and embedded structure. Each one is thoroughly discussed, complete with benefits, drawbacks, and examples. While passive charging techniques are straightforward to apply, it is apparent that they are not the best way to optimize charging speed while minimizing degradation. When sounds arise, they are unable to retain a high level of resilience. The active charging methodology will increase the battery's performance as well as the charger's efficiency. The cost function incorporates several battery status variables, which is critical for developing a battery model-based charging process. Active charging controllers are implemented utilizing a variety of best practices that have been outlined. According to the findings, MPC-based strategies are the most popular and promising ways for determining the optimal charging problem. The current MPC technique can only be used in linear systems, which means that the full-order battery model must be adequately simplified before being simulated. As a result, the control's precision may be compromised. In future research, the following features of optimum charging techniques should be addressed: (1) To account for the overall aging effects of the charging process; (2) To assess the precision of the embedded battery model under the proposed charging current; (3) To update battery parameters in real-time; (4) To investigate the cost of implementing the proposed charging algorithm in BMS hardware; (5) To propose universal optimal charging strategies; (6) To provide light on how to charge a battery efficiently at low temperatures.

References

Abdel-Monem M, Trad K, Omar N, Hegazy O, van den Bossche P, van Mierlo J (2017) Influence analysis of static and dynamic fast-charging current profiles on ageing performance of commercial lithium ion batteries. Energy 120:179–191

Agubra V, Fergus J (2013) Lithium ion battery anode aging mechanisms. Materials 6(4):1310–1325

Amanor-Boadu JM, Guiseppi-Elie A, Sanchez-Sinencio E (2018) Search for optimal pulse charging parameters for Li-ion polymer batteries using Taguchi orthogonal arrays. IEEE Trans Ind Electron 65(11):8982–8992

Amanor-Boadu JM, Abouzied MA, Sánchez-Sinencio E (2018) An efficient and fast li-ion battery charging system using energy harvesting or conventional sources. IEEE Trans Ind Electron 65(9):7383–7394

Andre D, Appel C, Soczka-Guth T, Sauer DU (2013) Advanced mathematical methods of SOC and SOH estimation for lithium ion batteries. J Power Sources 224:20–27

Anseán D, González M, Viera JC, García VM, Blanco C, Valledor M (2013) Fast charging technique for high power lithium iron phosphate batteries: a cycle life analysis. J Power Sources 239:9–15

Anseán D et al (2016) Fast charging technique for high power $LiFePO_4$ batteries: a mechanistic analysis of aging. J Power Sources 321:201–209

Appiah WA, Park J, Byun S, Ryou M-H, Lee YM (2016) A mathematical model for cyclic aging of spinel $LiMn_2O_4$/graphite lithium ion cells. J Electrochem Soci 163(13):A2757–A2767

Ashwin TR, McGordon A, Jennings PA (2017) Electrochemical modelling of Li-ion battery pack with constant voltage cycling. J Power Sources 341:327–339

Ashwin TR, McGordon A, Widanage WD, Jennings PA (2017) Modified electrochemical parameter estimation of NCR18650BD battery using implicit finite volume method. J Power Sources 341:387–395

Baghdadi I, Briat O, Delétage J-Y, Gyan P, Vinassa J-M (2016) Lithium battery aging model based on Dakin's degradation approach. J Power Sources 325:273–285

Bahiraei F, Ghalkhani M, Fartaj A, Nazri G-A (2017) A pseudo 3D electrochemical-thermal modeling and analysis of a lithium ion battery for electric vehicle thermal management applications. Appl Therm Eng 125:904–918

Barillas JK, Li J, Günther C, Danzer MA (2015) A comparative study and validation of state estimation algorithms for Li-ion batteries in battery management systems. Appl Energy 155:455–462

Barré A, Deguilhem B, Grolleau S, Gérard M, Suard F, Riu D (2013) A review on lithium ion battery ageing mechanisms and estimations for automotive applications. J Power Sources 241:680–689

Börner M et al (2017) Correlation of aging and thermal stability of commercial 18650-type lithium ion batteries. J Power Sources 342:382–392

de las Casas C, Li W (2012) A review of application of carbon nanotubes for lithium ion battery anode material. J Power Sources 208:74–85

Cai L, White RE (2010) An efficient electrochemical–thermal model for a lithium ion cell by using the proper orthogonal decomposition method. J Electrochem Soc 157(11):A1188

Campestrini C, Kosch S, Jossen A (2017) Influence of change in open circuit voltage on the state of charge estimation with an extended Kalman filter. J. Energy Storage 12:149–156

Castaing R et al (2014) Degradation diagnosis of aged $Li_4Ti_5O1_2$/$LiFePO_4$ batteries. J Power Sources 267:744–752

Chandrasekaran R (2014) Quantification of bottlenecks to fast charging of lithium ion-insertion cells for electric vehicles. J Power Sources 271:622–632

Chaturvedi NA, Klein R, Christensen J, Ahmed J, Kojic A (2010) Algorithms for advanced battery-management systems. IEEE Control Syst Mag 30(3):49–68

Chen LR (2007) A design of an optimal battery pulse charge system by frequency-varied technique. IEEE Trans Ind Electron 54(1):398–405

Chen L-R, Wu S-L, Shieh D-T, Chen T-R (2013) Sinusoidal-ripple-current charging strategy and optimal charging frequency study for li-ion batteries. IEEE Trans Ind Electron 60(1):88–97

Chen C, Shang F, Salameh M, Krishnamurthy M (2018) Challenges and advancements in fast charging solutions for EVs: a technological review. In: Proceedings ITEC, Long Beach, CA, USA, June 2018, pp 695–701

Cheng KWE, Divakar BP, Wu H, Kai D, Ho HF (2011) Batterymanagement system (BMS) and SOC development for electrical vehicles. IEEE Trans Veh Technol 60(1):76–88

Cheng M-W, Lee Y-S, Liu M, Sun C-C (2015) State-of-charge estimation with aging effect and correction for lithium ion battery. IET Elect Syst Transp 5(2):70–76

Choe S-Y, Li X, Xiao M (2013) Fast charging method based on estimation of ion concentrations using a reduced order of electrochemical thermal model for lithium ion polymer battery. In: Proceedings WEVA, Barcelona, Spain, pp 782–792

Docimo DJ, Fathy HK Analysis and control of charge and temperature imbalance within a lithium ion battery pack. IEEE Trans Control Syst Technol to be published

Dubarry M et al (2011) Identifying battery aging mechanisms in large format Li ion cells. J Power Sources 196(7):3420–3425

Edouard C, Petit M, Forgez C, Bernard J, Revel R (2016) Parameter sensitivity analysis of a simplified electrochemical and thermal model for Li-ion batteries aging. J Power Sources 325:482–494

Etacheri V, Marom R, Elazari R, Salitra G, Aurbach D (2011) Challenges in the development of advanced Li-ion batteries: a review. Energy Environ Sci 4(9):3243–3262

Fadaee M, Radzi MAM (2012) Multi-objective optimization of a standalone hybrid renewable energy system by using evolutionary algorithms: a review. Renew Sustain Energy Rev 16(5):3364–3369

Fang H, Wang Y, Chen J (2017) Health-aware and user-involved battery charging management for electric vehicles: Linear quadratic strategies. IEEE Trans Control Syst Technol 25(3):911–923

Fu R, Choe S-Y, Agubra V, Fergus J (2015) Development of a physics-based degradation model for lithium ion polymer batteries considering side reactions. J Power Sources 278:506–521

Gao Y, Zhang X, Cheng Q, Guo B, Yang J (2019) Classification and review of the charging strategies for commercial lithium-ion batteries. IEEE Access 7:43511–43524

Guo M, Sikha G, White RE (2011) Single-particle model for a lithium ion cell: thermal behavior. J Electrochem Soc 158(2):A122–A132

Han X, Ouyang M, Lu L, Li J (2015) Simplification of physics-based electrochemical model for lithium ion battery on electric vehicle. Part I: diffusion simplification and single particle model. J Power Sources 278:802–813

Hsieh G-C, Chen L-R, Huang K-S (2011) Fuzzy-controlled Li-ion battery charge system with active state-of-charge controller. IEEE Trans Ind Electron 48(3):585–593

Hu X, Li S, Peng H (2012) A comparative study of equivalent circuit models for Li-ion batteries. J Power Sources 198:359–367

Hu X, Li S, Peng H, Sun F (2013) Charging time and loss optimization for LiNMC and $LiFePO_4$ batteries based on equivalent circuit models. J Power Sources 239:449–457

Hu X, Cao D, Egardt B (2018) Condition monitoring in advanced battery management systems: moving horizon estimation using a reduced electrochemical model. IEEE/ASME Trans Mechatronics 23(1):167–178

Hussein A-H, Batarseh I (2011) A review of charging algorithms for nickel and lithium battery chargers. IEEE Trans Veh Technol 60(3):830–838

Ito A, Kawashima A, Suzuki T, Inagaki S, Yamaguchi T, Zhou Z (2018) Model predictive charging control of in-vehicle batteries for home energy management based on vehicle state prediction. IEEE Trans Control Syst Technol 26(1):51–64

Jaguemont J, Boulon L, Dubé Y (2016) A comprehensive review of lithium ion batteries used in hybrid and electric vehicles at cold temperatures. Appl Energy 164:99–114

Jannesar MR, Sedighi A, Savaghebi M, Guerrero JM (2018) Optimal placement, sizing, and daily charge/discharge of battery energy storage in low voltage distribution network with high photovoltaic penetration. Appl Energy 226:957–966

Jin X et al (2017) Physically-based reduced-order capacity loss model for graphite anodes in Li-ion battery cell. J Power Sources 342:750–761

Jokar A, Rajabloo B, Désilets M, Lacroix M (2016) Review of simplified pseudo-two-dimensional models of lithium ion batteries. J Power Sources 327:44–55

Keyser M et al (2017) 'Enabling fast charging—battery thermal considerations. J Power Sources 367:228–236

Khan AB, Choi W (2018) Optimal charge pattern for the high performance multistage constant current charge method for the Li-ion batteries. IEEE Trans Energy Convers 33(3):1132–1140

Kujundžić G, Ileš Š, Matuško J, Vašak M (2017) Optimal charging of valve-regulated lead-acid batteries based on model predictive control. Appl Energy 187:189–202

Lee JL, Chemistruck A, Plett GL (2012) One-dimensional physics-based reduced-order model of lithium ion dynamics. J Power Sources 220:430–448

Li Z, Huang J, Liaw BY, Metzler V, Zhang J (2014) A review of lithium deposition in lithium ion and lithium metal secondary batteries. J Power Sources 254:168–182
Lin X, Hao X, Liu Z, Jia W (2018) Health-conscious fast charging of Li-ion batteries via a single particle model with aging mechanisms. J Power Sources 400:305–316
Liu Y-H, Luo Y-F (2010) Search for an optimal rapid-charging pattern for Li-ion batteries using the Taguchi approach. IEEE Trans Ind Electron 57(12):3963–3971
Liu Y-H, Hsieh C-H, Luo Y-F (2011) Search for an optimal five-step charging pattern for Li-ion batteries using consecutive orthogonal arrays. IEEE Trans Energy Convers 26(2):654–661
Liu K, Li K, Zhang C (2017) Constrained generalized predictive control of battery charging process based on a coupled thermoelectric model. J Power Sources 347:145–158
Liu K, Li K, Yang Z, Zhang C, Deng J (2017) An advanced Lithium ion battery optimal charging strategy based on a coupled thermoelectric model. Electrochim Acta 225:330–344
Liu J, Li G, Fathy HK (2017) An extended differential flatness approach for the health-conscious nonlinear model predictive control of lithium ion batteries. IEEE Trans Control Syst Technol 25(5):1882–1889
Lu L, Han X, Li J, Hua J, Ouyang M (2013) A review on the key issues for lithium ion battery management in electric vehicles. J Power Sources 226:272–288
Lu B, Song Y, Zhang J (2016) Selection of charge methods for lithium ion batteries by considering diffusion induced stress and charge time. J Power Sources 320:104–110
Mayne DQ (2014) Model predictive control: recent developments and future promise. Automatica 50(12):2967–2986
Meintz A et al (2017) 'Enabling fast charging—vehicle considerations. J Power Sources 367:216–227
Moura SJ, Argomedo FB, Klein R, Mirtabatabaei A, Krstic M (2017) Battery state estimation for a single particle model with electrolyte dynamics. IEEE Trans Control Syst Technol 25(2):453–468
Naznin F (2013) State of charge (SOC) governed fast charging method for lithium-based batteries. In: Proceedings COMSOL conference, Bangalore, pp 1–7
Notten PHL, Veld JHGOH, van Beek JRG (2005) Boostcharging Li-ion batteries: a challenging new charging concept. J Power Sources 145(1):89–94
Nuhic A, Terzimehic T, Soczka-Guth T, Buchholz M, Dietmayer K (2013) Health diagnosis and remaining useful life prognostics of lithium ion batteries using data-driven methods. J Power Sources 239:680–688
Parvini Y, Vahidi A, Fayazi SA (2018) Heuristic versus optimal charging of supercapacitors, lithium ion, and lead-acid batteries: an efficiency point of view. IEEE Trans Control Syst Technol 26(1):167–180
Patnaik L, Praneeth AVJS, Williamson SS (2019) A closed-loop constant-temperature constant-voltage charging technique to reduce charge time of lithium ion batteries. IEEE Trans Ind Electron 66(2):1059–1067
Perez HE, Hu X, Dey S, Moura SJ (2017) Optimal charging of Li-ion batteries with coupled electro-thermal-aging dynamics. IEEE Trans Veh Technol 66(9):7761–7770
Perez HE, Dey S, Hu X, Moura SJ (2017) Optimal charging of Li-ion batteries via a single particle model with electrolyte and thermal dynamics. J Electrochem Soc 164(7):A1679–A1687
Pop V, Bergveld HJ, Regtien PPL, Veld JHGOH, Danilov D, Notten PHL (2007) Battery aging and its influence on the electromotive force. J Electrochem Soc 154(8):A744–A750
Pramanik S, Anwar S (2016) Electrochemical model-based charge optimization for lithium ion batteries. J Power Sources 313:164–177
Purushothaman BK, Landau U (2006) Rapid charging of lithium ion batteries using pulsed currents a theoretical analysis. J Elctrochem Soc 153(3):A533–A542
Purushothaman BK, Morrison PW, Landau U (2005) Reducing masstransport limitations by application of special pulsed current modes. J Elctrochem Soc 152(4):J33–J39
Remmlinger J, Tippmann S, Buchholz M, Dietmayer K (2014) Low-temperature charging of lithium ion cells part II: model reduction and application. J Power Sources 254:268–276

Serhan HA, Ahmed EM (2018) Effect of the different charging techniques on battery life-time: review. In: Proceedings ITCE, Aswan, Egypt, Feb 2018, pp 421–426
Smith K, Wang C-Y (2006) Solid-state diffusion limitations on pulse operation of a lithium ion cell for hybrid electric vehicles. J Power Sources 161(1):628–639
Smith KA, Rahn CD, Wang C-Y (2010) Model-based electrochemical estimation and constraint management for pulse operation of lithium ion batteries. IEEE Trans Control Syst Technol 18(3):654–663
Subramanian VR, Diwakar VD, Tapriyal D (2005) Efficient macro-micro scale coupled modeling of batteries. J Electrochem Soc 152(10):A2002–A2008
Sun X-H, Yamamoto T, Morikawa T (2016) Fast-charging station choice behavior among battery electric vehicle users. Transp Res D, Trans Environ 46:26–39
Tanim TR, Rahn CD (2015) Aging formula for lithium ion batteries with solid electrolyte interphase layer growth. J Power Sources 294:239–247
Tippmann S, Walper D, Balboa L, Spier B, Bessler WG (2014) Low-temperature charging of lithium ion cells part I: electrochemical modeling and experimental investigation of degradation behavior. J Power Sources 252:305–316
Trippe AE, Arunachala R, Massier T, Jossen A, Hamacher T (2014) Charging optimization of battery electric vehicles including cycle battery aging. In: Proceedings IEEE PES innovation smart grid technology Europe, Istanbul, Turkey, October 2014, pp 1–6
Vašak M, Kujundžić G (2018) A battery management system for efficient adherence to energy exchange commands under longevity constraints. IEEE Trans Ind Appl 54(4):3019–3033
Vazquez S et al (2014) Model predictive control: a review of its applications in power electronics. IEEE Ind Electron Mag 8(1):16–31
Vazquez S, Rodriguez J, Rivera M, Franquelo LG, Norambuena M (2017) Model predictive control for power converters and drives: advances and trends. IEEE Trans Ind Electron 64(2):935–947
Verma P, Maire P, Novák P (2010) A review of the features and analyses of the solid electrolyte interphase in Li-ion batteries. Electrochim Acta 55(22):6332–6341
Vo TT, Chen X, Shen W, Kapoor A (2015) New charging strategy for lithium ion batteries based on the integration of Taguchi method and state of charge estimation. J Power Sources 273:413–422
Waag W, Sauer DU (2013) Adaptive estimation of the electromotive force of the lithium ion battery after current interruption for an accurate state-of-charge and capacity determination. Appl Energy 111(4):416–427
Waldmann T, Hogg B-I, Wohlfahrt-Mehrens M (2018) Li plating as unwanted side reaction in commercial Li-ion cells—a review. J Power Sources 384:107–124
Wei Z, Zou C, Leng F, Soong BH, Tseng K-J (2018) Online model identification and state-of-charge estimate for lithium ion battery with a recursive total least squares-based observer. IEEE Trans Ind Electron 65(2):1336–1346
Xia G, Cao L, Bi G (2017) A review on battery thermal management in electric vehicle application. J Power Sources 367:90–105
Xiong R, Yu Q, Wang LY, Lin C (2017) A novel method to obtain the open-circuit voltage for the state of charge of lithium ion batteries in electric vehicles by using H infinity filter. Appl Energy 207:346–353
Xiong R, Zhang Y, He H, Zhou X, Pecht M (2018) A double-scale, particle-filtering, energy state prediction algorithm for lithium-ion batteries. IEEE Trans Ind Electron 65(2):1526–1538
Yang X-G, Leng Y, Zhang G, Ge S, Wang C-Y (2017) Modeling of lithium plating induced aging of lithium ion batteries: transition from linear to nonlinear aging. J Power Sources 360:28–40
Yang XG, Zhang G, Ge S, Wang CY (2018) Fast charging of lithium ion batteries at all temperatures. Proc Nat Acad Sci USA 115(28):7266–7271
Ye Y, Shi Y, Tay AAO (2012) Electro-thermal cycle life model for lithium iron phosphate battery. J Power Sources 217:509–518
Zhang C, Jiang J, Gao Y, Zhang W, Liu Q, Hu X (2017) Charging optimization in lithium ion batteries based on temperature rise and charge time. Appl Energy 194:569–577

Zhou X, Bernstein DS, Stein JL, Ersal T (2017) Battery state of health monitoring by estimation of side reaction current density via retrospective-cost subsystem identification. J Dyn Syst Meas Control 139(9):091007

Zou Y, Hu X, Ma H, Li SE (2015) Combined state of charge and state of health estimation over lithium ion battery cell cycle lifespan for electric vehicles. J Power Sour 273:793–803

Zou C, Manzie C, Nešić D, Kallapur AG (2016) Multi-time-scale observer design for state-of-charge and state-of-health of a lithium ion battery. J Power Sources 335:121–130

Zou C, Hu X, Wei Z, Tang X (2017) Electrothermal dynamics-conscious lithium ion battery cell-level charging management via state-monitored predictive control. Energy 141:250–259

Zou C, Hu X, Wei Z, Wik T, Egardt B (2018) Electrochemical estimation and control for Lithium ion battery health-aware fast charging. IEEE Trans Ind Electron 65(8):6635–6645

Zou C, Manzie C, Nešić D (2018) Model predictive control for lithiumion battery optimal charging. IEEE/ASME Trans Mechatron 23(2):947–957

Chapter 11
The Future of EV Wireless Power Transfer Technologies

Since WPT was commercialized in the 1980s, WPT has entered a rapid development period in recent years (Covic and Boys 2013). Thirty years later, especially facing the shortage of global resources and environmental declines, WPT has achieved unprecedented focus, the development of that could be a key to solve the tricky problem (Rim and Mi 2017; Hui et al. 2014). The future is prospering, a summary of trends of WPT development is presented in Fig. 11.1, and more details are discussed as follows:

1. High efficiency

   Essentially, WPT is an energy transmitting device, efficiency is one of the basic criteria to describe the system's performance (Zhang et al. 2019). In general, when it comes to efficiency, the term "system efficiency" is often involved. According to the internationally recognized definition, the system efficiency is always measured as the output power of the dc-load divided by that of dc-input at the inverter (Deng et al. 2017).

   Higher efficiency means better system performance, which is also the development trend of WPT. However, compared to plug-in charging, the system efficiency of the WPT system is 5 ~ 6% lower (Mayordomo et al. 2013). Especially when the distance between the onboard receiving coil (secondary side coil) and the ground transmitting coil (primary side coil) is far or offset, the system efficiency is lower.

   Keep the high efficiency not only in alignment but also misalignment is the issue that would be done for some time to come.
2. High transmitted power

   Among the evaluation criteria of WPT, one of the most significant criteria is transmitted power capability (Kazmierkowski and Moradewicz 2012; Sample et al. 2011). The lower transmitted power, the longer charging time equipment need, in another word, consumers spend much more time on charging, it does

X. Zhang et al., *Wireless Power Transfer Technologies for Electric Vehicles*, Key Technologies on New Energy Vehicles,
https://doi.org/10.1007/978-981-16-8348-0_11

Fig. 11.1 Trends of EV wireless power transfer technologies

not affect relieving range anxiety. Otherwise, the low transmitted power limited the wide application of WPT.

To gain higher transmitted power, there is still much room for progress.

Considering the resonant inverter, the application of new semiconductors could be a way to out of the dilemma. A new inverter with an enhancement-mode gallium nitride(eGaN) was proposed to further develop the system power ability even in the MHz frequency band (Choi et al. 2018). Despite the single-phase inverter, the multiphase parallel resonant inverter could also be helpful to obtain a high-power level possibly (Deng et al. 2017).

3. System reliability

   The complexity of the system and the changeable environment determine that the reliability of WPT is the top priority (Mayordomo et al. 2013; Botsford and Szczepanek 2009). As mentioned in the previous chapter, the system is composed of multiple parts, which means whatever parts happen to fault will lead the whole system to run out of control, breaks down, and cause unpredictable safety hazards. With the trend of high transmitted power, parts of the WPT system are more likely to break down. In addition, the system is not always indoor, such as dynamic charging, which is always exposed outdoors (Zhang et al. 2019). This means that a changeable environmental should also be considered as system reliability inevitably.

4. Interoperability

Due to the particularity of the WPT system, it is hard to establish a standard that fits all devices. Take the WPT system for EVs as an example, the battery capacity of each model and each manufacturer is different, and the location of the battery pack of cars is also different, which is precisely the most critical part during designing a WPT system. Table 11.1 shows the Power battery parameters of three commercial EVs.

Table 11.1 Power battery parameters of four commercial EVs

Model	Rated voltage of battery pack (V)	Battery pack capacity (kWh)	Pure electric mileage (km)
Tesla model S	360	100	570
Toyota PRIUS (plug-in)	346	4.4	22
BMW mini E	380	35	250

Obviously, the large-scale application of WPT needs to consider the match between different coils and topologies, different power levels, and different models produced by different manufacturers.

Since the commercialization of wireless charging, three different standards organizations have been formed, A4WP, PMA, and WPC proposing QI standards (Zhang et al. 2019). A4WP standard is based on resonant technology while PMA and QI are based on inductive charging. Absolutely, the three various standards make it hard to be compatible with each other. And the incompatibility hinders the development of WPT to a certain extent.

To accelerate the promotion of WPT, it's inevitable to unify the three organizations and publish a comprehensive standard to fits all devices and all over the world. In 2014, A4WP took action firstly that it merged PMA. This is undoubtedly a propellant for the process of wireless charging.

Certainly, the compatibility of different standards is an unchangeable trend of WPT. But it takes time.

5. Improvement of ecosystem

WPT is not an independent technology, it should be the basic technology of construction of ecosystem of intelligent EVs. And the structure of the ecosystem is depicted in Fig. 11.2.

As mentioned above, the performance of WPT is influenced greatly by positioning accuracy. Part of improving the capacity of misalignment, another choice is combining the system with driverless and automatic parking technology, as shown in Fig. 11.3.

We could call the combination an ecosystem of intelligent EVs. The three technologies promote each other, driverless and automatic parking's high-precision positioning improves the performance of the WPT system, while the WPT completes the driverless and automatic parking.

What's more, Internet companies are devoted to layout their products based on the largest personal mobile terminal, automobiles. At the same time, the mature supporting facilities are an essential part of the ecosystem, such as the Wireless charging parking lot.

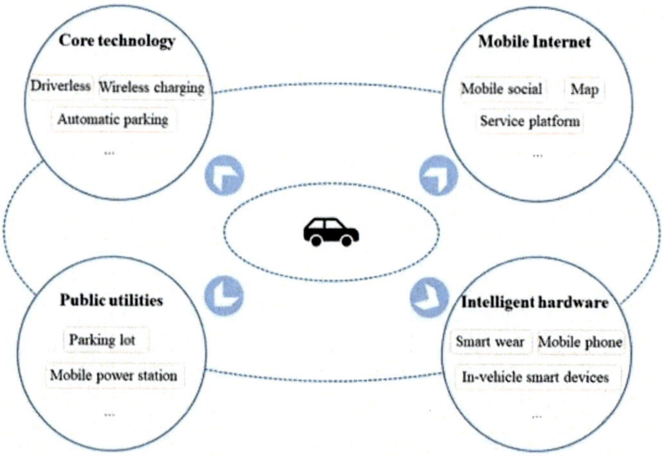

Fig. 11.2 Ecosystem of intelligent EVs

Fig. 11.3 Automatic parking for precise positioning. *Source* https://www.sohu.com/a/240269775_455835

References

Botsford C, Szczepanek A (2009) Fast charging vs. slow charging: pros and cons for the new age of electric vehicles. In: Presented at the 24th electric vehicle symposium, Stavanger, Norway, May 2009

Choi J, Tsukiyama D, Tsuruda Y, Davila JMR (2018) High-frequency, high-power resonant inverter with eGaN FET for wireless power transfer. IEEE Trans Power Electron 33(3):1890–1896

Covic GA, Boys JT (2013) Inductive power transfer. Proc IEEE 101(6):1–14

Deng QJ, Liu JT, Czarkowski D, Hu WS, Zhou H (2017) An inductive power transfer system supplied by a multiphase parallel inverter. IEEE Trans Ind Electron 64(9):7039–7048

Hui SYR, Zhong W, Lee CK (2014) A critical review of recent progress in mid-range wireless power transfer. IEEE Trans Power Electron 29(9):4500–4511

Kazmierkowski MP, Moradewicz AJ (2012) Unplugged but connected: review of contactless energy transfer systems. IEEE Ind Electron Mag 6(4):47–55

Mayordomo I, Dräger T, Spies P, Bernhard J, Pflaum A (2013) An overview of technical challenges and advances of inductive wireless power transmission. Proc IEEE 101(6):1302–1311

Rim CT, Mi C (2017) Wireless power transfer for electric vehicles and mobile devices. Wiley-IEEE Press, Hoboken, NJ, USA

Sample AP, Meyer DT, Smith JR (2011) Analysis, experimental results, and range adaptation of magnetically coupled resonators for wireless power transfer. IEEE Trans Ind Electron 58(2):544–554

Zhang Z, Pang H, Georgiadis A, Cecati C (2019) Wireless power transfer—an overview. IEEE Trans Industr Electron 66(2):1044–1058